MIGRANT PESTS: PROGRESS, PROBLEMS AND POTENTIALITIES

MIGRANT PESTS: PROGRESS, PROBLEMS AND POTENTIALITIES

PROCEEDINGS OF
A ROYAL SOCIETY DISCUSSION MEETING
HELD ON 6 AND 7 DECEMBER 1989

ORGANIZED AND EDITED BY
THE LATE R. C. RAINEY, F.R.S., K. A. BROWNING, F.R.S.,
R. A. CHEKE, AND MISS MARGARET J. HAGGIS

LONDON
THE ROYAL SOCIETY
1990

Printed in Great Britain for the Royal Society
by the
University Press, Cambridge

First published in *Philosophical Transactions of the Royal Society of London*,
series B, volume 328 (no. 1251), pages 515–755

♾ The text paper used in this publication meets the minimum requirements of American National Standard for Information Sciences Permanence of Paper for Printed Library Materials, ANSI Z39.48-1984.

Copyright

British Library Cataloguing in Publication Data

Migrant pests: a discussion organized and edited by the late R. C. Rainey, F.R.S.,
K. A. Browning, F.R.S., R. A. Cheke and Margaret Haggis.
1. Pests: migrant insects
I. Rainey, R. C. (Reginald Charles) 1913–1990. II. Royal Society
632.7

ISBN 0-85403-414-5

Published by the Royal Society
6 Carlton House Terrace, London SW1Y 5AG

PREFACE

Reg Rainey, the inspiration behind this book, died on 18 January 1990. This book stands as a tribute to his lifelong commitment to the problem of migrant pests. It was he that conceived and meticulously planned the Royal Society Discussion Meeting on 6 and 7 December 1989 on which this book is based. For this preface, I could hardly do better than quote from Reg Rainey's letter in November 1987 to the Hooke Committee of the Royal Society whose job it is to select worthy topics for Discussion Meetings.

To explain the title of the meeting, Migrant pests: progress, problems and potentialities, Rainey wrote: 'Progress is illustrated', he said 'by recent epidemiological evidence from the World Health Organization Onchocerciasis Control Programme in West Africa; at the half-way stage of this 20-year programme an estimated three million children were in consequence growing up free of river-blindness following the degree of success achieved in dealing with particular problems of long-range re-invasion by the black-fly vectors and resistance to the particular insecticides initially used'.

'Problems' he continued, 'are currently illustrated both by a 2000 km move of Desert Locust swarms in Autumn 1987 from Sudan, Chad and Niger across the Sahara to Morocco and northern Algeria, and by the exceptional severity of infestations of this species in sensitive areas of northern Ethiopia. This has evoked an unprecedented international response of swarm spraying … (and) a current annual expenditure of the order of a hundred million dollars on the control of half a dozen species of locusts in Africa alone.' He then went on to stress 'the key problem' of 'the extended periods, of months at a time, when all contact has been lost with what could subsequently be recognised as the main populations of a species'.

'Potentialities for meeting this problem', he asserted 'are represented in particular by the fact that all components are now available (and extensively field-tested) for a complete airborne system for seeking, detecting, assessing and if appropriate attacking, in flight, the crucially important missing concentrations of migrant pests, exploiting the manner in which their displacements and distribution have been found to be dominated by the wind-systems in which they fly…'

So it was that, after some hard work by two of the organizers, Dr R. A. Cheke and Miss Margaret J. Haggis, an international gathering of speakers, embracing research and operational as well as biological and physical interests, came together in London in December 1989. This book records their lectures together with the lively discussions that ensued under the chairmanship of Professor R. J. V. Joyce, Dr L. G. Goodman and myself.

April 1990   K. A. Browning, F.R.S.

CONTENTS

CONTENTS

CONTENTS

PROGRESS IN THE INTERNATIONAL CONTROL OF ANOTHER MAJOR MIGRANT PEST: THE BLACKFLY VECTOR OF RIVER-BLINDNESS

ORGANIZATIONS CONCERNED WITH PEST CONTROL

There are many organizations concerned with research and practical control of animal pests, on both national and international scales. A number of these are referred to throughout the proceedings of this symposium by abbreviations, given below.

C.O.P.R.   Centre for Overseas Pest Research, U.K. Ministry of Overseas Development. College House, Wrights Lane, London W8 5SJ, U.K.

CILSS   Comité Inter-États pour la Lutte contre la Sécheresse dans le Sahel.

D.L.I.S.   Desert Locust Information Service.

DLCO-EA   Desert Locust Control Organization for Eastern Africa, P.O. Box 4255, Addis Ababa, Ethiopia.

FAO   Food and Agriculture Organization of the United Nations. Rome, Italy.

IRLCOCSA   International Red Locust Control Organization for Central and Southern Africa. [From 1970.] Mbala, Zambia.

IRLCS   International Red Locust Control Service. [Covered all of southern Africa from Abercorn, Northern Rhodesia (now Mbala, Zambia) until 1970.]

CONTENTS

OCCGE   Organisation de Coordination et de Coopération pour la lutte contre les Grandes Epidémies en Afrique de l'Ouest.
[Upper Volta.]

OCLALAV   Organisation Commune de Lutte Antiacridienne et de Lutte Antiaviare.
B.P. 1066, Dakar, Senegal.

OCP   Onchocerciasis Control Programme.

OICMA   Organisation Internationale centre le Criquet Migrateur Africain.
B.P. 136, Bamako, Mali.

OMS   Organisation Mondiale de la Sante.

ORSTOM   Office de la Recherche Scientifique et Technique Outre-Mer.
[France.]

Phil. Trans. R. Soc. Lond. B **328**, 519–521 (1990)
Printed in Great Britain

Dedication

Dr Reginald Charles Rainey, O.B.E., F.R.S.

These Transactions are dedicated to Dr Reginald Rainey, the Organizer of this Discussion Meeting on Migrant Pests. Reg died on January 18th 1990 after an illness which had severely restricted his physical activities, satisfied, I believe, that his meeting had been successful, lively and well attended, and that every possible action had been taken to ensure that the dead-lines of the Transactions' Editors were being met.

Reg spoke to me as early as January 1987 about his wish to follow the 1977 meeting on Migrant Pests (*Trans. R. Soc. Lond.* B **287**) by a further one to highlight progress, problems and potentialities. This was at a time when his previous robust health was beginning to decline and when he was much engaged with his book, *Migration and Meteorology* (published, happily, before he died and on display at the Meeting). Despite this preoccupation and his deteriorating health, Reg completed heroically all the correspondence relating to the meeting, did more than his fair share of preparing our joint paper, and finalized and proof-read his book, though largely confined to his chair and handicapped by cataracts on both eyes for much of the time. I had many telephone conversations with him during this period and his cheerfulness, clarity of thought and quickness of response disguised from me the seriousness of his condition. It was only towards the end of the year, when Reg's messages were sometimes conveyed by Margaret, his wife, that I became aware that only Reg's determination and dedication had made it possible for all the preparations for the Meeting to be completed.

Reg was a graduate and post-graduate of Imperial College, London, and of the Imperial College of Tropical Agriculture in Trinidad. As a student, he was a keen glider. His first appointment as an Entomologist at the Empire Cotton Growing Corporation's Research Station in Barberton, South Africa, in 1938, engaged him in research into cotton insect pests which were active flyers, and also gave him the opportunity of exploring and publishing accounts of the thermal up-currents by which he was able to soar to heights of over 1300 m from a ridge just outside Pretoria. When the war came, and being rejected as a pilot because of impaired eyesight, Reg joined the South African Air Force as a Meteorologist and served in this capacity throughout the African and Italian campaigns. As a meteorologist, Reg was far more at home with physics than are most biologists. After the war, Dr D. Gunn, then the Senior Scientist at the Anti-Locust Research Centre in London, discovered Reg's dual expertise and Reg needed little persuasion to transfer from cotton pest research to the study of a group of insects, the Acrididae, of some members of which it might be said the air was their habitat. Reg was probably the first to estimate how much time locusts spent in flight and recognize how the air on which they flew determined where they fed, where they reproduced and where they sheltered. The Desert Locust, in particular, fascinated him. This species inhabits a zone from West Africa to eastern India, which can be hostile to all life for months or years on end, and yet populations locate and colonize those parts when the imminence of rain promises to provide the conditions necessary for survival and reproduction. Careful field observations by Reg and his colleagues, accompanied by novel methods, showed that, despite the apparent purposefulness of swarm migration, in fact, flight was usually intermittent and the direction effectively random. Reg demonstrated that the displacement of such swarms must inevitably be downwind and this would lead to their accumulation in zones of wind-convergence, that is in zones where more air is entering than leaving on the horizontal plane, the excess air being displaced upwards, providing the first requirement for rain. He published his convergence hypothesis in *Nature* in 1951.

The initial major test of the hypothesis was undertaken retrospectively under the auspices of the World Meteorological Organization on all available Desert Locust data from all countries for 1954–55, together with relevant meteorological data. This analysis led to the conclusion that the movements and distribution of Desert Locusts on scales from 10^2 to 10^3 km^2 were not only correlated with, but largely determined by the corresponding low-level wind fields. Desert Locust populations were in almost continuous movement over a total area three times the size of Europe at speeds of up to several thousand kilometres per month, and locusts flew on those winds which, by and large, led to their survival. The second major test of the hypothesis was undertaken using current locust and synoptic data of the whole of the Desert Locust distribution area during 1960–67 to provide interpretations, forecasts and warnings to all countries concerned, in the form of communications of the International Desert Locust Information Service which Reg set up and ran on behalf of the Food and Agricultural Organization of the United Nations. The results were independently assessed by a specialist FAO Committee who recorded 'the very satisfactory results obtained in the verification of the Desert Locust Information Service predictions'.

While the possible limitations of the circumstantial evidence for the convergence hypothesis were fully offset by the number of observations considered, Reg was anxious to secure the smaller-scale direct evidence of the relation of the density and structure of populations of airborne insects and the wind fields in which they flew. It was Reg who initiated the radar studies of the late Dr Glen Schaefer which have been so successfully followed up by Dr J. Riley. This series of radar observations established that wind convergence determined not only the destination of airborne insect populations, but also gave rise to concentrations determining their density and structure. These conclusions could not have been reached without the massive contributions made by Reg. He and his colleague, Miss Margaret Haggis, flew some hundreds of hours in seven countries in Africa and North America in aircraft fitted with Doppler wind-finding equipment, which permitted them to measure quantitatively the convergence which concentrated airborne insects and to locate with unprecedented accuracy the separation of wind-discontinuities, usually only a few kilometres, where insects accumulated and towards which insects, for example moths, moved from both sides with ground speeds much in excess of their own airspeed. This work included studies on the Spruce budworm moth on behalf of the Canadian Forestry Service, on the African armyworm moth on behalf of the East African Agricultural and Forestry Research Organization and on the blackfly vector of Onchocerciasis (river blindness), on behalf of the World Health Organization. They stimulated parallel work in the U.S.A., Australia and China, and established beyond doubt a view previously expressed by many entomologists that we had grossly underestimated the flight activity of most major pests.

Reg's work has profound implications for the understanding of insect population dynamics. It emphasizes that the spatial redistribution of pests which spend much time in flight is often more important than their total numbers and is characteristically dominated by the structured atmospheric systems of winds and weather. Moreover, this redistribution includes not only long range displacement of populations but also massive changes in density under the influence of convergent winds. The arrival of such concentrations, if not already at damaging levels, represents an enormous potential increase in birth-rate which may overwhelm locally occurring controlling agents giving rise to pest outbreak. The very important role of flight behaviour in insect population dynamics can be fully appreciated only when sufficient account is taken of the dynamics of the atmospheric environment.

Reg's profound curiosity regarding the interaction of insects and weather was matched only by his concern for his colleagues and co-workers. He was swift to recognize their contributions and to give them every possible help and encouragement. His memory and his work will have a continuing influence on entomological thought.

R. J. V. JOYCE
12 February 1990

Phil. Trans. R. Soc. Lond. B **328**, 523–524 (1990)
Printed in Great Britain

BIOLOGICAL FACTORS

Introductory remarks

By R. J. V. JOYCE

Cranfield Institute of Technology, Bedfordshire MK43 0AL, U.K.

(*Formerly Director, DLCO-EA and AARU*)

The Chair for this opening Session should, of course, have been taken by Dr Rainey, but you will be aware that, unhappily, Reg is not well enough to attend. This is very sad as Reg is the instigator and inspirer of this meeting and has spent at least two years in preparing for it with all his accustomed attention to detail. We are grateful to Reg and his co-organizers, Dr K. A. Browning, Dr R. A. Cheke and Miss M. J. Haggis that, despite illness, the meeting is taking place as planned.

It is an honour to have been invited by Reg to chair this session, an honour that I felt able to accept only because it gives me an opportunity to pay a tribute to Reg in the presence of so many of his old friends and colleagues and of so many distinguished scientists.

The meeting itself is a tribute to Reg, integrating, as it does the biological and physical sciences for a better understanding of the phenomenon of migration, Reg's unique contribution. From his student days, Reg was a keen glider – and what activity demands a more intimate knowledge of the atmosphere than gliding? At the outbreak of war, he was a Cotton Entomologist with the old Empire Cotton Growers Corporation at its research station in Barberton. He would have joined the Airforce there as a pilot, but was prevented from doing so because of poor eye-sight. Instead, he joined the Royal South African Airforce as a Meteorologist. In this capacity he served throughout the war with the South African forces in Kenya, Somalia, the western Desert and Italy. His war-time duties carried great responsibilities that involved providing vital information to aircrew embarking on difficult and dangerous missions. It was this experience, perhaps, which shaped his approach to locust research later, Reg always being acutely aware that his research had to be directed primarily to provide information and help to field staff in their job of control. It was also perhaps his war-time experience that generated the respect that he always had for aircrew, as is evidenced by the presence today of many whose primary task has been the practical execution of locust control rather than planning or research.

Reg combines to a remarkable degree single mindedness in his approach to research, with an intellectual integrity that allows him to explore and understand other possible approaches. He is always among the first to see merit in ideas and concepts, the development of which might bring rewards to their originator and to science. Through this he has stimulated and encouraged more research workers, particularly in 'developing countries', than any other research worker I know.

It is a happy coincidence that his book *Migration and meteorology* (Oxford University Press 1989) has just been completed and in time to be displayed at this meeting. I believe that it will be on sale in the New Year. This authoritative account will be of immense value to all workers in this field.

We shall sadly miss Reg at this discussion meeting and I am sure you will all join with me in sending him our greetings and best wishes for a speedy recovery.

Dr Rainey was to have chaired this session and wrote the following opening remarks.

Welcome to all participants and guests. I would like to make just two points.

First, the function of such Society meetings as this is to secure an authoritative exchange of differing opinions on matters of major scientific and practical importance. On Desert Locusts, Skaf directs attention to one such outstanding question – the significance of low-density populations, crucial to the understanding of the species as a whole, as well as of the biological feasibility of the airborne radar system.

I would now only direct attention to the authoritative presentation of both the still conflicting points of view (including my own!) in the Society's *Philosophical Transactions* published record of the previous meeting on Migrant Pests (Gunn & Rainey 1979). They are not to be confused with my own new book *Migration and meteorology* (Oxford University Press 1989), the very first copies of which incredibly appeared only last week – now also on display here.

At the previous meeting it was claimed that 'none of the earlier cases of gregarization appeared indeed to have been *proved* to play a major part in the overall development of the plague, since substantial gregarious populations were already present elsewhere in each of these cases; and gregarization has often occurred without upsurge.' I would now only urge that any more recent evidence to the contrary be placed on record in detail.

My second point is on the intractable problem of dieldrin. On an entirely different environmental point I have recently been unexpectedly impressed by the objectivity of some of the information material now published by non-official organizations such as Friends of the Earth. Might it now be too far fetched to suggest attempting to interest such an organization in dieldrin in the hope of eliciting a similarly entirely independent report on the evidence?

Reference

Gunn, D. L. & Rainey, R. C. (eds) 1979 *Strategy and tactics of control of migrant pests* (*Phil. Trans. R. Soc. Lond.* B **287**), pp. 245–488. London: The Royal Society.

6 December 1989   R. C. RAINEY, F.R.S.

Phil. Trans. R. Soc. Lond. B **328**, 525–538 (1990)

Printed in Great Britain

525

The Desert Locust: an international challenge

By R. Skaf, G. B. Popov† and J. Roffey

*Food and Agriculture Organization of the United Nations, via delle Terme di Caracalla,
Rome 00100, Italy*

A new plague of the Desert Locust *Schistocerca gregaria* started in 1986; it developed quickly in 1987 in the Sahelian countries and reached northwest Africa at the end of 1987. It expanded in 1988 in north Africa, the Sahel, the Sudan, the Near East, southwest Asia, and in October 1988, swarms crossed the Atlantic to the Caribbean. The plague declined dramatically in the past quarter of 1988, and by March 1989 the plague was over. A study of the latest known upsurges provides more support to the theory that a build-up of locusts arises from initially low-density populations rather than from the persistence of undetected swarming populations. The decline of the recent plague was probably attributable to the cumulative effect of control operations combined with natural environmental factors. Prevention of new plagues of the Desert Locust will be much more difficult as the restriction on the use of dieldrin poses major technical, logistic and financial problems.

1. Introduction

There has been a series of important events involving locusts since the 1977 Royal Society Discussion Meeting on migrant pests (Gunn & Rainey 1979). A short-lived plague of the Desert Locust *Schistocerca gregaria* (Fôrsk.) occurred in 1976–79, which has already been summarized (Roffey 1979, 1982). Other smaller upsurges occurred in Niger in 1980 (Castel 1982) and Indo-Pakistan in 1983 (Food and Agriculture Organization (FAO), Desert Locust Situation Summary and Forecast nos. 59–62). These were successfully checked by control operations. In 1986 there was a further upsurge in Desert Locust populations, which led to the plague which is the main subject of this article.

The period 1985–1987 also witnessed upsurges of all four of the other main locust species in Africa. The African Migratory Locust *Locusta migratoria* (R. & F.) gregarized in several parts of Africa: Sudan, Eastern and southern Africa, Madagascar and most recently in Chad during 1989. The Brown Locust *Locustana pardalina* (Wlk.) erupted in the Karoo region of the Republic of South Africa and invaded Botswana. The Red Locust *Nomadacris septemfasciata* (Serv.), after years of deep recession, bred successfully in several outbreak areas from which several swarms escaped. Tree Locusts *Anacridium melanorhodon* (Thnbg.) became significant pests in Sudan and Chad. In southwest Asia the Moroccan Locust *Dociostaurus maroccanus* (Thnbg.) continued to infest the border areas of Iran and Iraq and a new plague has developed since 1987 in the northern provinces of Afghanistan. Simultaneously, very important upsurges of the Senegalese grasshopper, *Oedaleus senegalensis* (Krauss), and other species of grasshoppers occurred in the Sahel. The American continent also experienced important events. The Brazilian Locust, identified by Carbonell (1988) as *Rhammatocerus schistocercoides* (Rehn), which had never been

† Present address: 129a Hammersmith Grove, London W5 0NJ, U.K.

[7]

recorded as a gregarious species in the past, started to swarm in 1985 in Mato Grosso State in Brazil. It has since invaded several other States, infesting an area of some 35 million ha†. The central American Locust, *Schistocerca piceifrons* (Wlk.) has again swarmed in the region since 1985 and emergency control operations had to be organized in Costa Rica in 1986–87, in the Gulf of Fonseca (Nicaragua, Honduras, El Salvador) in 1988–89 and a number of small fully gregarized swarms developed in the Yucatan Peninsula in Mexico.

2. Desert Locust recessions and plagues: an introductory synopsis

The following is a brief summary of the salient features of Desert Locust recessions and plagues to provide a background to our interpretation of events during the 1986–89 plague. Fuller accounts are given by Pedgley (1981); Roffey *et al.* (1970); Steedman (1988); Waloff (1966).

The similarities and differences between the timing and distribution of breeding of Desert Locust populations during recessions and plagues are clearly revealed by a comparison of the monthly degree-square frequency maps for the major 25-year (1939–1963) plague period (prepared by the Anti-Locust Research Centre, unpublished) and those for the 20-year recession period 1964–1984 (less 1968) (FAO, unpublished). The main similarity is, naturally, that breeding follows the onset of seasonal rainfall, but because of the reduced mobility of non-swarming adults, there are large areas where gregarious breeding occurs but not low density breeding e.g. in northwest Africa, the countries to the north of the Arabian peninsula, north-central India, southern Somalia and East Africa. In the Central Region even the spring breeding grounds in the interior of Arabia, the monsoon breeding grounds in the interior of Sudan and the short-rains breeding grounds in the Somali peninsula remain almost completely clear of adults and of breeding except during major upsurges, when they soon become involved.

High-frequency breeding areas, where locusts are more numerous and occur more often than elsewhere, also exist within the recession area. Thus, in the Central Region, Desert Locust populations most frequently occur during recessions in the coastal areas around the Red Sea and Gulf of Aden, where there are overlapping rainfall regimes and where breeding beginning in one season often continues into the next, and sometimes lasts for many months. The coastal areas thus constitute a major reservoir within which locust populations survive and breed, producing two to three generations annually. Breeding occurs principally from July to March, but occasionally it may last until April or even June, with the production of an extra generation. Other high-frequency recession breeding areas are found in the Western Region, where they are associated with the major geomorphological relief and drainage features in the southern and western Sahara, while in the Eastern Region they correspond to the drier and warmer, low-lying part of the spring and the central part of the summer breeding grounds. To a large extent, these differences can be ascribed to the biological and behavioural differences between plague and recession populations.

There are marked differences in the flight behaviour of swarms characteristic of plagues, which migrate by day, and the mainly non-swarming recession populations which fly principally at night (Roffey & Popov 1968). The intensive gregariousness of individuals in swarms frequently results in swarms overflying habitats that are suitable for non-swarming populations, while the latter are prevented from reaching areas attained by swarms because of

† 1 hectare = 10^4 m².

low night temperatures. As a result, the area which can be reached by swarms, the invasion area, is almost double that of the recession area. The latter corresponds to the great Sindo-Saharan desert where most of the locust habitats receive less than 100 mm of rain annually.

The Desert Locust is not wholly resistant to drought at any stage of its development and has no built-in mechanism, such as an egg diapause, to protect it from extreme dessication. When females lay, water must be available in the soil, in sufficient quantities to ensure both the development of the eggs and the growth of vegetation to sustain the resulting hoppers and adults. The Desert Locust survives in its arid environment by moving from one area where rain has fallen to another, which may be several thousand kilometres distant.

It follows that the chances of survival and the resulting numbers of locusts are highest where sequences of sufficient rainfall are most frequent and reliable, where direct rain is enhanced by run-off and flooding, and especially where certain combinations of soils and vegetation create particularly suitable habitats. Occasionally, when such sequences of successful breeding span several successive generations, population build-up of an order of magnitude may occur at each generation. The resultant insects can aggregate and gregarize in particular habitats leading to hopper band and swarm formation. In addition, wind fields and mesoscale convergence provide transportation and at times the concentration of locusts. Thus Uvarov's concept of outbreaks of *L. migratoria* arising through a process of phase changes in particular outbreak areas (Uvarov 1921) also holds true for the Desert Locust, but in a more dynamic context, involving complexes of temporary breeding habitats linked by successive displacements.

The present strategy of plague prevention which evolved during the 1960s consists essentially of conducting surveys in seasonal breeding areas and controlling any gregarious or significant gregarizing populations. A second school of thought favours delaying control operations until most of the populations are in bands and swarms. Although the first is the recommended strategy, in practice the second is often implemented (Bennett 1976). Its weakness is that a major population build-up may occur in areas inaccessible to control; a phenomenon successively observed in Eritrea, Chad, Mauritania and Western Sahara during the 1986–89 plague upsurge.

3. The Desert Locust plague 1986–89

The upsurge of the Desert Locust in 1985–86 in Africa and Arabia rapidly attained plague proportions at the end of 1987 when locusts invaded northwest Africa from the Sahel countries. In 1988 all the Sahelian belt from Mauritania to Ethiopia, northwest Africa, the Arabian Peninsula and the western part of southwest Asia were affected (see figures 1 and 2; Skaf 1990.)

The period from the summer of 1985 to the summer of 1988 was marked by repeated successful breeding and gregarization by the Desert Locust in the central and western regions of its invasion area. The plague developed because suitably timed rains fell in areas which were reached rapidly by increasingly gregarious populations which, in 1987–88, allowed almost continuous breeding in west and northwest Africa. Because of the inaccessibility of certain areas (successively Eritrea in 1986, Chad in 1987, northern Mauritania and Western Sahara in late 1987 and early 1988), and the restrictions on the use of persistent pesticides, the locust situation was not brought under control in spite of very large scale operations. However, control continued through 1988, but it was not until the last quarter of 1988 and the first quarter of 1989 that a combination of adverse weather and control led to a dramatic decline of the plague.

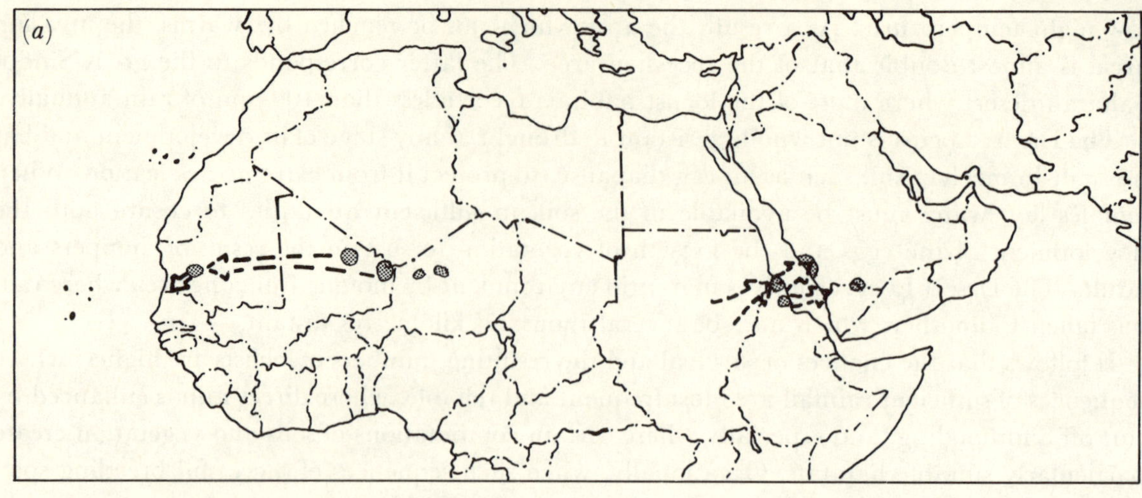

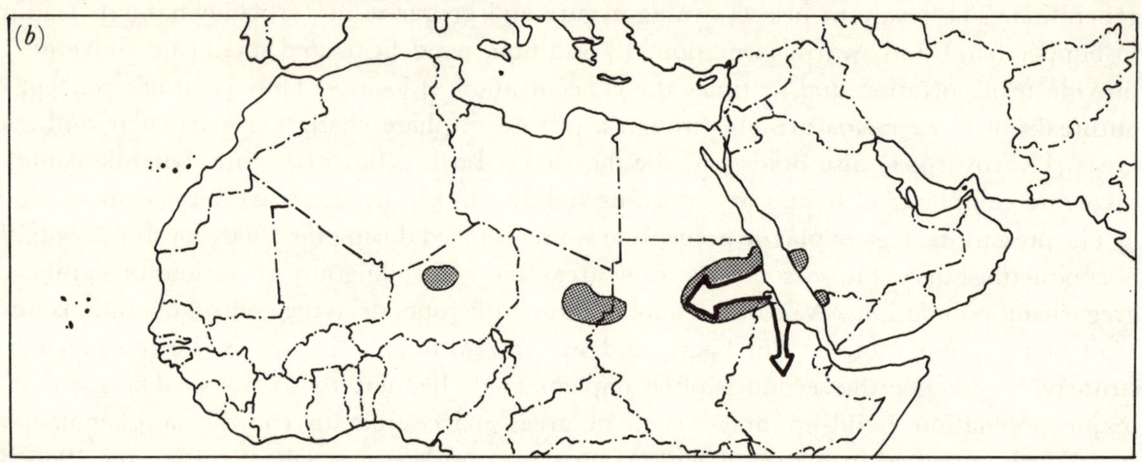

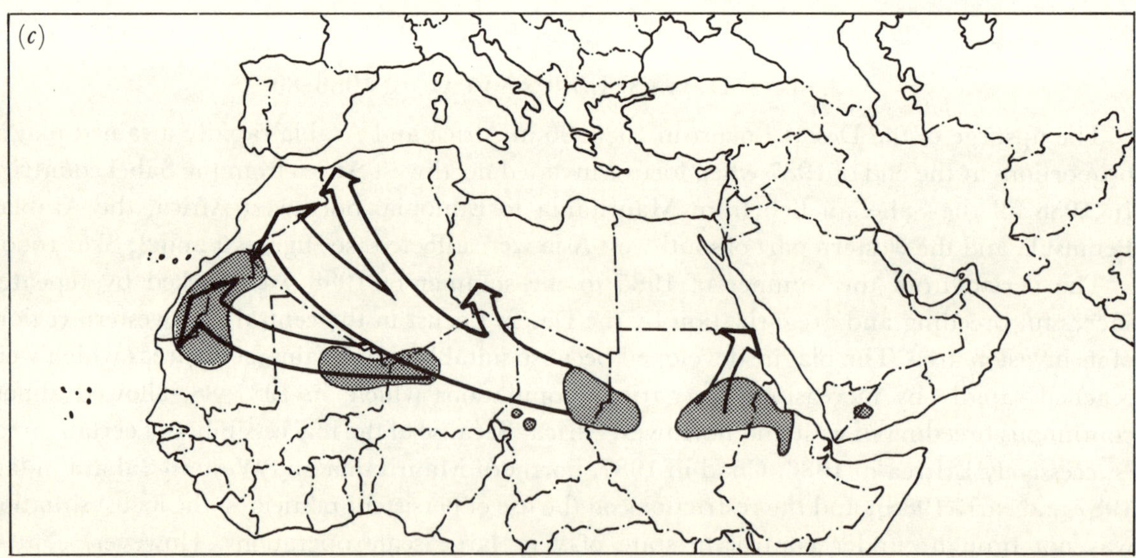

FIGURE 1. (*a*) 1985 campaign. (*b*) 1986–87 Winter-Spring breeding. (*c*) 1987 Summer breeding.
(breeding; ⇒, migration; ⇢, possible migration.)

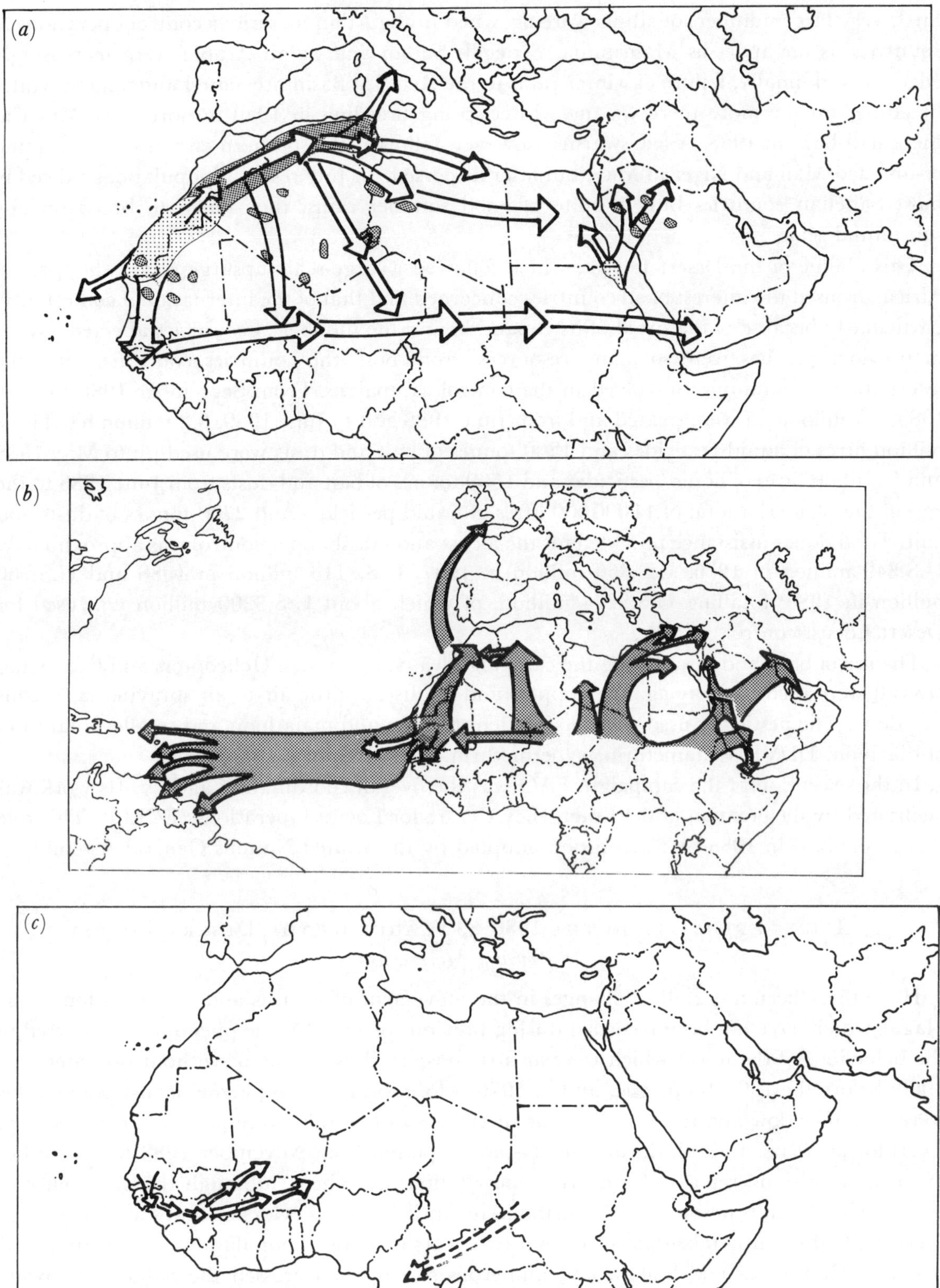

FIGURE 2. (*a*) Winter-Spring breeding 1987–88. (*b*) Movements of swarms of the first summer generation (June–September 1988). (*c*) Movements of swarms of the second summer generation (October 1988–May 1989). (◓, breeding Jan.–Mar.; ◓, breeding Apr.–Jul.; ⟹, migrations; ⇢, possible migrations.)

First, very large numbers of adults were drowned in the Atlantic, serious control operations in countries as far apart as Mauritania, Sénégal, Sudan and Saudi Arabia were increasingly effective and, finally, failure of winter rains in the Red Sea Basin prevented anticipated winter breeding. Furthermore there was no winter-spring breeding in 1989 in northwest Africa as there had been in 1988. A few swarms, however, followed the southern circuit in Guinea and re-invaded Mali and Niger in May, but later dispersed. In July residual populations existed in most Sahelian countries but, in spite of good summer rains, they failed to breed on any substantial scale.

This plague of the Desert Locust, which followed a large-scale upsurge of grasshoppers in Africa, aroused the interest of all countries concerned and that of the international community, particularly because of the speed of its expansion and the number of countries affected. Never in previous plagues were so many resources, from both the countries concerned and the international community, involved in the control campaigns. From September 1986 to May 1988, 7.6 million ha were treated and from June 1988 and to June 1989, 9.2 million ha. Three million litres of liquid pesticides and 1200 tonnes of bait and dusts were used up to May 1988 and 8 million litres of liquid pesticides and 1500 tonnes of bait and dusts from June 1988 to the end of the plague: a total of 11 000 000 litres of liquid pesticides and 2700 tonnes of dusts and bait. Total donor assistance provided for the locust and grasshopper control was approximately U.S.$49 million in 1986, U.S.$50 million in 1987, U.S.$115 million in 1988 and U.S.$60 million in 1989, totalling U.S.$274 million, of which about U.S.$200 million was used for Desert Locust control.

The use of baits and dust during the campaign was very limited. Helicopters and fixed-wing aircraft were used to spray against hoppers and adults but little air-to-air spraying of swarms was done. The pesticides used were mainly fenitrothion and malathion and smaller quantities of diazinon, DDVP, deltamethrin, cypermethrin, dursban and carbaryl.

In the execution of the campaign, FAO was the overall coordinating agency. Its work was facilitated by the creation of the Emergency Centre for Locust Operations (ECLO). This role was recognized in 1988 by a resolution adopted by the United Nations General Assembly.

4. Issues arising from the 1986–89 plague of the Desert Locust
(a) Swarm Movements

Have there been any radical changes in the movement of swarms and the evolution of the plague which have not been recorded during previous plagues? Some phenomena occurred in the behaviour of the insect which were at first considered as unique or without precedent. In 1966, before the 1967–68 plague, and in 1976 before the 1977–78 plague, locust populations were extremely low and this was also true in 1984–85 when they were possibly at their lowest levels for 50 years. Indeed, during the 24-month period from November 1983 to November 1985 there were no reports of gregarious populations anywhere, although the occurrence of pink coloured adults in southeast Mauritania in April 1985 was evidence that some gregarious breeding had recently taken place. Within two years gregarious populations were widespread between Mauritania and Sudan, and numerous swarms had crossed the Sahara to invade Morocco and northern Algeria. In spring 1988, the southward movement of swarms produced during winter-spring breeding in northwest Africa began very early: southern Mauritania, Sénégal and the Cape Verde islands were reached in late March; later, swarms moved eastwards through Mali and Niger reaching Chad and Sudan in late May and Ethiopia and the Yemen Arab Republic in July; such a major and prolonged easterly movement is without recorded precedent. Even more spectacular was the movement of locusts in October 1988 out over the Atlantic Ocean, leading to repeated invasions of the Cape Verde Islands and a large scale invasion of the eastern Caribbean and northern coastal areas of South America on a front of some 1500 km. Again, such a crossing of the Atlantic Ocean was without historical precedent. There were also unprecedented migrations, although on a smaller scale, to Italy (March, May), Greece (May), Turkey, Syria and Lebanon from the west (December). Other migrations were not exceptional and have been recorded previously (Donnelly 1947; Rainey 1963; Waloff 1966; Pedgley 1981).

(b) *Plague commencement*

The start of the plague can only be attributed, like the short-lived 1978 plague (Roffey 1979, 1982) to the progressive and rapid build-up and gregarization of initially solitary-living populations as a result of successive and successful breeding. Indeed the transformation from 1984 to 1986–88 provides better evidence for such processes in the generation of plagues than did earlier upsurges. It does not lend support to an alternative hypothesis to explain the generation of plagues which requires the continuity of major populations and invokes a prolonged loss of contact with swarms to account for gaps in their recorded occurrences (Rainey & Betts 1979). The controversy is far from being academic; it has operational, practical and strategic implications. The acceptance of the 'prolonged loss of contact' hypothesis presupposes the establishment of a costly fleet of aircraft equipped with sophisticated equipment and operating, probably with little return, all the year. Also, it will discourage the preventative control strategy which we are trying to strengthen and/or introduce in all the countries concerned.

(c) *Control tactics*

Where and how could the developments of the 1986–89 Desert Locust plague have been most effectively dealt with? There are three possible explanations for the developments that occurred:

(i) A lack of preparation for large-scale control operations which resulted in inadequate control of the early infestations in Mauritania (late 1985) and the Central Region in late 1985, and early 1986.

(ii) The decreasing use of chlorinated hydrocarbons (including dieldrin) and their restriction when most locusts in the invasion area were concentrated in relatively small areas: August 1987 in northeastern Chad and northwest Sudan; November 1987 to March 1988 in northern Mauritania and Western Sahara.

(iii) The lack of security which limited surveys and control operations in Ethiopia in 1986, Chad in 1987 and Western Sahara in 1988.

These events have already been described (Skaf 1988). In spite of predictions and warnings, opportunities to check the plague in Chad and then in Mauritania by the use of dieldrin in barrier spraying were missed.

(d) *Methods of pesticide application*

Only limited quantities of baits and dusts were used during the campaigns. Most operations were conducted with liquid pesticides including both ultra low volume (ULV) and emulsifiable concentrate (EC) formulations.

However, until the spring of 1988 several countries were still using a high volume application technique at the rate of 25 l of diluted formulations per hectare! Their conversion to ULV spraying required modification and training, which were executed at remarkable speed. Also, in spite of the very large-scale aerial control operations covering millions of hectares, several countries and donors ignored the experience with ULV aerial application gained in previous plagues, and applied direct spraying with narrow swaths even with large aircraft. It was only after the end of the plague that a joint US-FAO technical consultation meeting, held on 6 and 7 November 1989, agreed key principles of ULV aerial application for Desert Locust control (US-FAO 1989).

As no successful government aerial unit exists in Africa, most aerial operations were undertaken by private companies using different techniques. The dissolution of l'Organisation Internationale Contre le Criquet Migrateur Africain (OICMA), the disappearance of l'Organisation Commune de Lutte Antiacridienne et de Lutte Antiaviare (OCLALAV) as an operational regional organization, the decline of the operational power of the Desert Locust Control Organization for Eastern Africa (DLCO-EA), and the scarcity of suitable aircraft and trained pilots able to undertake air-to-air spraying have exacerbated the situation.

(e) The use of dieldrin

The use of dieldrin was very limited at the start of the plague and, later, strict limitations on its use were imposed. This gave rise to controversies and high-level discussions at many meetings during the campaigns. Technicians were almost unanimously in favour of its use under special circumstances to stop the plague, backed by Gunn's arguments (Gunn 1979), which remain valid. As a result of the technicians' insistence, a special meeting was organized in October 1988 in Rome on the use and hazards of dieldrin in Desert Locust control (FAO 1988c). Widely divergent opinions were expressed on the future use of dieldrin in Desert Locust control, ranging from a complete ban and the destruction of stocks to recommending its use against hopper infestations in remote areas, under technical guidance. It was suggested that 'dieldrin is a pesticide of the past', not on account of its hazards, but 'as it is no longer produced'. It was noted that the use of alternative pesticides for Desert Locust control would result in substantially higher control costs and opportunities for a rapid reduction of the current plague would be missed. Only existing stocks needed to be considered so it was agreed that: FAO would make a detailed inventory of existing stocks; a study would be undertaken of the area where Desert Locusts would have to be controlled and these areas would be classified according to the potential environmental risk that application of dieldrin might entail; on the basis of this information, a use or destruction plan for all or part of the dieldrin stocks would be prepared.

(f) The size of the 1986–89 plague

Although unprecedentedly large-scale control operations were undertaken during 1987 and 1988, there is little evidence to support the view, often expressed during the campaign, that the plague was the most serious ever recorded. In fact not only was no systematic plotting of infestations done, except in Morocco, Algeria and at OCLALAV in 1988, but there were repeated occasions when estimates of the area claimed to be infested were inflated. No quantitative assessment was done of the real number, size and density of swarms and/or of adult populations at the lower densities. The only quantitative studies on hopper infestations were by an FAO/OCLALAV/CIDA team in early September 1988 followed by an FAO/multi donor evaluation study in late September 1988, both in southeastern Mauritania (FAO 1988 *a*); and by G. B. Popov (unpublished results) in Mali and Niger in July–August 1988 in the Sahelian belt.

In July 1988 there were 40000 ha of swarms in Mauritania, 30000 ha in northwestern Mali and 50000 ha in Niger. By September, a gross area of 2750000 ha within southeastern Mauritania contained infestations. In general, the area actually infested was 5% (137500 ha). Hopper bands in Mauritania varied in size between 0.1–400 ha with averages of 8 ha in the centre and 15–140 ha in the south and southeast, respectively. The average number of hopper bands varied between 0.5 km^{-2} in some blocks, with mainly small bands, and 0.1 per km^2 in other blocks with mainly larger bands. The total hopper population was potentially able to produce 5×10^{11} adults, i.e. 1 million ha of new swarms in October. Something like this actually happened; for in addition to the many millions which flew out over the Atlantic, numerous large swarms invaded Morocco, where the 3000000 ha treated represented repeated applications to prevent invasions of crops, while others went southwards into Sénégal, Gambia and Guinea Bissau. Similarly, in Niger, escapes from the 60000 ha area originally infested gave rise later to 600000 ha of adult infestations, which required control. In Sudan over one million hectares were also treated.

(g) The decline of the Desert Locust plague in 1988–89; control operations or natural causes?

In spite of problems posed by the assessment of the effects of control measures on the development of the overall Desert Locust situation, as against the effect of natural causes, it is clear that throughout 1988 control operations were conducted in the affected countries on unprecedented scales. Of particular impact were those campaigns waged against swarms in Morocco from October to December 1988, when most known swarms in northwest Africa were located, against hopper infestations in Sénégal and Gambia and elsewhere in West Africa, and in Saudi Arabia against the invading swarms in late 1988 and early 1989.

In addition to efficient control operations, natural factors cannot be ignored. Of particular impact was the loss of swarms over the Atlantic and the failure of breeding in the Central Region and northwest Africa in the winter of 1988–89.

The reasons for the failure of the 'southern circuit' swarms to lead to substantial breeding in the Sahelian countries in the summer of 1989 are not known. The apparent loss of fecundity by the locusts could, perhaps, be ascribed to the effects of cumulative sublethal doses of pesticides or the cumulative effect on the environment of continuous control operations undertaken since the start of the plague.

Other likely contributing factors include: mortality factors throughout the period (attacks by man, birds and other predators, disease, adverse climatic conditions); fragmentation of swarms (as was observed in southern Mali and Burkina Faso); the possible loss of a reproductive ability postulated for 'southern circuit' swarms (Pedgley 1981).

5. CONTROL STRATEGY AND ALTERNATIVES

The intensity of the 1986–89 Desert Locust plague, the amount of external assistance involved (in excess of U.S.\$200 million) and the interest that it aroused have confirmed the necessity of improving control strategies and considering new alternative approaches to the prevention of new plagues. For its part FAO organized a series of expert meetings in 1988–89, dealing with various aspects of the problem (FAO 1988b, 1989).

As stressed by Bennett (1976), a great deal of improvement is needed in the organization of campaigns once an upsurge has started, to concentrate control operations when and where large locust populations are concentrated for protracted periods of time (strategic areas). Such concentrations have frequently occurred in Morocco and northern Somalia; other potential target areas such as the lakes region of Mali have been identified in the Sahel. Radar and aircraft properly equipped with doppler radar could be useful in the strategic areas at appropriate times, but we have to recognize the limits of radar's use and its feasibility because of the frequent distribution of locusts in remote and inaccessible areas and the short period when adult control can be undertaken. On the other hand, we support the use of radar to detect night flying adults at the early stage of plague periods.

Consideration of previous upsurges suggests that a critical situation may arise very rapidly, leading to a locust population in the range of 10^9 or 10^{10} individuals within an area of $20\,000$ km^2, comprised of a large number of small bands (covering 1–5 % of a band zone, defined as an area containing bands separated from another band zone by at least a few kilometres). They may be very difficult to locate. They are also very costly to control, for the simple reason that band zones will have to be covered when using non-persistent pesticides and all insecticides applied to the gaps between bands will be wasted.

Although the merit of treating the insects themselves rather than the band zone containing them appears obvious, we find in practice that most attacks on locusts can only be directed at a treatment of the band zones for achieving the work in the time available; a band zone may remain extant for about six weeks, but a swarm may pass in a day. Given the difficulties of finding more than a small proportion of the small bands, except in inhabited areas, and of controlling flying swarms by air-to-air spraying, it is difficult to envisage the prevention of the spread of plagues without:

(i) the rehabilitation of dieldrin for barrier spraying;

(ii) the organization of large-scale hopper band zone spraying, using 50 times the necessary quantity of non-persistent pesticides and unnecessarily polluting the environment;

(iii) the organization of systematic tracking and spraying of swarms in flight as was widely practiced in the 1950s, often efficiently (because airborne locusts collect a higher proportion of the spray than settled targets) and at acceptable cost, or against settled swarms as in 1988 at a very high cost.

6. THE INTERNATIONAL CHALLENGE; POSSIBILITIES OF PREVENTING AND CHECKING FUTURE PLAGUES OF THE DESERT LOCUST

It is to be hoped that the implementation of a preventive control strategy involving the strengthening and/or creation of national locust control units in all the countries concerned, during recession periods, and the establishment of adequate regional task forces able to intervene and check initial upsurges, will prevent important upsurges and new plagues in the future. The recent large-scale campaigns against grasshoppers and locusts undertaken since 1985 in Africa and the Near East, generously supported by the international community, have resulted in the better equipping of national units and a better understanding by governments of the need for preventive control. However, as the use of dieldrin is limited, an economical reduction of numbers by barrier spraying will not be possible. Consequently, for preventive control to be successful, more equipment, manpower and pesticides than were used during the 1962–1985 period will be needed, as non-persistent insecticides will require the finding and spraying of individual concentrations or bands of hoppers. For future planning the following factors must be borne in mind: the decreasing role of regional locust control organizations and their replacement by the national plant protection services; the economic difficulties of developing countries that cannot support the continuity required for preventive control; the depth of the involvement of international community assistance in controlling grasshoppers and locusts during the recent pest years, thereby creating a heavy reliance on foreign assistance and implying the necessity of commitment by donors to support preventive control costs over a long period; the persistence of political problems within the recession area of the Desert Locust; the time needed to promote new alternative pesticides and control strategies.

Therefore, to be realistic, we cannot deny the likelihood of new upsurges and even new plagues in the decades to come. The challenge consists in understanding the problems involved and envisaging approaches and solutions able to check situations in time before explosions occur.

References

Bennett, L. V. 1976 The development and termination of the 1968 plague of the desert locust. *Bull. ent. Res.* **66**, 511–551.

Carbonell, C. S. 1988 *Rhammatocerus schistocercoides* (Rehn, 1906), especie prejudicial para la agricultura en la región centro oeste de Brasil (orth. Acrididae, Gomphocerinae). *Bol. do Meseu national Zool.* **318**.

Castel, J. M. 1982 La poussée du criquet pèlerin en Afrique de l'ouest en 1980 *in* FAO Stat. rech. acrid. sur le terrain. Sér. techniques No. AGP/DL/TS/23, pp. 75–107. Rome: FAO.

Donnelly, U. 1947 Seasonal breeding and migrations of the desert locust (*Schistocerca gregaria* Forsk.) in West and North-West Africa. *Anti-Locust Mem.* **3**.

FAO 1988*a* *Report of an FAO/Multidonor mission on the feasibility of using large aircraft for controlling the Desert Locust (with special reference to Mauritania), 21–30 September 1988.* (Restricted internal FAO report.)

FAO 1988*b* *Report of meeting on Desert Locust Research 'Defining future research priorities', Rome, 18–20 October 1988.*

FAO 1988*c* *Report of the Meeting on the use and hazards of dieldrin in Desert Locust control, Rome, 21 October 1988.*

FAO 1989 *Report of the FAO Research Advisory Panel, 2–5 May 1989, Rome, Italy.*

Gunn, D. L. 1979 Strategies, systems, value judgements, and dieldrin in control of locust hoppers. *Phil. Trans. R. Soc. Lond.* B **287**, 429–445.

Gunn, D. L. & Rainey, R. C. (eds) 1979 *Strategy and tactics of control of migrant pests.* London: The Royal Society.

Pedgley, D. E. (ed) 1981 *Desert Locust Forecasting Manual.* (vols. 1 and 2). London: Centre for Overseas Pest Research.

Rainey, R. C. 1963 Meteorology and the migration of Desert Locusts *Tech. Notes Wld met. Org.* **54**.

Rainey, R. C. & Betts, E. 1979 Continuity in major populations of migrant pests: the Desert Locust and the African armyworm. *Phil. Trans. R. Soc. Lond.* B **287**, 359–374.

Roffey, J. 1979 Developments in the Desert Locust situation during 1976–79. *Phil. Trans. R. Soc. Lond.* B **287**; 479–488.

Roffey, J. 1982 The Desert Locust upsurge and its termination 1977–79. FAO Des. Loc. Field Res. Stat. Tech. Series No. AFP/DL/TS/23, pp. 1–74, Rome: FAO.

Roffey, J. & Popov, G. 1968 Environmental and behavioural processes in a Desert Locust outbreak. *Nature, Lond.* **219**, 446–450.

Roffey, J., Popov, G. B. & Hemming, C. F. 1970 Outbreaks and recession populations of the Desert Locust. *Bull. ent. Res.* **59**, 675–680.

Skaf, R. 1988 A story of a Disaster: why locust plagues are still possible. *Disasters* **12**, 122–126.

Skaf, R. 1990 The development of a new plague of the Desert Locust *Schistocerca gregaria* (Fôrskal) (orth. Acrididae) 1985–1989. *Proceedings of the fifth international meeting of the orthopterists' society, 17–20 July 1989, Valsain (Segovia), Spain.* (In the press.)

Steedman, A. (ed) 1988 *Locust handbook* (2nd edn). London: Overseas Development National Resources Institute.

US-FAO 1989 Key principles of ULV aerial application for Desert Locust control. *US-FAO technical consultative meeting. New Mexico State University, 6–7 November 1989.*

Uvarov, B. P. 1921 A revision of the genus *Locusta*, L (= *Pachytylus*, Fieb.) with a new theory as to the periodicity and migrations of locusts. *Bull. ent. Res.* **12**, 135–163.

Waloff, Z. 1966 The upsurges and recessions of the Desert Locust plague: an historical survey. *Anti-Locust Mem.* **8**.

Discussion

G. B. Popov (*FAO, Rome, Italy*). From what was said by Dr Skaf, I would like to highlight the following: the degree square monthly frequency maps prepared by FAO for the recession period 1963–1984 (excluding 1968, a plague year), when compared with those prepared by the Anti-Locust Research Centre for the 1939–1962 plague period, show marked differences; of particular significance is the fact that the major high-frequency spring breeding grounds in northern Africa and the Near East are not reached by recession populations. This is evidently because they fly by night and are thus subject to lower temperatures and different atmospheric conditions than the swarms that fly by day. The implications are far-reaching; we know from Waloff (1966), who studied the dynamics of plague upsurges and declines in great detail, that as a rule, major upsurges have followed a successful double-generation spring breeding, (let us point out *en passant* that this was yet again the case during the last plague upsurge, which found its impetus following 1987–88 winter-spring breeding, despite the massive control of over 5 million hectares). It follows that if we could prevent locusts from reaching the spring breeding grounds in numbers, this would help towards plague prevention.

It seems that we are now yet again in a recession period: what strategy should we adopt to prevent a recurrence of plagues in the future? The majority of us probably agree that plagues start from a successful build-up and gregarization of initially non-swarming low-density locust populations, a process that is furthered by sequences of favourable rainfalls in time and in space. There are basically two schools of thought as to how this process can be prevented from reaching plague proportions; the first advocates survey of the recession breeding areas and control of any populations that begin to gregarise. The second (on the grounds that many minor outbreaks are abortive and that swarms represent the most cost-effective control targets), would delay control operations until most of the populations are in swarms. At present the former is the strategy, but in practice, the second is often implemented.

Taking western Africa, of which I have first-hand field experience, as an example, the first strategy was effectively practised there by the regional locust control organization (OCLALAV) until the decline of that organization in the late seventies. During the 1966–81 sixteen-year period, control of gregarizing populations was conducted during 11 years, the scale varying from one to tens of thousands of hectares. This strategy aimed at culling gregarious and high-density gregarizing populations, leaving sometimes fairly large areas of low-density populations uncontrolled. But the latter, by virtue of their solitarious nocturnal flight behaviour failed to reach the high-frequency spring-breeding grounds, and remained confined to the recession area with the prospect of having to survive the adverse winter desert conditions. During the 1966–82 period this strategy effectively prevented locusts from reaching the spring breeding grounds in any numbers, except on two notable occasions: the first in October 1968, when swarms originating in the Central Region traversed the continent in little more than a week, reaching southern Morocco, where they were effectively controlled; and the second, in 1980, by which time OCLALAV had ceased to be an effective control organization and had failed to prevent swarm formation in the Mali and Niger outbreak areas and their escape to Algeria. There is surely a lesson to be learnt here.

The weakness of the second strategy, which advocates delaying control until swarms are formed, is that major population build-up may occur in areas where no control is possible. The 1986–88 plague upsurge provides a good example of this.

Reference

Waloff, Z. 1966 The upsurges and recessions of the Desert Locust plague: an historical survey. *Anti-Locust Mem.* **8**.

J. ROFFEY (*FAO, Rome, Italy*). I would like to draw attention to two issues: the suddenness with which the Desert Locust numbers increased and the qualitative changes associated with these increases. The first upsurge, which with hindsight we can see led to the plague, occurred at the end of the summer of 1985, the year when the drought of 1982–84 ended. In 1985 there were good rains which triggered off big increases of grasshoppers in the Sahel, of *Locusta migratoria* in Sudan and the first significant increases in Desert Locust numbers at both ends of the Sahel: in Mauritania and in the Red Sea Basin. Then in 1986, 1987 and 1988 there were good rains over much of the area infested with Desert Locusts. So by October 1988, with two generations during the winter-spring and two in the summer, there had been 12 or 13 generations and the populations had risen by about five orders of magnitude. There was a multiplication rate of about three per generation but in some areas it was much higher and in others lower. Some control was conducted against all these generations, but it was not until the last quarter of 1988 that such measures had a significant impact on the largest infestations. Dr Skaf has pointed out many of the reasons for this, which included the lack of security in Eritrea, in northern Chad, western Sudan, and the Western Sahara, and technical deficiencies in the application of the insecticide and the limitations imposed on the use of dieldrin. Equally important, however, were qualitative changes in the populations. These started off with the switch from night flying to day flying in late 1985, but by 1988 the locusts became much more vigorous than we expected standard gregarious locusts to be, as evidenced by a number of remarkable migrations. These included the eastward one from West Africa across to the Yemen during the late spring and early summer of 1988, and the migrations across the Atlantic Ocean to the Caribbean, with repeated invasions of the Cape Verde Islands, in October and November 1988. But there were also numbers of migrations northwards to the U.K., Italy, Greece and later in the year to Turkey, Syria and Lebanon from the southwest, that is, across the Mediterranean Sea. If you have insects whose performance can greatly exceed those of the normal animal, this means that forecasting during upsurge situations is going to be particularly difficult: it will be much more difficult to predict plague upsurges than it has been over the last two or three decades.

R. S. SCORER (*Imperial College, London, U.K.*). I well remember the occasion when Dr Rainey came to the Meteorology Department to try out his ideas about locust migration on me and my colleague Frank Ludlam. When we had spoken for a bit he suddenly stopped; and when we indicated that we awaited the controversial points he had warned us to expect, he said 'that's all'. We completely agreed with his thesis as being in accord with our ideas about air movements. The point of this remark is to show that I still (35 years later) do not understand the stubbornness of doubts among many non-physically oriented thinkers to recognizing the primacy of air motion for an airborne species.

I do not agree with Dr Rainey's suggestion in the opening remarks that a campaigning organization would be appropriate to make an assessment of the case for and against the use of dieldrin. An economist would probably be better equipped to assess the overall advantage of various anti-pest strategies. What we need is a widely supported case for dieldrin from a variety of scientific disciplines, clearly showing the quantitative advantages.

J. HEWITT (20 *Hartington Road, Chiswick, London, U.K.*) The real problem facing operations of any kind during years of recession is that governments, while responding to crises, never anticipate them.

Phil. Trans. R. Soc. Lond. B **328**, 539–553 (1990)

Printed in Great Britain

539

A migrant pest in the Sahel: the Senegalese grasshopper *Oedaleus senegalensis*

By R. A. Cheke

Overseas Development Natural Resources Institute, Central Avenue, Chatham Maritime, Chatham, Kent ME4 4TB, U.K.

The Senegalese grasshopper *Oedaleus senegalensis* is periodically a major pest of millet and other crops of subsistence agriculture in the Sahel zone of West Africa. Aspects of the species' biology are described. Eggs can survive several seasons and adults sometimes migrate up to 350 km per night, adaptations that contribute to the species' success in semi-arid areas. Evidence for migrations, both northwards with the S.W. monsoon and southwards with N.E. harmattan winds, is reviewed with particular reference to studies in Mali and Niger. Control strategies such as monitoring migrations and egg-laying to predict the sites of future outbreaks, possibly up to three years after heavy infestations, are considered.

1. Introduction

The Senegalese grasshopper *Oedaleus senegalensis* (Krauss) occurs in semi-arid grasslands in Africa, the Middle East and the Indian sub-continent (Batten 1969; Ritchie 1981; Popov 1988). The species is the most important grasshopper pest in the Sahelian zone of West Africa. This was not widely recognized until 1974 when, together with other grasshopper species, it infested 3500×10^3 ha† in West Africa and was responsible for the loss of 368000 tonnes of agricultural production (Bernardi 1986). *O. senegalensis* has recurred as a major pest, notably in 1975, 1977, 1985, 1986 and 1989. During 1986, insecticides were sprayed from aircraft and by ground applications on 3385500 ha of grasshopper-infested zones between Sénégal and Chad (FAO 1987a). The most intensive spraying, involving four DC-7 airplanes in Sénégal, Mali, Mauritania and the Gambia at a cost of 2700000 U.S. Dollars, was reported to have restricted crop losses to 5% (Walsh 1986), but this may be an over-optimistic estimate (N. D. Jago, personal communication).

With a preference for sandy soils, ideal for its egg-laying (see, for example, Cheke *et al.* 1980b), *O. senegalensis* feeds preferentially on grasses such as *Cenchrus biflorus* and *Aristida mutabilis* in savannas where these plants predominate. However, *O. senegalensis* is catholic in its choice of grass (Boys 1978) and often ravages millet and other crops grown by subsistence farmers (Cheke *et al.* 1980a; Centre for Overseas Pest Research (C.O.P.R.) 1982). Both nymphs and adults will consume the leaves and seed heads of millet plants, particularly when the latter are in the milky stage of grain formation. Attacks on seedlings, usually early in the season, may force farmers to sow again.

O. senegalensis can disperse over distances of up to at least 350 km during one night (Riley & Reynolds 1979) and is capable of surviving in the egg stage for at least three years (Fishpool & Cheke 1983). These adaptations for life in dry environments contribute to its success and pest status. This paper summarizes what is known of the species' ecology with emphasis on migrations within the Sahelian zone of West Africa and implications for control strategies.

† 1 hectare = 10^4 m².

[21]

2. Life cycle in West Africa

(a) Eggs

In West Africa nymphs can be found between March and November and adult *O. senegalensis* occur between April and December (Fishpool & Popov 1984). The intervening dry season is spent underground in the egg stage, with up to 45 and an average of about 25 eggs per pod (Launois-Luong 1979; Cheke *et al.* 1980*b*). Commonly, females lay twice, rarely three times, with five or six days between ovipositions (Launois-Luong 1979). The pods are usually aggregated, with densities of up to 13 pods m^{-2} (Popov 1980). The eggs may be attacked by a variety of predators, particularly larvae of Diptera (Bombyliidae) and Coleoptera (Histeridae, Meloidae and Tenebrionidae) (Cheke *et al.* 1980*b*; Popov 1980).

The time taken for the eggs to develop is complex, depending on the time of year, whether the eggs are in diapause or not and the rainfall. Not all the eggs in a pod will hatch with the first rains of the season. Laboratory experiments on the hatching of field-laid eggs showed clearly that the times taken to hatch eggs within a particular pod were highly variable, even when the whole pod was subjected to the same temperature (constant 30 °C) and humidity conditions. The shortest interval between a pod being moistened and the hatching of the first egg was 11 days (Cheke *et al.* 1980*a*), but the longest was 901 days (Fishpool & Cheke 1983). Hatching was protracted in all the pods studied, with a maximum of 1214 days (3 years and 4 months) elapsing between the hatching of the first and last eggs from the same pod (Fishpool & Cheke 1983). Thus, some eggs may not hatch until three wet seasons after they were first laid. There is also field evidence of viability after as long as five years (Saraiva 1962).

The pods used in the hatching experiments (Cheke *et al.* 1980*a*; Fishpool & Cheke 1983) had all been laid at the end of a wet season, but those laid at other times of the year may not include any diapausing eggs (Popov 1980). Nevertheless, these may also survive, quiescent, for three months in the absence of rain (Launois & Launois-Luong 1988). It is thought that decreasing day-length from September onwards, initiates the laying of diapausing eggs (Popov 1980). The variability in egg-laying and hatching behaviours is not only an adaptation for survival in habitats where the rainfall is erratic and patchy, but is also an insurance against isolated showers that are not followed by enough rain to maintain food supplies for the hoppers.

(b) Nymphs and adults

The nymphs can be serious pests, especially early in the wet season when they attack millet seedlings. There are five hopper instars, taking about three weeks to reach adulthood after eclosion, but as short a span as 15 days has been recorded in Mali in 1978 (Fishpool 1982) and in 1986 (N. D. Jago, personal communication). Accelerated rates of development contribute to increases in the intrinsic rate of natural increase of Desert Locusts *Schistocerca gregaria* (Forskål) (Cheke 1978) and may also play an important part in *O. senegalensis* outbreaks.

There have been reports of gregarious behaviour by *O. senegalensis* nymphs, including marching bands (Joyce 1952; Descamps 1953; Batten 1969; Popov 1988) whose members appear darker than solitary nymphs (Popov 1988; N. D. Jago, personal communication), but there is no unequivocal evidence for a phase change other than behavioural. Although both nymphs and adults can occur as either green or brown forms and a melanic form occurs in the Cape Verde Islands (Ritchie 1978), the proportion of each is seemingly under environmental control. Figure 1 shows how the percentage of brown adults at Watagouna (15° 15′ N,

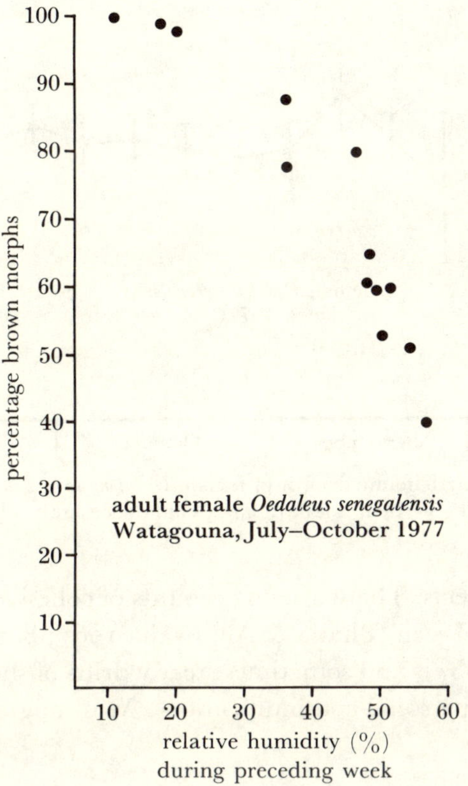

FIGURE 1. The percentage of brown morphs among collections of adult female *Oedaleus senegalensis* (sample sizes ranging from 9 to 1052) at Watagouna, Mali, during July–November 1977, in relation to the average relative humidity in the week preceding the collection (the averages of daily readings taken between 17h00. and 18h30.).

00° 43′ E) in Mali, during the 1977 season was inversely related to the humidity in the preceding week. This result, using field data, concurs with Rowell (1971) who said in his review of the subject that 'there is general agreement that humidity is the most important single factor in predisposing experimental populations in favour of the green morph.' However, it contrasts with a laboratory study in which only light and not temperature or humidity affected the colour of *O. senegalensis* (Abushama & El Khider 1973). Morphometric evidence for phase changes is also lacking. There was very little variation in the tegmen:femur ratio of adult females at Niamey, Niger (13° 31′ N, 02° 07′ E) during August–November 1977 (figure 2). Similar results were found for both tegmen:femur and tegmen:caput ratios at Danga (14° 31′ N, 01° 56′ E) in Niger in August–September 1974 during a serious outbreak of *O. senegalensis* (J. M. Ritchie, personal communication). However, Bhatia & Ahluwalia (1967) found a shift from a tegmen:femur ratio of 1.68 in an area where no concentrated breeding occurred to one of 1.73 where there was such breeding, but only 47 insects were measured.

After the final moult adult females take about ten days to reach maturity and become capable of egg-laying (Launois 1978; Fishpool 1982). There is evidence that if they are going to disperse most do so before maturing (Jago 1979; Riley & Reynolds 1983). Cheke *et al.* (1980*a*) found that females caught in light traps were more than three days old but many were six or more days old. Such samples consisted mostly of young adults, with immature ovaries, or females that had just laid, whereas samples netted in adjacent fields had more mixed age

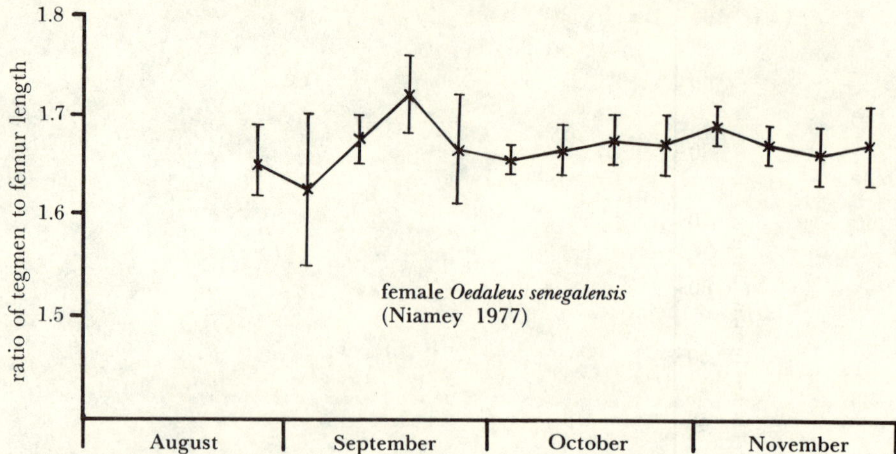

FIGURE 2. The ratio of tegmen length:femur length of female *Oedaleus senegalensis* caught at Niamey during August–November 1977. The vertical lines are 95% confidence limits about the mean values.

structures including gravid insects. There are no accounts of cohesive adult swarms, but flights of loose aggregations have been seen (Bhatia & Ahluwalia 1967; Batten 1969; Ahluwalia *et al.* 1976; Popov 1976; Lecoq 1978*a*), and sometimes steady drifts of short flights in one direction have been noted (N. D. Jago, personal communication). Most migrations probably take place at night.

3. MIGRATION

(a) *Evidence for northward movements with S.W. monsoon winds*

The evidence for northward movements, by which I mean any displacements north of the source, including those with N.E. or N.W. biases, is more circumstantial than is the case for southward migrations. Northward movements would be expected when the Inter-Tropical Convergence Zone (ITCZ) is north of the grasshoppers and the prevailing winds are southerlies or south-westerlies. Southward migrations including those with S.W. or S.E. biases, will tend to occur when the ITCZ is south of the insects and the prevailing winds are the dry, warm, generally N.E. 'Harmattan' winds. These are, of course, generalizations and events will often be determined by smaller scale phenomena such as storms and other disturbances.

There are three possible reasons for the paucity of data on northward movements. One is that they may not occur regularly. Another is that few intensive studies have been conducted at the beginning of the rainy seasons, when southerly winds predominate. Finally, adult populations will tend to be smaller and less easy to detect at this time. However Lecoq (1978*b*) noted that a population of up to 360 adults per hectare at Saria (12° 53′ N, 02° 19′ W) in Burkina Faso, on 17 June 1976 had all but disappeared three days later (with no evidence for a sudden mass mortality) and he inferred that they had emigrated north with the advancing rains. N. D. Jago (personal communication) observed a mass influx of grasshoppers, including *O. senegalensis*, at Mourdiah in Mali (14° 28′ N, 07° 28′ W) on 29–30 June 1986. The first heavy rains of the year (37 mm) also fell during the same night and there were no substantial populations north of Mourdiah, suggesting a movement from the south. The grasshoppers remained, at a density of 5 m^{-2}, for three days and left abruptly on 3 July during a night of further rainstorms. A similar northward movement, of about 50 km, was noted at Hombori (15° 16′ N, 01° 40′ W) in Mali during June 1985 (Popov 1988).

Reynolds & Riley (1988), by using radar, documented a north-eastward movement of grasshoppers, including *O. senegalensis*, on 10–11 October 1978 near Gao (16° 16′ N, 00° 03′ W) in Mali. This occurred when the ITCZ was north of the insects, and the prevailing winds were south-westerlies. Similar events were recorded on other nights (D. R. Reynolds, personal communication).

(b) *Evidence for southward movements with NE Harmattan winds*

In October 1974, there was evidence for a southward movement in Nigeria. Between 5 and 15 October, there were high densities of adult *O. senegalensis* at Maiduguri (11° 53′ N, 13° 16′ E) with a maximum noted on the night of 10 October. On 12 October they began to be recorded at latitude 11° 15′ N, 250 km further south (Popov 1976).

Lecoq (1978*c*) recorded an influx into his Saria study site in October 1975, which culminated with densities of 1000 adults per hectare. He concluded that these were immigrants as no soft-bodied juveniles were noted, only grasshoppers with hard integuments, and catches at light were abundant.

During the last week of September 1977 there were reports of heavy *O. senegalensis* infestations in the Maradi (13° 28′ N, 07° 06′ E) and Zinder (13° 48′ N, 08° 59′ E) regions of southern Niger and in northern Nigeria (Cheke *et al.* 1980*a*). At the time a line of light traps, supplemented by ground surveys, was maintained along the length of the Niger river from Gao, in Mali, southeast through Niger to Malanville (11° 52′ N, 03° 24′ E) in Benin. At one of these sites, Niamey, the trap catch reached 4416 *O. senegalensis* on the night of 3 October, when there was no interference from moonlight. This was the first catch of the season in excess of 1000 and presaged an influx of unprecedented proportions, culminating in a peak of 15030 on 13 October. Figure 3 shows the position of the ITCZ at mid-day on 1, 2, 3 and 4 October in relation to records of high-density occurrences of adult *O. senegalensis* during the period 2–5 October and areas where the species was known to cause damage to millet (*Pennisetum americanum*). These data provide good circumstantial evidence of grasshopper movements in relation to those of the ITCZ. Figure 4 shows a possible reason for the mass movement: during September the rainfall had been half the average in the most heavily infested areas, rendering the habitats inimical to *O. senegalensis*. The drought conditions, by forcing the grasshoppers out of desiccating grasslands into comparatively lush millet fields, may also have accounted for the crop damage.

The strongest evidence for *O. senegalensis* migrations is provided by the radar studies of Riley & Reynolds (1979, 1983). In the earlier paper they describe how they observed targets with wingbeat frequency 'signatures' corresponding to those of *O. senegalensis* flying S.W. over the 'middle Niger' area of Mali. Back trajectories showed that the insects were often travelling at least 50 km in a night and sometimes as far as 350 km. In 1978, two radars were deployed in the Gao area of Mali. On the night of 21 October a dense concentration of targets, with the characteristics typical of *O. senegalensis*, overflew one radar and was then detected only two hours later by the second radar positioned 100 km further south (Riley & Reynolds 1983).

Ritchie (1978) documented catches at sea up to at least 100 miles off the west African coast and concluded that in nearly all cases the insects had been blown offshore by easterly winds.

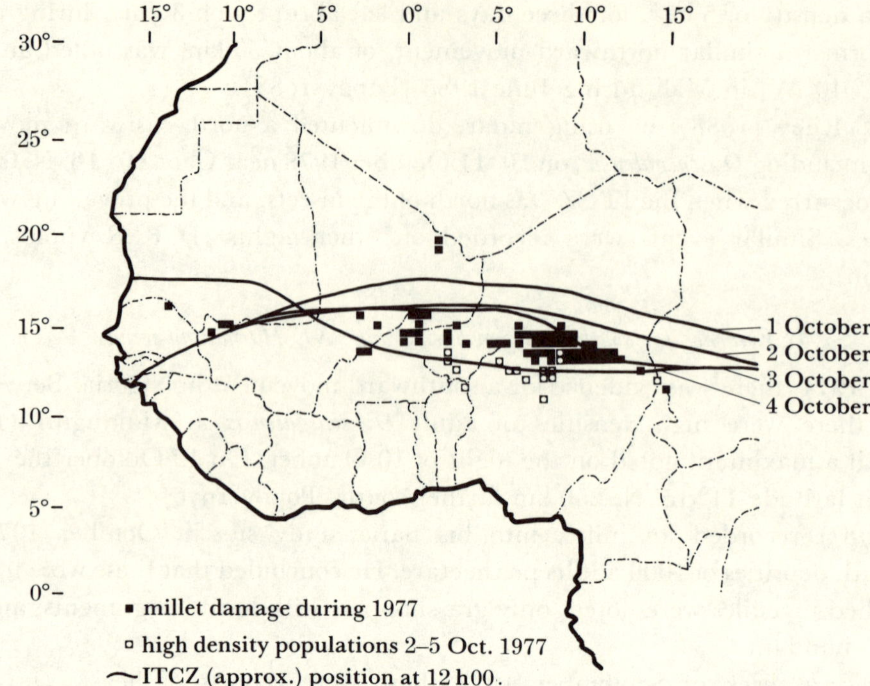

FIGURE 3. The position of the Inter-Tropical Convergence Zone at mid-day on 1, 2, 3 and 4 October 1977 and the occurrence of *Oedaleus senegalensis* in the region. Closed symbols, sites where millet was damaged by *Oedaleus senegalensis* during 1977; open symbols, sites where high density populations were suddenly recorded during 2–5 October 1977.

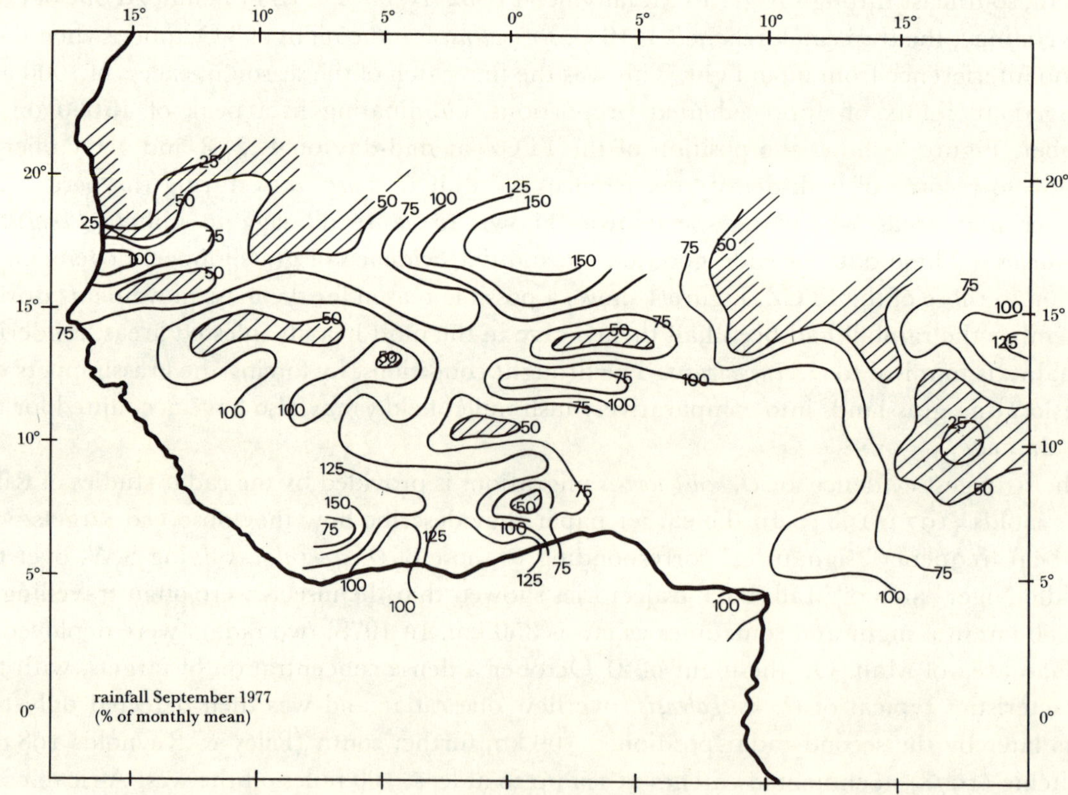

FIGURE 4. The rainfall in West Africa during September 1977 expressed as a percentage of the September monthly mean. (The shaded areas highlight zones with less than 50%.)

In exceptional circumstances *O. senegalensis* may be blown as far as the Cape Verde Islands, a distance of 600 km or more.

4. SIGNIFICANCE OF MIGRATIONS

Given a period of about six weeks to complete a life cycle in the wet season, three or four generations may occur in one season. Launois (1979), incorporating these figures, proposed a scheme whereby each of three 'generations' were associated with different ecological zones, according to rainfall and the movements of the ITCZ. He suggested that there was an 'area of initial multiplication' between the annual rainfall isohyets of 1000 and 750 mm where development occurs at the beginning (April and May) and end (November) of the wet season. Emigrants from the April and May populations fly north to the 750 to 500 mm isohyet belt (the 'transitional area of multiplication') where, during June and July, they mix with locally bred insects derived from eggs laid in that region in the previous season. These in turn provide more northward migrants which reinforce populations hatching in the 500 to 250 mm rainfall belt (the 'northern area of multiplication'), during August and September. Finally, there is a southward migration from August to October through all three zones, thus completing the cycle.

As a simple summary of aspects of the species' phenology, Launois' scheme has been followed by others (see, for example, Popov 1988) but it does not allow for events such as: (i) extensive hatching in any zone at any time, if rain falls on eggs laid in previous seasons: (ii) the vagaries of the ITCZ and mesoscale wind convergences that may lead to movements in directions other than those predicted and, (iii) up to three generations following each other successively in the same site or zone, as shown by Fishpool (1982) at Watagouna. Indeed Fishpool noted that in 1978, taking account of all his study area, there were four generations.

Predictive algorithms derived from Launois' scheme, which use indices to effect changes at ten-day intervals, exist (Arnaud *et al.* 1982; Launois 1979, 1984; Bernardi 1986; Launois & Launois-Luong 1988; Lecoq 1989). Environments are classified into up to 36 categories on the bases of photoperiod (day-length more or less than 12 h), mean temperature, humidity (ratio of actual to potential evapotranspiration) and a 0–5 index of vegetation turgidity. The responses of *O. senegalensis* eggs, nymphs or adults within a particular environment are scored according to a set of scaled indices representing speed of development, survival and 'success' (defined as the product of the other two indices). The score for a particular geographical area will depend on: (i) its environment category: (ii) a combination of current and historical values of rainfall, temperature, and potential evapotranspiration and (iii) a mean index for the *O. senegalensis* populations at the end of the preceding dry season. Migration is represented by populations being carried in or out of a zone dependent on whether the environment is or is not favourable to the grasshoppers becoming sedentary, according to a scale from 0 (no displacement) to 5 (maximum flight activity). Finally the simulation plots maps of areas with different degrees of risk in terms of densities of egg-pods, nymphs or adults. The densities are scaled from 1 to 5. For nymphs 1 is a negligible density (less than 0.1 per m^{-2}) and 5 is high (more than 30 per m^{-2}). Thus the simulations do not use quantitative data on the grasshoppers' biology, and the only meteorological information that they include (other than that embodied in the definition of environment categories) are details of rainfall, temperature and potential evapotranspiration. Nevertheless, it is claimed that predictions are correct in 70–80 % of cases when attempts have been made at validation (Launois & Launois-Luong 1988).

The accuracy of the computer simulations would be improved if they assessed more critically the effects of local meteorological conditions on grasshopper migrations, instead of depending, as they appear to do, on the notion of a shift in the species' 'centre of gravity' in line with movements of the ITCZ as each season progresses. Furthermore, they omit important biological attributes: for instance, Arnaud *et al.* (1982) ignored eggs surviving more than one season on the grounds that the experiments of Cheke *et al.* (1980a) were unrealistic. Arnaud *et al.* (1982) argued that it is normal to have an extended hatching period with a constant temperature of 30 °C, as the breaking of diapause with time is always more progressive and less synchronous than diapause broken by cold spells. However, many of the eggs studied by Cheke *et al.* (1980a), including one which did not hatch until 413 days after its first wetting, were collected in February 1978 and so would have experienced cold spells in the field. Furthermore, Venkatesh *et al.* (1972) conducted experiments at different temperatures and concluded that water was responsible for the termination of *O. senegalensis* diapause, irrespective of the temperature at which the eggs were kept. Launois & Launois-Luong (1988) acknowledged that diapausing eggs can survive a year or more in the complete absence of rain, but ignore the possibility that they may not hatch even when rained on: the laboratory egg experiments showed that even with an abundance of water, some eggs may not hatch for many months or years. The diapause has had applied importance: Saraiva (1962) described how there was a major outbreak on Boa Vista in the Cape Verde Islands in 1948 after five Senegalese grasshopper-free years; and an upsurge in Mali in 1985 and 1986 was ascribed to the accumulation of eggs over successive years (Steedman 1988). Although the contribution of diapausing eggs to outbreaks requires evaluation, it should not be ignored as it may be as significant as migration.

5. CONTROL

As both nymphs and adult *O. senegalensis* can damage crops, the ideal control strategy would be the elimination of nymphal populations before the grasshopper numbers build up excessively and long-distance movements take place. However, this would require a rapid response on an extensive scale. An alternative would be to predict where extensive hatching will take place on the basis of the knowledge of previous seasons' laying activities and current rainfall patterns. This approach has already been recommended (FAO 1987b) and is current practice in some areas (FAO 1987a). Even if influxes of grasshoppers at the end of seasons do not cause any damage, knowledge of the migrations of *O. senegalensis* is important so that places where large numbers of grasshoppers may have laid can be surveyed to aid the plans for subsequent control interventions.

It is not yet clear if high populations of *O. senegalensis* always require control, for strategic purposes, or whether they can be tolerated under some circumstances. Most of the mass migrations occur in October at the end of the season and in some cases this is after crops have already been harvested, so control of such populations could only be justified as a strategic measure to minimize egg-laying. There is some evidence, including the events of 1977 described above, to suggest that the grasshoppers move out of their preferred pastures into cultivated land when the former become excessively dry. It may also be only under such circumstances, when drought conditions have forced them to seek more amenable areas to feed or oviposit, that the grasshoppers migrate. Thus, there may be times when although high populations are thriving in grasslands they do not need controlling as they are not threatening crops. If this is often the case, then a good strategy would be to issue farmers with facilities such as ultra low volume spraying equipment or supplies of insecticide-impregnated bran for them to use as necessary, rather than the organization of mass spraying campaigns involving aircraft. However, when massive upsurges occur it is most unlikely that such small-scale control efforts could stem the tide of invading grasshoppers.

The possibilities for identifying aerial target populations, once migrations have started, by airborne radar are discussed elsewhere in this volume (Pedgley, this symposium; Riley & Reynolds, this symposium).

To date, control has relied on ground and aerial applications of insecticides. Henry *et al.* (1985*a*) reported that in trials of applying the microsporidian parasite *Nosema locustae* Canning, *O. senegalensis* took up the spores and is probably susceptible to infection by *Nosema*. Other more virulent pathogens have also been isolated from *O. senegalensis* (Henry *et al.* 1985*b*) and so there are possible alternatives to insecticide use, which merit further trials.

I thank Dr L. D. C. Fishpool, Dr N. D. Jago, Dr D. R. Reynolds and Dr J. M. Ritchie for their helpful discussions and suggestions for improving the manuscript. Collection of the 1977 field data depended on the Centre for Overseas Pest Research/Organisation Commune de Lutte Antiacridienne et de Lutte Antiaviaire (COPR/OCLALAV) Grasshopper project team, especially Dr L. D. C. Fishpool. I also thank Professor O. S. Bindra for the data from Nigeria.

References

Abushama, F. T. & El Khider, E. T. M. 1973 Colour change in the grasshopper *Oedaleus senegalensis* Krauss (Fam. Acrididae). *Z. angew. Ent.* **72**, 415–421.

Ahluwalia, P. J. S., Sikka, H. L. & Venkatesh, M. V. 1976 Behaviour of swarms of *Oedaleus senegalensis* Krauss Orthoptera: Acrididae-subfamily-Oedipodinae. *Indian J. Entomol.* **38**, 114–117.

Arnaud, M., Forest, F. & Launois, M. 1982 Automatisation d'un modèle écologique original propre à *Oedaleus senegalensis* (Krauss 1877) (Orthoptera, Acrididae). *Agron. trop.* **37**, 159–171.

Batten, A. 1969 The Senegalese grasshopper *Oedaleus senegalensis* Krauss. *J. appl. Ecol.* **6**, 27–45.

Bernardi, M. 1986 Le problème des sauteriaux. In Compte-Rendu du Séminaire International du Projet CILSS de Lutte Integrée, Niamey (Niger) 6–13 December 1984, pp. 43–57.

Bhatia, D. R. & Ahluwalia, P. J. S. 1967 *Oedaleus senegalensis* Krauss (Orthoptera: Acrididae-subfamily Oedipodinae) plague in Rajasthan (India). *Pl. Prot. Bull., New Delhi* **18**, 8–12.

Boys, H. A. 1978 Food selection by *Oedaleus senegalensis* (Acrididae: Orthoptera) in grassland and millet fields. *Entomologia expl. appl.* **24**, 278–286.

Cheke, R. A. 1978 Theoretical rates of increase of gregarious and solitarious populations of the desert locust. *Oecologia* **35**, 161–171.

Cheke, R. A., Fishpool, L. D. C. & Forrest, G. A. 1980*a* *Oedaleus senegalensis* (Krauss) (Orthoptera: Acrididae: Oedipodinae): an account of the 1977 outbreak in West Africa and notes on eclosion under laboratory conditions. *Acrida* **9**, 107–132.

Cheke, R. A., Fishpool, L. D. C. & Ritchie, J. M. 1980*b* An ecological study of the egg-pods of *Oedaleus senegalensis* (Krauss) (Orthoptera: Acrididae). *J. nat. Hist.* **14**, 363–371.

C.O.P.R. 1982 *The locust and grasshopper agricultural manual.* London: Centre for Overseas Pest Research.

Descamps, M. 1953 Observations relatives au criquet migrateur Africain et à quelques autres espèces d'Acrididae du nord Cameroun. *Agron. Trop.* **6**, 576–614.

F.A.O. 1987*a* Migrant Pest Newsletter (nos. 51–52), 1987. Rome: FAO.

F.A.O. 1987*b* *FAO guidelines for egg-pod surveys with special reference to the Senegalese grasshopper Oedaleus senegalensis* (Kr.), *1987.* Rome: FAO.

Fishpool, L. D. C. 1982 The ecology of the West African grasshopper fauna of the Sahel and sudan savanna with special reference to *Oedaleus senegalensis* (Krauss 1877). Ph.D. thesis, University of London.

Fishpool, L. D. C. & Cheke, R. A. 1983 Protracted eclosion and viability of *Oedaleus senegalensis* (Krauss) eggs (Orthoptera, Acrididae). *Entomologists mon. Mag.* **119**, 215–219.

Fishpool, L. D. C. & Popov, G. B. 1984 The grasshopper faunas of the savannas of Mali, Niger, Benin and Togo. *Bull. Inst. fondam. Afr. noire* **43**, 275–410.

Henry, J. E., Fowler, J. L., Wilson, M. C. & Onsager, J. A. 1985*a* Infection of West African grasshoppers with *Nosema locustae* Canning (Protozoa: Microsporida: Nosematidae). *Trop. Pest Mgmt*, **31**, 144–147.

Henry, J. E., Wilson, M. C., Oma, E. A. & Fowler, J. L. 1985*b* Pathogenic micro-organisms isolated from West African grasshoppers (Orthoptera: Acrididae). *Trop. Pest Mgmt*, **31**, 192–195.

Jago, N. D. 1979 Light trap sampling of the grasshopper *Oedaleus senegalensis* (Krauss, 1877) (Acrididae; Oedipodinae) and other species in West Africa: a critique. *Proc. 2nd. Trienn. Meet. (21–25 July 1979)*, pp. 165–198. Pan American Acridological Society, Bozeman, Montana, U.S.A.

Joyce, R. J. V. 1952 The ecology of grasshoppers in East Central Sudan. *Anti-Locust Bull.* **11**.

Launois, M. 1978 Modélisation écologique et simulation opérationelle en Acridologie. Application à *Oedaleus senegalensis* (Krauss, 1877). *Ministère de la Cooperation, Paris*.

Launois, M. 1979 An ecological model for the study of the grasshopper *Oedaleus senegalensis* in west Africa. *Phil. Trans. R. Soc. Lond.* B **287**, 345–355.

Launois, M. 1984 Les bio-modèles à géométrie variable appliqués à la surveillance des criquets ravageurs. *Agron. trop.* **39**, 269–274.

Launois, M. & Launois-Luong, M. H. 1988 *Oedaleus senegalensis* (Krauss, 1877) sauteriau ravageur du Sahel. Collection Acridologie Opérationelle. 3. *Ministère des Affaires Etrangères des Pays-Bas, La Haye et CIRAD/PRIFAS, Montpellier*.

Launois-Luong, M. H. 1979 Etude de la production des oeufs d'*Oedaleus senegalensis* (Krauss) au Niger (Région de Maradi). *Bull. Inst. fondam. Afr. noire* **41**, 128–148.

Lecoq, M. 1978*a* Le problème sauteriaux en Afrique soudano-sahélienne. *Agron. trop.* **33**, 241–257.

Lecoq, M. 1978*b* Les déplacements par vol à grande distance chez les Acridiens des zones Sahélienne et soudanienne en Afrique de l'Ouest. *C. r. Acad-Sci., Paris* (D) **286**, 419–422.

Lecoq, M. 1978*c* Biologie et dynamique d' un peuplement acridien de zone soudanienne en Afrique de l' Ouest (Orthoptera, Acrididae). *Annls. Soc. ent. Fr.* (N.S.) **14**, 603–681.

Lecoq, M. 1989 Les biomodèles en Acridologie et leurs applications opérationnelles. In *Report of the meeting on desert locust research: defining future research priorities*, pp. 82–99. Rome: FAO.

Popov, G. B. 1976 The 1974 outbreak of grasshoppers in Western Africa. In *Report of the Sahel Crop Pest Management Conference, held at the invitation of the Agency for International Development, United States Department of State, Washington DC, USA, December 11–12, 1974* (ed. R. F. Smith & D. E. Schlegel), pp. 35–43. Berkeley: University of California (for United States Agency for International Development).

Popov, G. B. 1980 Studies on oviposition, egg development and mortality in *Oedaleus senegalensis* (Krauss), (Orthoptera, Acridoidea) in the Sahel. Centre for Overseas Pest Research. *Misc. Rep.* **53**.

Popov, G. B. 1988 Sahelian grasshoppers. *ODNRI Bull.* **5**.

Reynolds, D. R. & Riley, J. R. 1988 A migration of grasshoppers, particularly *Diabolocatantops axillaris* (Thunberg) (Orthoptera: Acrididae), in the West African Sahel. *Bull. ent. Res.* **78**, 251–271.

Riley, J. R. & Reynolds, D. R. 1979 Radar-based studies of the migratory flight of grasshoppers in the middle Niger area of Mali. *Proc. R. Soc. Lond.* B **204**, 67–82.

Riley, J. R. & Reynolds, D. R. 1983 A long-range migration of grasshoppers observed in the Sahelian zone of Mali by two radars. *J. Anim. Ecol.* **52**, 167–183.

Ritchie, J. M. 1978 Melanism in *Oedaleus senegalensis* and other Oedipodines (Orthoptera, Acrididae). *J. nat. Hist.* **12**, 153–162.

Ritchie, J. M. 1981 A taxonomic revision of the genus *Oedaleus* Fieber (Orthoptera: Acrididae). *Bull. Br. Mus. nat. Hist.* (Ent.) **42**, 83–183.

Rowell, C. H. F. 1971 The variable coloration of the acridoid grasshoppers. *Adv. Insect. Physiol.* **8**, 146–198.

Saraiva, A. C. 1962 Plague locusts – *Oedaleus senegalensis* (Krauss) and *Schistocerca gregaria* (Fôrskal) in the Cape Verde Islands. (In Portuguese with English summary.) *Estudos agron.* **3**, 61–89.

Steedman, A. (ed). 1988 Locust handbook (2nd. edn). London: ODNRI.

Venkatesh, M. V., Ahluwalia, P. J. S. & Harjai, S. C. 1972 Influence of rainfall on the egg diapause of *Oedaleus senegalensis* Krauss (Orthoptera, Acrididae). *Pl. Prot. Bull., New Delhi* **23**, 20–27.

Walsh, J. 1986 Grasshopper control program successful. *Science, Wash.* **234**, 815–816.

Discussion

N. D. JAGO (*Overseas Development Natural Resources Institute, Chatham, U.K.*). The following comments reinforce the excellent paper by Dr Cheke. Two subject areas are of interest, the first biological, the second agro-economic and logistic and focuses on the estimated crop loss caused by *Oedaleus senegalensis* and other economically important Sahelian acridids.

ODNRI is well placed to comment because the Overseas Development Administration (ODA) has supported an ODNRI, farmer level, millet crop protection project in northwest

Mali since 1985. As leader of the programme, I have had a unique opportunity to observe fluctuations in grasshopper pest species during the period, starting with the massive outbreaks of *O. senegalensis* and *Kraussaria angulifera* (Krauss) in 1985 and 1986 and culminating in the truly remarkable and uniquely extensive outbreaks of no less than nine species in 1989. In rank order of pest status these were:

1. *Oedaleus senegalensis*, 2. *K. angulifera*, 3. *Hieroglyphus daganensis* (Krauss), 4. *Diabolocatantops axillaris* (Thunberg), 5. *Cataloipus cymbiferus†* (Krauss), 6. *Kraussella amabile†* (Krauss), 7. *Ornithacris cavroisi†* (Finot), 8. *Cryptocatantops haemorrhoidalis* (Krauss), and 9. *Aiolopus simulatrix‡* (Walker).

Biological observations: migration

An example of conclusive evidence of northward movement occurred on 29–30 June 1986. Early and precocious rains between 22–27 May at latitude 13° 30′ N had caused hatching of *O. senegalensis* in an east-west belt of over 100000 ha, with hoppers at a density of 5 m^{-2}. These fledged some three weeks later. On 29–30 June a massive overnight arrival of young adults, associated with 37 mm of rain, took place at Madina Kagoro, a village at 14° 20′ N, just south of the project base at Mourdiah (14° 28′ N, 07° 28′ W), in an area which had hitherto received only light rain (7 mm). The fields showed some millet germination, but no previous hatching of *O. senegalensis*. The project mobilized a field survey and discovered an area of cultivated millet of some 200 ha covered with young adult insects, at an estimated density of 5 m^{-2}. Three days later, and before control could be mobilized, the insects disappeared during a night on which heavy rain fell. The Mourdiah region was once again empty.

The heavy rain led to wholesale hatching over the Mourdiah region. Eventually, intensive survey showed that 320 km^2 was occupied by hoppers, which formed marching bands and exhibited a tendency to grey-brown/black cuticular colour, in spite of verdent and abundant wild vegetation. Of the potential cultivated millet area, 25% was destroyed despite repeated attempts at replanting. Fledging took place at the end of July, and all the *O. senegalensis* left the Mourdiah area between 28 July and 30 August, leaving devastation behind them. From then till late September, areas south of 14° 50′ N were virtually free of the species.

Evidence for massive southward movement of *O. senegalensis* was obtained in September 1989. Between 5 August and 9 September, project areas south of latitude 15° 00′ N were completely overwhelmed by 6 major grasshopper pest species (see species 2–5 listed above). At this stage *O. senegalensis* was not included. After 11 September, north of 15° 00′ N, the fields and savanna were found to be filled with *O. senegalensis*, as fifth instar hoppers and young adults. By 14 September crop damage had started to accelerate and young adults were seen to be making local diurnal movements. On 19 September a massive influx of young adults occurred overnight at Nara (15° 10′ N, 07° 17′ W).

On 21 September a southward daylight displacement took place at a village 7 km west of Dilli (15° 00′ N, 07° 40′ W) (J. Legg, personal communication). The passage of flying adults took 1.5 h and started at 11h30. The movement took place on a front of at least 7 km, because it passed simultaneously through Dilli. At Mourdiah, 70 km further south and in the area devoid of the species since early August, young adults started coming to the light trap in small numbers during 17–18 September, numbers rising to gigantic proportions on the nights of

† Species that became major pests in Mali for the first time in 1989.
‡ Usually major pests in crops on heavy clay soils, hence of less importance to millet than to sorghum.

26–29 September and declining to 3 October. This set of observations represents evidence for a southward displacement of 70 km in about five days.

Observations on crop damage

Grasshoppers afford a much greater long-term and chronic menace to agriculture than do the classic locusts. The outbreaks of 1986 saw at least 2850 km² of the Nara area infested in mid-September with a mixture of *Kraussaria angulifera* and *Oedaleus senegalensis*. North of 15° 00′ N populations were almost exclusively *O. senegalensis*. Hoppers and adults formed densities of 90000–160000 ha⁻¹ in the south, but *O. senegalensis* hoppers were commonly found at 170000–400000 ha⁻¹ in the north. The scale of crop loss was probably on the same scale as that experienced in 1989, though over a smaller area. In addition, the USAID-sponsored aerial control came too late to protect northwestern Mali, though their helicopters were later able to tackle a part of an 800 km² area infested north of 15° 00′ N.

The crop loss incurred in the 10000 km² of infestations in northwestern Mali in August–October 1989 has been carefully studied by the ODA Project. This is the first time that a detailed assessment of crop loss due to grasshoppers has been made. Analysis of results is not yet complete, but a preliminary study shows that 5.7 % of the farmer population lost 70–90 % of their millet (mean loss 85 %) (C. Lock, personal communication). These losses occurred in the 13 worst-hit villages under study. In a separate study at Mamaribougou Ferribé (14° 56′ N, 07° 46′ W) by J. Legg (personal communication), results showed that farmers lost 24 % of the millet crop to grasshoppers. In the same fields the millet head miner *Heliocheilus albipunctella* Joannis caterpillar had fallen to very low levels compared with 1988, so that very little crop loss can be attributed to that pest. C. West (personal communication) estimates that the 280 pilot farmers over the region as a whole (latitudes 14° 00′ N to 15° 30′ N) lost 18 % of the millet to grasshoppers, confirming the impression that losses were greater toward the Mauritania border and north of latitude 15° 00′ N.

Concerning estimates of crop loss (see, for example, Walsh 1986), firstly, the area lost to agriculture because of early season attack is usually ignored. Secondly, harvest losses are more serious than they first appear because they are percentage losses on a harvest which, without grasshopper attack, is already inadequate for the needs of the family for one year. Thus, farmer families can be roughly divided into 90 % of small to medium size and poorer families (less than 29 members) and 10 % of richer, larger families (29 or more members). Lock *et al.* (1988, unpublished report) have shown that poorer families produce only 7.7 to 8.6 months supply in years of good rains, the equivalent for larger families being 11.2. I have calculated that taken overall, the harvest will be around 8.5 months supply without major grasshopper attack, but roughly 7.0 months supply when good rains are accompanied by massive losses generated by grasshoppers. This is on a scale which, if repeated for a number of years, could not be compensated for by the economic strategies currently open to the farmers. Last year (1988), for example, C. Lock (personal communication) discovered that 10 villages in the Dilli 'arrondisement' had incurred 70–95 % millet crop losses due principally to grasshoppers, with an estimated value to the average farm family of between 150000 and 200000 F.CFA (£300–£400). As a result, up to a quarter of families had left their village in search of millet and work elsewhere.

The value of crops lost interacts directly with the problem of the cost of the treatments available for grasshopper control. ODA, for example, is currently determined that any method of crop-yield improvement or crop-loss reduction should, as far as possible, pay for itself. The benefit:cost ratio could be greatly improved by avoiding waste. Pesticides are very expensive, yet for much of the time, when used against grasshoppers such as *O. senegalensis* attacking subsistence level agriculture, they are ineptly applied. Difficulties of aircraft navigation contribute to failure, even though the technology exists to avoid this, and personnel, including pilots, are often inexperienced. Electronic and other equipment needs to be made more suitable for field use in the tropics. This also applies to pesticide formulations for aerial and ground applications.

J. M. RITCHIE (*Overseas Development Natural Resources Institute, Chatham, U.K.*). Dr Cheke has rightly drawn attention to the anecdotal nature of much of our information on the biology of *Oedaleus senegalensis*. By analogy to phase changes in locusts we accept that nymphal colouration changes do occur in *Oedaleus* in response to crowding. However, research is needed to clarify the effect of crowding on the biology of *Oedaleus* and, in particular, the conditions that cause band formation and the development of the characteristic brown and orange colour of 'gregarious' nymphs. Similarly, we know little of the degree of interaction between adults in dense populations and the extent to which their mobility and their fertility are affected by density as they are, for example, in the Desert Locust.

Further morphometric research is needed, employing multivariate techniques, to analyse a range of character measurements by using large samples from populations of known density, if we are to establish whether *O. senegalensis* merely shows allometry in certain characters in relation to overall size of the individual or whether, independent of size, there is a genuine separation into two distinct morphological phases, as in the major locust species.

Dr Cheke has emphasized that advance warning of potentially large nymphal populations at the start of the season can be provided by egg pod surveys. These surveys are greatly speeded up and improved by soliciting the observations of the farmers themselves on the extent and position of egg-laying by females towards the end of the previous rainy season.

Information from such surveys needs to be combined with new knowledge on the degree of hatching response of *Oedaleus* eggs to differing amounts of rainfall, building on the work of Dr Cheke and his colleagues. Improved coordination of rainfall data collection and processing in the Sahel would show likely areas for major concentration of egg hatching year by year. Early indications of rainfall can be obtained by the use of satellite imagery to detect changes in cloud temperature associated with precipitation and subsequent greenness resulting from the flush of new vegetation. Currently, ODNRI is planning to carry out further work on the inception and breaking of diapause in *Oedaleus*. The results of this study will then be combined with meteorological and egg-pod survey information to improve the forecasting of outbreaks.

The use of various pathogenic microorganisms to control *Oedaleus* has been touched on by Dr Cheke. The Commonwealth Agricultural Bureau International Institute of Biological Control is beginning a programme of research on pathogenic fungi, especially *Beauveria* and *Metarhizium*, which is targeted against the Desert Locust, but which might be equally applicable to *Oedaleus*. These pathogens are not released into the environment in the hope that they will become permanently established. Some strains are probably already present in areas occupied by the target insects. Instead, the aim would be to augment the natural level of infection by application of pathogens as biopesticides, sprayed onto the target as droplets in a

R. A. CHEKE

carrier oil or mixed with baits. Pathogens therefore suffer from the same logistic problems of movement and storage as conventional pesticides, with the added problem of increased sensitivity to sunlight and high temperatures.

In *Oedaleus* most damage is done by final instar nymphs and young adults. The use of biocontrol will therefore require very precise monitoring of population build-up at the start of the rains so as to target young first-generation nymphs while they are still relatively immobile and before they cause damage. The delayed knock-down of pathogens compared with pesticides may make them initially unattractive to control personnel and farmers who are used to being able to verify the efficacy of treatment immediately.

L. D. C. FISHPOOL (*Overseas Development Natural Resources Institute, Chatham, U.K.*). I agree with Dr Cheke that existing models of the phenology of *Oedaleus senegalensis* populations fail to pay sufficient regard to the joint phenomena of egg quiescence and egg diapause. The picture of progressive northward displacements early in the wet season in response to the advancing rains, resulting in a more or less complete evacuation of the southerly source areas as the wet season advances, to be followed by complementary north to south return movements at the end of the rains, abandoning those opportunistically exploited more northerly latitudes as conditions cease to be favourable, is too simplistic. The implied quasi-total desertion by *O. senegalensis* of parts of its range at different times of the wet season is, I believe, more apparent than real. Significant populations of *O. senegalensis* eggs almost certainly occur throughout its range at all times of the year. While these populations may not be very large in relative terms, they are none the less important and need to be considered when attempting to interpret and describe the life cycle of *O. senegalensis*.

One consequence of them is that they call into question the concept of strategic control of *O. senegalensis*. Thus control of early season adult or nymphal populations in the south with a view to preventing or significantly reducing their subsequent recolonization of more northerly latitudes is unrealistic.

Given the vagaries of rainfall in the region in timing, amount and distribution, a life-cycle strategy involving an extended, but variable egg diapause by part of the population, makes sound evolutionary sense. It is almost literally a case of not putting all your eggs in one basket.

R. C. RAINEY, F.R.S. (*Elmslea, Old Risborough Road, Stoke Mandeville, Bucks. U.K.*). Our own immediate interest in *O. senegalensis* is the evidence from detailed field observations of a direct effect of the Inter-tropical Convergence Zone on the distribution and structure of grasshopper concentrations. The evidence from radar offsets the absence of such observations on the Desert Locust from the Sahel region. During periods of massive grasshopper flights, the Airborne Radar (ABR) system might be a useful supplement to other monitoring methods.

P. DARLING (*Agricultural Economist, 46a Ophir Road, Bournemouth, U.K.*). From what I have seen of semi-arid Africa, pastoralism is the most important means of economic livelihood; yet, in nearly all reports of locust and grasshopper damage (including this talk), the emphasis is firmly centred on damage to crops belonging to sedentary farmers. As far as I know, very little or no research has been done to assess the impact of grasshoppers on forage and browse. Some implications of this are that research on migratory pests is creating, confirming or conforming to the excessive attention paid by many national governments to the needs of cultivators at the expense of pastoralists. Surely, those conducting research on migratory pests should seriously consider expanding their terms of reference or work towards redressing this imbalance in some way? Spraying rangelands may be a more effective means of achieving control – it covers outbreak areas more thoroughly and it is easier to spray effective lines across extensive rangeland than limited cropland (though more widespread spraying may expedite the treadmill effect of strains becoming more resistant to sprays). Many past economic cost-benefit assessments of control measures in semi-arid areas must be totally inadequate for, by excluding the economic (and environmental) costs and benefits of control measures on pasturelands, they are often omitting the paramount form of economic (and environmental) activity in the areas being studied.

R. A. CHEKE. There have been studies showing that over-grazing enhances grasshopper populations in the tropics (Merton 1959; Roffey 1970; Parihar 1981) and others with equivocal results (Amatobi *et al.* 1988) but, like Dr Darling, I know of no work devoted to the effects of grasshoppers or locusts on pastoralists in Africa. Such investigations would be difficult since a *sine qua non* of nomadic life is to migrate away from poor pastures, be they induced by insects or by drought. Nevertheless, I agree that studies would be worthwhile as both N. D. Jago (personal communication) and I have heard reports of grasshoppers destroying forage for cattle and camels. The extent of the depredations that grasshoppers inflict on grasslands, which we have both witnessed, must influence the well-being of shifting herds. Odiyo (1979) refers to a case where an infestation of armyworms (*Spodoptera exempta* (Walk.)) covering 65 km^{-2} of the Athi plains in Kenya, at a mean density of 28 sixth-instar larvae per square metre, were feeding at a rate equivalent to that of about 8000 cattle!

Regarding the spraying of rangelands, I urge caution as this might exacerbate grasshopper outbreaks by destroying too many of their natural enemies.

References

Amatobi, C. I., Apeji, S. A. & Oyidi, O. 1988 Effects of farming practices on populations of two grasshopper pests (*Kraussaria angulifera* Krauss and *Oedaleus senegalensis* (Krauss) (Orthoptera: Acrididae) in northern Nigeria. *Trop. Pest. Mgmt.* **34**, 173–179.

Merton, L. F. H. 1959 Studies in the ecology of the Moroccan Locust (*Dociostaurus maroccanus* Thunberg) in Cyprus. *Anti-Locust Bull.* **34**.

Odiyo, P. O. 1979 Forecasting infestations of a migrant pest: the African armyworm *Spodoptera exempta* (Walk.). *Phil. Trans. R. Soc. Lond.* B **287**, 403–413.

Parihar, D. R. 1981 Effects of over grazing on grasshopper population in the grasslands of Rajasthan desert. *Ann. Arid Zone* **20**, 291–294.

Roffey, J. 1970 *The effects of changing land use on locusts and grasshoppers* (ed. C. F. Hemming & T. H. C. Taylor), pp. 199–204. *Proc. Internat. Study Conf.* London: Centre for Overseas Pest Research.

Phil. Trans. R. Soc. Lond. B **328**, 555–569 (1990)

Printed in Great Britain

Progress and developments in forecasting outbreaks of the African armyworm, a migrant moth

By P. O. Odiyo

Desert Locust Control Organization for Eastern Africa and Kenya Agricultural Research Institute, P.O. Box 30023, Nairobi, Kenya

This paper describes the development of the forecasting service since 1977 by highlighting the results of various attempts that have been directed at elucidating the origins of primary populations of *Spodoptera exempta* over eastern Africa. The dates between the last outbreak of each season and the first outbreak of the next season (off-season period) provided a mean number of days for estimating dates of potential onset of primary outbreaks. The development of computer-based data processing routines (database management) and of an expert system for forecasting the occurrence of outbreaks in East Africa are described, as examples of decision-support tools in pest management.

1. Introduction

The African armyworm *Spodoptera exempta* (Walk.) (Lepidoptera: Noctuidae) is a major pest of cereal crops, sugar cane and pasture grasses in Africa south of the Sahara and in southwestern Arabia. At times it causes damage comparable with that of locusts: for example, 5% loss of maize, wheat and rice in Kenya and Tanzania alone can be worth some U.S. \$50 million and on rangeland, over one square kilometre, an infestation at 100 larvae per square metre can consume as much vegetation as 100 head of cattle (Odiyo 1979); in the Yemen Arab Republic up to 11 000 ha† of intensively cultivated areas were treated with chemicals in 1974 and another 60 000 ha in 1984, at considerable cost (Moharram & Nasseh 1989).

The moths migrate at night for tens or even hundreds of kilometres from their emergence site to the next breeding area (Rose *et al.* 1985), on winds commonly over 3–4 m s^{-1} (11–14 km h^{-1}). This is likely to bring them into zones of wind convergence (and therefore areas of potential rainfall) such as the African Rift Convergence Zone (ARCZ), or the Inter-Tropical Convergence Zone (ITCZ) (Brown *et al.* 1969; Haggis 1971, 1979). At the beginning of the rains, the localized wind convergence associated with storm outflows directly affects the distribution of moths and subsequent outbreaks (Tucker & Pedgley 1983). In the arrival area, females mate before taking up water or dew, ovipositing the next night and this leads to the development of new armyworm outbreaks (Page 1988).

The geographical distribution of outbreaks follows a seasonal pattern, with contemporary but different populations, one generally moving southwards from November or December from Malawi, Mozambique, Zambia or Zimbabwe across southern Africa, and the other moving northwards from Tanzania to affect the whole of East Africa (Uganda, Kenya and Tanzania) and countries to the north as far as the Yemen Arab Republic (Brown *et al.* 1969; Blair & Catling 1974; Haggis 1984, 1986). In East Africa, outbreak seasons usually cover the period November–December till May–June, while June–July to October–November represents the

† 1 hectare = 10^4 m^2.

'off-season' period, when no outbreaks are reported; there are often infestations in Ethiopia, Somalia and southwestern Arabia during these months (Brown *et al.* 1969; Haggis 1984, 1986).

In some years there is evidence that the early outbreaks in southwestern Tanzania were produced by moths originating in Zambia and brought eastwards by a movement of the ARCZ (Odiyo, in Rainey 1979; Rainey & Joyce 1990). More frequently, the first outbreaks of the season in East Africa are reported in east-central Tanzania and eastern Kenya (Odiyo 1979, 1981; Tucker *et al.* 1982). The sources of the moths that produce them was long debated (Rose 1979); the current view is that they generally come from scattered populations that survive the dry season in favoured localities in eastern Kenya and eastern Tanzania, probably near the coast (Tucker 1984*b*), although in some years there is evidence that they were produced by moths originating in neighbouring countries (Brown *et al.* 1969; Tucker 1984*b*; Rainey & Joyce, this symposium).

Outbreaks for which the parent moths cannot be traced to previously known outbreaks are now termed 'primary outbreaks' (Rose 1984; Rose *et al.* 1987; Pedgley *et al.* 1989); these outbreaks are usually small and scattered. Moths that develop from the 'primary' sources spread out over increasing areas with subsequent generations, producing series of secondary outbreaks (Odiyo 1975, 1981; Pedgley *et al.* 1989); as these are often more serious and expensive to control (Rose 1984), the concept of 'strategic control' has been proposed (Rose *et al.* 1987) and is being evaluated.

In the past, the sudden appearance of outbreaks could take farmers by surprise and the larvae were often in their late instars before control could be organized (Brown & Odiyo 1968). Forecasting where and when infestations might occur was seen as a key contribution to overcoming many of the problems of timely control (Brown 1970). The development in East Africa of the armyworm forecasting service from the trial period of 1969–70 to the full-scale permanent service since the 1970–71 season has been reported (Odiyo 1972, 1979; Betts 1976). While much progress has been made in understanding moth migration and behaviour, the main gap in knowledge remains the carry-over from one season to the next. Studies which could facilitate timely prediction and early detection of primary outbreaks could therefore greatly improve the capacity for effective control of *S. exempta* larvae in eastern Africa.

This paper briefly reviews progress in forecasting since 1977 (Odiyo 1979; Rose 1979), and introduces the data base and expert system, which are being developed as potentially useful forecasting decision-support tools.

2. Advances in forecasting

Early evidence of an association between armyworm outbreaks and storms (Rose & Law 1976) and of severe outbreaks following drought (Hattingh 1941; Brown 1962) has been followed up in studies on East African data.

Tucker & Pedgley (1983) studied the most likely egg-laying dates at 53 outbreak sites in East Africa in the 1973–74 and 1974–75 seasons in relation to the occurrence of rainstorms (20 mm or more a day), and found that in these seasons mass egg-laying by moths was more closely associated with the occurrence of rainstorms during the drier period of the season (January–March) than in April–May, when the long rains usually occur. Real-time monitoring of the distribution of very cold clouds by Meteosat satellite imagery is now being used for the third season to provide evidence of convective weather activity for pin-pointing potential breeding areas early in the season.

From studies of records of outbreaks in eight areas of Kenya and Tanzania and rainfall in October–December in 21 seasons (1960–61 to 1980–81), Tucker (1984*a*) also found an inverse relationship between the numbers of armyworm outbreaks and early season rainfall. This tendency for a bad armyworm season to follow low rainfall in October–December, and *vice versa*, has offered the possibility for a seasonal forecast to be issued early in the calendar year.

Monthly summaries of armyworm occurrences in administrative units (Districts and Provinces) in Tanzania, Kenya and Uganda for 20 years (1960–1979), showed that the onset and durations of infestations in each country tend to follow certain geographical patterns, which can be monitored through the use of insect traps and data from Agricultural Extension services (Odiyo 1984). To compare events occurring between the off-season months of July–October and the start of outbreak seasons from November–December, Odiyo (1981) studied occurrences of *S. exempta* larvae for 19 years (1961–1979) and records of moth catches between southern Kenya and east-central Tanzania (1–8° S and 34–41° E) for 16 years (1963–1978) and concluded that moths emerging from low density and scattered larvae in October–November in southern Kenya and northern Tanzania had probably been spreading and breeding over increasing areas of Kenya and Tanzania.

In studying the redistribution of armyworm populations in East Africa during the 1973–74 and 1974–75 seasons, Tucker *et al.* (1982) compared wind trajectories with known outbreaks before and after nights of sudden increase in moth catch, and concluded that some outbreaks, especially early in the season, were probably derived from unreported sources, which in these seasons were most likely to have been in eastern Kenya or eastern Tanzania.

Areas where the first outbreaks of the season most often occur had already been identified (Odiyo 1979). Backtracks calculated for the oviposition periods of first outbreaks in 14 seasons between 1965–66 and 1981–82 showed that the parent moths generally came from the eastern parts of Kenya and Tanzania, to generate infestations in Kenya and in eastern and southeastern Tanzania, though for outbreaks in Kenya and in southwestern Tanzania sometimes there may also have been immigration from neighbouring countries (Tucker 1984*b*). Networks of pheromone traps are now set up annually in the suspect source areas to monitor for increases in moth numbers several weeks before the anticipated start of the outbreak season.

3. Use of historical data

Patterns in the incidence of armyworm outbreaks have been analysed, as indicated above, from which 'rules of thumb' have been derived to assist with forecast production by synthesizing the accumulated years of experience customarily exercised in manual forecasting. For example, historical data from Kenya and Tanzania were used to develop a system for forecasting the occurrence of primary outbreaks from the dates of secondary infestations at the end of the preceding season. Using the records archived at Muguga, Kenya, for the 25 seasons from 1962–63 to 1986–87, dates of the last outbreaks (secondary outbreaks) of each season were listed with all dates of the first outbreaks in the following season. The number of inter-season days (between successive seasons) was counted for each of the 24 years to obtain an annual mean figure. The mean figure was then added to the last date in each season to produce 'predicted' dates for the onset of potential primary outbreaks. These were tabulated against

TABLE 1. PREDICTED DATES OF FIRST OCCURRENCE OF *S. EXEMPTA* LARVAE IN TANZANIA BASED ON THE MEAN NUMBER OF DAYS (247) BETWEEN DATES OF THE LAST (SECONDARY) AND THE NEXT (PRIMARY) OUTBREAKS OF THE SEASONS

outbreak season	date first outbreak predicted to occur	date first outbreak reported	location of first outbreak by district	difference in days between prediction (2) and occurrence (3)	last date larvae reported in this season
1962–63	—	—	—	—	06/04/63
1963–64	12/12/63	20/12/63	Kilosa	+8	31/01/64
1964–65	04/10/64	06/01/65	Kilosa	+94	03/03/65
1965–66	05/11/65	10/12/65	Kilosa	+35	03/05/66
1966–67	05/01/67	26/12/66	Kilosa	−10	10/05/67
1967–68	12/01/68	07/01/68	Kisarawe	−5	16/04/68
1968–69	19/12/68	04/11/68	Mbeya	−45	01/04/69
1969–70	04/12/69	22/12/69	Kilosa	+18	21/05/70
1970–71	23/01/71	28/11/70	Tunduru	−56	13/02/71
1971–72	18/10/71	28/12/71	Morogoro	+71	04/04/72
1972–73	07/12/72	06/05/73	Kisarawe	+150	30/05/73
1973–74	01/02/74	01/01/74	Lindi	−30	15/06/74
1974–75	14/02/75	03/01/75	Ulanga	−42	03/05/75
1975–76	05/01/76	01/01/76	Kilosa	−3	04/05/76
1976–77	06/01/77	21/12/76	Mpwapwa	−16	16/04/77
1977–78	19/12/77	01/01/78	Masasi	+13	12/02/78
1978–79	17/10/78	28/11/78	Mbozi	+42	19/04/79
1979–80	22/12/79	Nov. 1979	Kilimanjaro	?	27/05/80
1980–81	29/01/81	11/12/80	Kilosa/Pare	−49	27/04/81
1981–82	01/01/82	21/11/81	Nachingwea	−42	26/05/82
1982–83	28/01/83	26/12/82	Kilosa	−33	?
1983–84	?	15/12/83	Mpwapwa	?	19/05/84
1984–85	21/01/85	26/11/84	Kilimanjaro	−56	14/04/85
1985–86	17/12/85	16/12/85	Kilimanjaro	−1	23/03/86
1986–87	27/11/86	17/11/86	Kilosa/Mbeya	−10	?

the actual dates of first outbreaks of the next season, to give the difference in days ($\pm$) between actual and predicted dates. The results are summarized in tables 1 and 2.

In this analysis of Tanzanian dates, 12 out of 23 predictions (52%) were confirmed between 1 and 35 days (within about a month) from the dates of actual outbreaks, 6 (26%) of them within 10 days. Accurate 'predictions' were made for all 13 Districts with primary outbreaks, all located in the areas outlined in Figure 2 of Odiyo (1979). For Kenya, only 4 out of 19 predictions (21%) were confirmed between 4 and 36 days from the actual dates of outbreaks, and accurate predictions were made for only Taita-Taveta and South Nyanza Districts, which border northeast and northwest Tanzania, respectively.

This system of prediction may complement existing expertise in forecasting early outbreaks in Tanzania, where first outbreaks of the season develop mainly in November, December and January (Odiyo 1981, 1984). In Kenya, however, outbreaks may start between October and February. The final outbreaks of the season have been recorded in Tanzania predominantly in April–May (range January–June), while those in Kenya were predominantly in June (range March–August).

TABLE 2. PREDICTED DATES OF FIRST OCCURRENCE OF *S. EXEMPTA* LARVAE IN KENYA BASED ON THE MEAN NUMBER OF DAYS (202) BETWEEN DATES OF THE LAST (SECONDARY) AND THE NEXT (PRIMARY) OUTBREAKS OF THE SEASONS

outbreak season	date first outbreak predicted to occur	date first outbreak reported	location of first outbreak by district	difference in days between prediction (2) and occurrence (3)	last date larvae reported in this season
1962–63	—	—	—	—	27/06/63
1963–64	13/01/64	?	—	—	?
1964–65	?	Feb. 1965	Narok	—	16/03/65
1965–66	02/10/65	mid Nov. 65	Laikipia	—	22/06/66
1966–67	08/01/67	05/04/67	Machakos	+87	12/06/67
1967–68	29/12/67	16/05/68	Laikipia	+138	16/05/68
1968–69	02/12/68	?	—	—	?
1969–70	?	22/12/69	Machakos	—	08/06/70
1970–71	25/12/70	04/11/70	Lamu	−51	28/06/71
1971–72	14/01/72	19/02/72	Taita-Taveta	+36	19/02/72
1972–73	06/09/72	30/12/72	Kisumu	+115	30/12/72
1973–74	18/07/73	late 02/74	South Nyanza	—	09/07/74
1974–75	25/01/75	26/02/75	Taita-Taveta	+32	11/08/75
1975–76	27/02/76	02/03/76	Taita-Taveta	+4	24/07/76
1976–77	09/02/77	10/12/76	Kajiado	−61	17/06/77
1977–78	03/01/78	20/02/78	South Nyanza	+48	20/02/78
1978–79	08/09/78	04/01/79	Kiambu	+118	21/04/79
1979–80	08/11/79	12/02/80	Machakos	+96	07/07/80
1980–81	24/01/81	22/10/80	Kwale	−107	07/07/81
1980–81	24/01/81	03/11/80	Meru	−82	07/07/81
1981–82	24/01/82	09/02/82	South Nyanza	+13	22/06/82
1982–83	08/01/83	20/10/82	Taita-Taveta	−80	20/10/82
1983–84	08/05/83	13/01/84	Machakos	+250	27/06/84
1984–85	31/01/85	05/10/84	Kitui	−118	20/05/85
1985–86	06/12/85	10/02/85	Taita/Machakos	−299	26/05/86
1986–87	12/12/86	18/10/86	Taita-Taveta	−55	12/06/87

4. COMPUTER DATABASE

The need to improve both the regional and national control strategies and to strengthen the Desert Locust Control Organization for Eastern Africa (DLCO-EA) armyworm forecasting service, requires that data from current and historical field records of armyworm occurrence, weather and vegetation be re-organized for rapid access and utilization. To do this, the warning system which was developed in East Africa in 1969 for forecasting seasonal occurrence of outbreaks has been computerized.

Reports of armyworm infestations vary greatly in quality and detail, coming from farmers, agricultural extension workers, trap operators, national pest control services or from the public or the press. Every effort is, however, made at the monitoring and forecasting centre to collate all the information, which is then recorded on a standard data sheet, mapped and archived at Muguga. The data sheet was specially designed to select key factors relating to outbreaks of larvae, and to arrange them in order of priority. Records of outbreaks of larvae for the period 1963–1989 have been transcribed onto the special data forms before being stored on computer. Historical records of trap catches of moths, being more uniform in format, have been entered without transcription to the dBASE IV system of database management.

A key part of the database development has been the design of a system specification to identify the set of operational requirements to be met by the system, and prepare precise and detailed statements of what the computer system is to do. The main components of the dBASE IV system control database maintenance and utilization. Software, code-named WORMBASE (Day 1987), has been developed to accommodate a range of data types and functions which are summarized in table 3.

TABLE 3. THE MAIN COMPONENTS AND FUNCTIONS OF THE DATABASE

data types	forecast types
light trap records	short term, e.g. weekly
outbreak records	medium term e.g. monthly
pheromone trap records	long term e.g. quarterly/seasonally
weather summaries	
data handling functions:	data analysis functions:
entering	data presentation
viewing	current
editing	historical
printing	data analysis
mapping	comparisons
summarizing	analogue searches
	forecast generation (expert system)
	forecast printing

The WORMBASE data entry programmes have been tested using imaginary data. This practical exercise proved valuable, enabling learning through trial and error by the operator, and suggesting corrective measures to the programmer. The tests showed the way the programme is likely to be used most of the time in East Africa, but also included ways that will not arise very often, but which may pose different operational problems. Grossly erroneous data (as might occur in mistakes) were deliberately entered as a test for safety checks, and logical problems were created to see if the system might assign a wrong number or inaccurately prompt 'No data' when some item was entered. Other practical benefits from these tests were that the system was thoroughly checked for typographical errors, and situations which might elicit 'problem error messages' on the screen; the wording on displays, menus etc. was checked for clarity of messages; menus were scrutinised for superfluous or incomplete items; the messages displayed in the status bars were verified, and extra help/information suggested which might be useful at any point for the benefit of the inexperienced user.

5. EXPERT SYSTEM

Expert systems are computer programmes that attempt to mimic human experts in identifying solutions to problems. They provide a framework for accessing quantitative information in databases and are particularly useful for problems that have qualitative components (Frenzel 1987; Waterman 1986). An expert system uses information; the reasoning is arranged in rules consisting of conditions and consequent actions. A rule can be a regulation or statement defining a particular attribute, such as the behaviour of a moth.

Expert system development for the armyworm database management programme encompasses both prescriptive problem solving, whereby a user answers a series of questions and receives a recommendation, and provision of information which is relevant to a problem to help the user make a better decision, by linking to the database for searching, analysis and presentation of historical information. Preliminary development of the more prescriptive components has utilized an expert system shell (Exsys 1985) which allows knowledge to be coded by using a rule-based format:

'IF ⟨conditions⟩ THEN ⟨conclusions⟩ ELSE ⟨other conclusions⟩'.

For example, IF moth influx in an area is primary/secondary AND the cereal crops are in a vulnerable stage AND the period is within the outbreak season AND the area has had outbreaks of larvae in the past AND weather elements (rain and wind) have been favourable THEN a warning of impending pest attack should be issued. Or, as another example, IF young maize, sugarcane or grazing is infested AND the larval density looks threatening AND natural enemies are not reducing the population AND the loss is likely to be economically unacceptable THEN control of the pest with insecticides is necessary ELSE IF older maize, sugarcane or grazing is infested AND only a few larvae are observed AND damage is unlikely to be serious THEN chemical control is not economical to apply.

In the above examples, the IF and the AND parts state the conditions, and the THEN part suggests necessary actions. The expert system may already have access to information concerning the conditions of a rule; if not, appropriate questions will be asked of the user. If all the conditions of a rule are satisfied, the rule is then selected, and the search continues from rule to rule until the programme reaches a final conclusion.

The armyworm forecasting problem is a suitable application for expert systems for a number of reasons. Firstly the problem can be well defined, as it involves describing the interactions between the key factors of time (of moth catch, larval infestation or rainfall), location (of trap site, outbreak or raingauge), size (of catch, outbreak or precipitation) and the history of such occurrences over the years (from the database). Also the problem is one that occurs frequently. Finally, the end-users are well defined (e.g. pest control services, farmers, government funding departments etc.).

The armyworm forecasting expert system was structured to address the problem of decision making by evaluating four components: the breeding potential of moths in current and recent catches at a trap representing an administrative district; the risks to be posed by different stages in the life of the insect (larvae, for current damage; pupae, as potential sources of new moths; moths, as potential parents for new larvae at near or distant sites); the degree of influence likely to be exerted on moths, larvae or pupae by the prevailing weather, especially rainfall, winds and temperature; and the lessons to be learned from the comparison of current with previous (analogous) population structures and levels.

The forecasting process, made explicit in the expert system, proceeds as a series of levels. An influx of moths intercepted at a trap site forms the first factor for making a decision in armyworm forecasting (Odiyo 1989). Forecasting considerations at level 1 are therefore based solely on current moth catches from the trap network in each country. The development of populations is assessed consecutively from week 1 at the anticipated start of the outbreak season to week 35 or more at the end of the season. Rule building at forecast level 2 would therefore include moth catches for weeks 1 and 2. The magnitude of moth catches, that is, peak numbers in a night, can vary according to locality, altitude, time of year, prevailing weather conditions and trap efficiency. Categories of peak moths catches (based on experience) may at certain times of the year be defined for high-catching stations, namely, Muguga in Kenya, Tengeru and Arusha in northern Tanzania, as:

1–100 moths per night = low catch (category 1),
101–500 moths per night = medium catch (category 2),
more than 500 moths per night = high catch (category 3),

and for low-catching stations, i.e. all others:

1–10 moths per night = low catch (category 1),
11–100 moths per night = medium catch (category 2),
more than 100 moths per night = high catch (category 3).

Consultation with the programme is structured to follow the usual sequence considered by the forecaster: (a) current peak moth catches from the light or pheromone trap network; (b) reports of current outbreaks of larvae in the country from pest control services; (c) synoptic weather summaries from the meteorological departments, and (d) access to the historical data from the archives (database). The computer will first ask the operator to identify the trap site and indicate whether this week's catch is known. The operator is then asked to enter the number of moths caught, whose value will be interpreted by the computer according to the in-built rules on moth catch category. Next, the preceding week's catch is entered as the actual catch or, if appropriate, as 'catch unknown'. The programme then moves to questions about conditions of known outbreaks during the current week, and so on, until all items are covered.

The final output at the end of the run (i.e. prediction and recommendation) is structured in the range:

1. 'Forecast impossible' when there is no information from the field on which to base a decision.

2. 'Forecast, none expected' when available reports are mostly of zero values.

3. 'Forecast low probability' when only minor infestations could result from known source areas.

4. 'Forecast medium probability' where larger populations of larvae might be discovered in the district.

5. 'Forecast high probability' when the occurrence of fresh outbreaks is imminent.

Examples of forecasts based on catches in week 1 are shown in table 4. For the following week, at forecast level 2, the relationship would be as shown in table 5. Output from the programme is in the form of text, maps, graphics and explanatory notes.

TABLE 4. RANGE OF FORECASTS POSSIBLE ON BASIS OF MOTH CATCHES IN FIRST WEEK OF SEASON

IF: catch category is	THEN: Forecast probability of outbreaks is
moth catch unknown	forecast impossible
moth catch zero	forecast none expected
moth catch category 1	forecast low probability
moth catch category 2	forecast medium probability
moth catch category 3	forecast high probability

TABLE 5. RANGE OF FORECASTS AVAILABLE ON BASIS OF MOTH CATCHES
IN FIRST AND SECOND WEEKS

IF catch in weeks 2 (current week) is	AND catch in week 1 (last week) was				
	unknown	zero	category 1	category 2	category 3
	THEN forecast probability of outbreaks (level 2) is				
unknown	impossible	none	low	low	medium
zero	none	none	low	low	medium
category 1	low	low	low	low	medium
category 2	low	low	medium	medium	high
category 3	medium	medium	high	high	high

6. DISCUSSION

Putter & Van der Graaf (1988), of the Food and Agriculture Organization of the United Nations, identified broad objectives about information for plant protection in terms of: the type of information required (e.g. text, map, photographs); the end-users who need the information (e.g. farmer, pest control officer); how the information is to be managed at the central office (e.g. as database, expert system, archive); why the information is needed (i.e. what purpose will it serve?) and what is the value i.e. cost-effectiveness of the information in the management/decision-making process? The development of the Armyworm Forecasting Service in East Africa has been in full awareness of these criteria, particularly in terms of how it has incorporated research findings and experience for improving its mandate. Areas of research have included biology, ecology, biometeorology and biogeography; management of the service at the national and regional levels (Odiyo 1985); and the economics of control (Brown 1970; Rose 1984; Iles 1986).

Operation of the service faces continuing challenges in several key areas: specific problems related to armyworm biology include the nocturnal behaviour of the moths, which necessitates the use and maintenance of light and pheromone traps for monitoring changes in moth distribution throughout the year, and the survival of larvae on very widely distributed host plants, which makes investigation and detection of low-density larvae expensive and practically impossible at the farm level. Management problems include detection of early larval instars, which requires massive inputs of labour, transport and the installation of monitoring devices, while reduced vigilance during the off-season, or years with few outbreaks can only be countered by adopting strategic monitoring policies. Economic problems hinge, for example, on the magnitude and frequency of attacks and therefore whether control of larvae on farms and grasslands should be based on strategic or crop protection control policies, especially where limited funds have to be disbursed on competing priorities at the farm or district level.

Encouragement for successful future development comes from the strengthening of national pest monitoring and forecasting services (in Kenya and Tanzania), continuing support for the regional monitoring and forecasting service under the auspices of the DLCO-EA, based in Nairobi, and the involvement in and support for collaborative research and development exemplified by the ODA/EEC/DLCO-EA Armyworm Project.

A well structured monitoring and forecasting system can provide various types of information, such as text which describes the general situation at a particular time and expected developments from the current combination of factors (Betts 1976; Odiyo 1979); a map showing the positions and intensities of current or recent outbreaks, with a short text explaining potential changes which might result from known situations; and short warnings delivered by the quickest possible means (e.g. telex, facsimile, radio) to reach end-users for immediate action.

To maintain and build on progress already achieved, training of personnel at all levels (e.g. trap operators, extension officers, country coordinators) is essential for effective collaboration and exchange of data and information needed for balanced decision-making.

The database and expert system described here, in combination with real-time satellite monitoring of rainfall, and rapid communication of reports of current distributions and concentrations of moths and larvae, especially from primary outbreak areas, will facilitate the formulation of appropriate pest reports and improve forecasting for armyworm outbreaks, for the application of strategic control of larvae.

Records of *S. exempta* kept at the National Agricultural Research Centre (NARC) of the Kenya Agricultural Research Institute (KARI), Muguga, which in turn came from routine data provided by the National Armyworm Coordinators/Pest Control Officers for Tanzania and Kenya in support of the Regional Armyworm Forecasting Services for eastern Africa, have formed the main base for this paper. I am therefore grateful to these officers and to former scientists in East and Southern Africa, especially E. Lewes of the Meteorological Department, Dagoretti (Nairobi). My grateful thanks also go to the Directors of NARC & KARI and the Director General, DLCO-EA, for support and permission to publish this paper; to Dr R. C. Rainey and Miss M. J. Haggis of the Centre for Overseas Pest Research (now Overseas Development Natural Resources Institute), U.K. for encouragement; Dr D. J. W. Rose of the EEC/ODA/DLCO-EA Armyworm Project in Nairobi and Dr R. Day of the Silwood Centre for Pest Management, U.K. for criticisms and useful comments.

REFERENCES

Betts, E. 1976 Forecasting infestations of tropical migrant pests: the Desert Locust and African armyworm. In *Insect Flight* (ed. R. C. Rainey) *Symp. R. Ent. Soc. Lond.* **7**), pp. 113–134. Oxford: Blackwell.

Blair, B. W. & Catling, H. D. 1974 Outbreaks of African armyworm, *Spodoptera exempta* (Walker) (Lepidoptera, Noctuidae), in Rhodesia, South Africa, Botswana and South-west Africa from February to April 1972. *Rhod. J. agric. Res.* **12**, 57–67.

Brown, E. S. 1962 *The African armyworm, Spodoptera exempta (Walker) (Lepidoptera, Noctuidae): a review of the literature.* (57 pages.) London: Commonwealth Institute of Entomology.

Brown, E. S. 1970 Control of the African armyworm, *Spodoptera exempta* (Walk). – an appreciation of the problem. *E. Afr. agric. For. J.* **35**, 237–245.

Brown, E. S., Betts, E. & Rainey, R. C. 1969 Seasonal changes in distribution of the African armyworm, *Spodoptera exempta* (Walk.) (Lep.: Noctuidae), with special reference to eastern African. *Bull. ent. Res.* **58**, 661–728.

Brown, E. S. & Odiyo, P. 1968 The rate of feeding of the African army worm *Spodoptera exempta* (Walk.) and its significance for control operations. *E. Afri. agric. For. J.* **33**, 245–256.

Day, R. K. 1987 Armyworm forecasting: *Report of a visit to Kenya, May-June, 1987.* T C Fellowship X158, Project A1736. Imperial College, Silwood Park, Ascot, U.K.

Exsys Incorporated, 1985 *Expert system development package users manual.* Albuquerque. New Mexico.

Frenzel, L. E. Jr. 1987 *Understanding expert systems.* Indiana: Haward W. Sams & Company. (A Division of Macmillan, Inc.)

Haggis, M. J. 1971 Light trap catches of *Spodoptera exempta* (Walk.) in relation to wind direction. *E. Afr. agric. For. J.* **37**, 100–108.

Haggis, M. J. 1979 African armyworm *Spodoptera exempta* (Walker) (Lepidoptera: Noctuidae) and wind convergence in the Kenya rift valley, May 1970. *E. Afr. agric. For. J.* **44**, 332–346.

Haggis, M. J. 1984 Distribution, frequency of attack and seasonal incidence of the African armyworm *Spodoptera exempta* (Walk.) (Lep., Noctuidae), with particular reference to Africa and southwestern Arabia. (116 pages.) *Rep. Trop. Dev. Res. Inst. Lond.* **L69**.

Haggis, M. J. 1986 Distribution of the African armyworm, *Spodoptera exempta* (Walker) (Lepidoptera: Noctuidae), and the frequency of larval outbreaks in Africa and Arabia. *Bull. ent. Res.* **76**, 151–170.

Hattingh, C. C. 1941 The biology and ecology of the army worm (*Laphygma exempta*) and its control in South Africa. (50 pages.) *Sci. Bull. Dep. Agric. For. Un. S. Afr.* **217**.

Iles, M. J. 1986 An economic assessment of the African armyworm, with special reference to Tanzania: Research proposals following a visit to Tanzania and Kenya, April–May, 1986. (26 pages.) *Rep. Trop. Dev. Res. Inst.* **R1380** (R).

Moharram, I. A. & Nasseh, O. M. 1989 Distribution and seasonal incidence of African armyworm *Spodoptera exempta* (Walker) (Noctiuidae-Lep) in Yemen Arab Republic. *Report of 3rd Arab Congress of Plant Protection, Al-Ain, United Emirates University, 5–9 December* 1988.

Odiyo, P. O. 1972 Reliability of the first full-scale (Armyworm) forecasting service. *Rec. Res. E. Afr. agric. For. Res. Org. 1971*, pp. 200–202.

Odiyo, P. O. 1975 The biology of armyworm outbreaks in East Africa. *Ken. entomol. Newsl.* **2**, 2–6.

Odiyo, P. O. 1979 Forecasting infestations of a migrant pest: the African armyworm, *Spodoptera exempta* (Walk.). *Phil. Trans. R. Soc. Lond.* B **287**, 403–413.

Odiyo, P. O. 1981 Development of the first outbreaks of the African armyworm, *Spodoptera exempta* (Walk) between Kenya and Tanzania during the 'off-season' months of July to December. *Insect Sci. Appl.* **1**, 305–318.

Odiyo, P. O. 1984 A guide to seasonal changes in the distribution of armyworm infestations in East Africa. *Insect Sci. Appl.* **5**, 107–119.

Odiyo, P. O. 1985 The development and operation of an armyworm forecasting service in East Africa. *Proceeding of the seminar on agrometeorology and pest forecasting, 24–28 June, 1985, CTA/TDRI, Fulmer Grange.* pp. 56–72.

Odiyo, P. O. 1989 An expert system for forecasting outbreaks of the African armyworm, *Spodoptera exempta* (Walker) (Lepidoptera: Noctuidae.). M.Phil. Thesis, University of Wales, Cardiff.

Page, W. W. 1988 Varying durations of arrested oocyte development in relation to migration in the African armyworm moth, *Spodoptera exempta* (Walker) (Lepidoptera: Noctuidae). *Bull. ent. Res.* **79**, 181–198.

Pedgley, D. E., Page, W. W., Mushi, A., Odiyo, P., Amisi, J., Dewhurst, C. F., Dunstan, W. R., Fishpool, L. D. C., Harvey, A. W., Megenasa, T. & Rose, D. J. W. 1989 Onset and spread of an African armyworm upsurge. *Ecol. ent.* **14**, 311–333.

Putter, C. A. J. & Van der Graff, N. A. 1988 Information needs in plant Protection. *Pl. Prot. Serv. FAO, Rome.*

Rainey, R. C. 1979 Control of the armyworm *Spodoptera exempta* in eastern Africa and southern Arabia: report of a mission to formulate an inter-regional project. *FAO-AGPP: Misc/32.*

Rose, D. J. W. 1979 The significance of low-density populations of the African armyworm, *Spodoptera exempta* (Walk). *Phil. Trans. R. Soc. Lond.* B **287**, 393–402.

Rose, D. J. W. 1984 Current views on the epidemiology of the African armyworm outbreaks and strategies for control. In *Working papers of the seminar on the African armyworm, Spodoptera exempta* (Walker) (Lepidoptera Noctuidae) 4–5 September 1984, *Trop. Dev. Res. Inst. Lond.* pp. 31–35.

Rose, D. J. W., Dewhurst, C. F., Page, W. W. & Fishpool, L. D. C. 1987 The role of migration in the life system of the African armyworm. *Spodoptera exempta. Insect Sci. Appl.* **8**, 561–569.

Rose, D. J. W. & Law, A. B. 1976 The synoptic weather in relation to an outbreak of the African armyworm, *Spodoptera exempta* (Wlk.). *J. ent. Soc. Sth. Afr.* **39**, 125–130.

Rose, D. J. W., Page, W. W., Dewhurst, C. F., Riley, J. R., Reynolds, D. R., Pedgley, D. E. & Tucker, M. R. 1985 Downwind migration of the African armyworm moth, *Spodoptera exempta*, studied by mark-and-capture and by radar. *Ecol. ent.* **10**, 299–313.

Tucker, M. R. 1984*a* Forecasting the severity of armyworm seasons in East Africa from early season rainfall. *Insect Sci. Appl.* **5**, 51–55.

Tucker, M. R. 1984*b* Possible sources of outbreaks of armyworm *Spodoptera exempta* (Walker) (Lepidoptera: Noctuidae) in East Africa at the beginning of the season. *Bull. ent. Res.* **74**, 599–607.

Tucker, M. R., Mwandoto, S. & Pedgley, D. E. 1982 Further evidence for windborne movement of armyworm moths, *Spodoptera exempta*, in East Africa. *Ecol. ent.* **7**, 463–473.

Tucker, M. R. & Pedgley, D. E. 1983 Rainfall and outbreaks of the African armyworm, *Spodoptera exempta* (Walker) (Lepidoptera: Noctuidae). *Bull. ent. Res.* **73**, 195–199.

Waterman, D. A. 1986 *A guide to expert systems.* Reading, M.A.: Addison-Wesley.

Discussion

R. DAY (*Imperial College, Silwood Park, Ascot, U.K.*). Mr Odiyo has raised the subject of so-called 'strategic control'. The idea of this approach is that by controlling all the primary armyworm outbreaks, or at least those which are critical in the sense that they lead to subsequent or secondary outbreaks, the extent and thus the cost of secondary outbreaks will be greatly reduced. The idea is certainly attractive, but there are a couple of possible problems with the approach that I think would merit closer investigation.

Strategic control is based on the theory that most of the moths causing secondary outbreaks have arisen from primary outbreaks. There is certainly good circumstantial evidence that secondary outbreaks are caused by moths that have come from the area of primary outbreaks, but what is difficult to show is that the majority of the moths causing secondary outbreaks have come from primary outbreaks themselves. If at the same time that immigrant moths cause primary outbreaks there are significant numbers that are not concentrated in the outbreak area, low density, but extensive synchronous populations could be established in the vicinity of the outbreak. These would emerge at roughly the same time as the moths from the primary outbreaks, and might appear to arise from the outbreak site, but would escape strategic control. The important question is thus what proportion of moths causing a secondary outbreak actually come from the primary outbreak site itself?

A second area of strategic control that needs to be considered is the problem of finding a significant proportion of primary or critical outbreaks. Primary outbreaks are often smaller than secondary outbreaks, and so more difficult to detect, particularly if they are in areas where human habitation is sparse. In this case the costs of monitoring and detecting the primary outbreaks become much greater. An economic analysis comparing strategic and tactical control (if that is the opposite), would be useful, so that the sensitivity of the method to the critical assumptions could be tested.

A second area of interest is Mr Odiyo's analysis of the number of days between the last outbreaks of one season and the first outbreaks of the following season. His examination of this statistic implies that there might be some suggestion that the outbreaks at the start of one season are related to those at the end of the previous season. This recalls the continuous gregarious populations debate, which I thought had been closed, at least for African armyworms.

What are the relative merits of forecasting by statistical methods and forecasting by mechanistic methods? Obviously there is overlap between the two, but my feeling is that the models and statistics that we use in forecasting should be based on mechanistic explanations or hypotheses, rather than simple statistical correlations. On the other hand, it would be folly not to use an inexplicable correlation if it provided a better forecast than a mechanistic model.

D. E. PEDGLEY (*ODNRI, Chatham, U.K.*). Regarding what Dr Day said on whether one can estimate the proportion of secondary outbreaks that come from primary outbreaks, one way is by examining a large number of armyworm seasons and backtracking, from wind-field maps, to make best estimates of where the parents came from which produced the secondary outbreak and compare these with the known primary outbreaks at the time when moths are likely to have taken to the air. This work is on-going, but as far as I can recall the majority of secondary outbreaks seem to result from moths which have come from primary outbreaks; I would not like to put a figure on it, but perhaps in two years time we will be able to answer the question.

A. G. GATEHOUSE (*University of Wales, Bangor, U.K.*). Dr R. Day expressed some doubt that the proportion of moths responsible for secondary outbreaks, which originate in primary outbreaks, is likely to be high. He suggested that many of them are likely to arise from undetected low-density populations in the same general area. This question is of crucial importance for the current approach to strategic control, which depends on the effective control of primary outbreaks as a means of interrupting the sequence of infestations developing from them.

We need some means of distinguishing moths from high- and low-density phase larvae, and methods based on possible biochemical differences are being assessed. However, I think there is reason to suppose that, in years when outbreaks are severe and frequent, the large majority of moths responsible for secondary infestations do indeed originate in primary or critical outbreaks (Rose *et al.* 1987). It is now well established that seasons of severe armyworm outbreaks in East Africa are associated with partial failure of the early or short rains. When this happens, suitable larval habitat, which is absolutely dependent on rainfall, is present only where rain storms have occurred and is therefore patchy and widely scattered. The only larval populations likely to be established under these circumstances are those resulting from the concentration of flying moths (originating from low-density populations in the dry season habitats) by intense convergent airflow associated with these storms, i.e. the primary outbreaks. In these seasons, the patchy distribution of rainfall means that suitable habitat to support low-density infestations of larval offspring of moths escaping such concentration is just not available.

There now seems little doubt that *Spodoptera exempta* breeds continuously through the dry season, surviving at low densities in areas where there is sufficient off-season rainfall or other sources of moisture to permit continued growth of its host grasses. Pheromone traps deployed in such areas, coastal regions of Kenya, for example, catch moths throughout the year but give no indication of a build-up of these populations towards the end of the dry season to levels expected if they are the sources of moths initiating the primary outbreaks. As the latter occur regularly on the eastern slopes of the first hills inland from the Kenya and Tanzania coasts and as back-tracking studies have shown, unequivocally, that the moths causing them originate from the east, these consistently low catches could be no more than a consequence of the problems inherent in monitoring the insects at extremely low population densities, a measure of the success of their 'low density strategy' (Gatehouse 1987). However, Mr Page of the Kenya National Armyworm Project has suggested that another mechanism may be involved.

Moth concentration leading to primary outbreaks very often occurs on the very first rains in these areas, irrespective of their precise timing. This requires the assumption, on the basis of current knowledge, that adequate numbers of moths leave the dry season coastal habitats every night at the end of the dry season and beginning of the rains, to account for the infestations following the first rainstorms. If this were happening, moths would also be available for concentration by rainstorms subsequent to the first ones but the evidence suggests that they are not. However, the apparent anomaly would be resolved if moths from the coastal habitats underwent a quiescence induced by the low humidity conditions that they encounter inland after flight, before the first rainfall. Even if this was of limited duration, it would allow accumulation of the insects over time in these regions, to be activated by environmental cues associated with the onset of the first rains and then concentrated by them to initiate the primary outbreaks. Diapause and quiescence, often prolonged, is known in adult noctuids (Oku 1983) and we are now looking into its possible occurrence in the African armyworm.

568 P. O. ODIYO

References

Gatehouse, A. G. 1987 Migration and low population density in armyworm (Lepidoptera: Noctuidae) life histories. In *Recent advances in research on tropical entomology* (ed. M. F. B. Chaudhury) *Insect Sci. Appl.* **8** 573–580.

Oku, T. 1983 Aestivation and migration in noctuid moths. In *Diapause and life cycle strategies in insects* (ed. V. K. Brown & I. Hodek), pp. 219–231. The Hague.

Rose, D. J. W., Dewhurst, C. F., Page, W. W. & Fishpool, L. D. C. 1987 The role of migration in the life system of the African armyworm *Spodoptera exempta*. In *Recent advances in research on tropical entomology* (ed. M. F. B. Chaudhury). *Insect Sci. Appl.* **8**, 561–569.

D. J. W. Rose (*EEC/ODA/DLCO-EA Armyworm Project, Nairobi, Kenya*). Twelve years have passed since Mr Odiyo presented his paper at the previous symposium on migrant pests, convened by the Royal Society in 1977 (Odiyo 1979). That paper described the Armyworm Forecast Service which had been developed for the three countries of the East African Community by using light traps to monitor moth populations. Since then, the forecast service has been extended to the Desert Locust Control Organisation for Eastern Africa (DLCO-EA) and the national armyworm coordinators in seven member countries, and much greater use has been made of moth pheromone traps in the national networks of each country for the regional forecast issued each week. In addition to this service, relatively dense networks of pheromone traps have been established in the most important primary outbreak areas in Kenya and Tanzania. These are monitored by the local extension officers and are used to locate the first outbreaks of the season so that these can be controlled as quickly as possible before moths spread to continue a succession of secondary outbreaks downwind from the original sites. These networks of pheromone traps, plus the satellite pictures of locations of storms likely to concentrate moths to cause outbreaks, are the most important sources of information for alerting crop protection officers and District Agricultural officers so that major control operations can be mounted. In addition, farmers and extension officers in these previously dry areas are warned to take heed of the first major rain storms of the season and to search for newly hatched caterpillars about a week later. Coloured and illustrated charts showing the stages of armyworm which may be found on specified days after the storms have been prepared for distribution to farms (W. W. Page, personal communication).

The extension of the forecast service with Mr Odiyo as Regional Armyworm Forecast Officer for DLCO-EA, the development of the data bank on computer and the introduction of 'expert system' software for programming forecasts are all exciting developments, which show great promise for the future. Seasonal forecasts of severity of armyworm outbreak years are already possible (Tucker 1984; A. W. Harvey, unpublished). The investigations of possibilities for producing long-term forecasts of the time and place of likely occurrence of primary outbreaks are still at a preliminary stage and no doubt will need to be modified as more data become available. Nevertheless, this possibility must be fully investigated, as prior warning increases the chance of implementing control operations before damage is done.

References

Odiyo, P. O. 1979 Forecasting infestations of a migrant pest: the African armyworm, *Spodoptera exempta* (Walk.). *Phil. Trans. R. Soc. Lond.* B **287**, 403–413.

Tucker, M. R. 1984 Forecasting the severity of armyworm seasons in East Africa from early season rainfall. *Insect Sci. Appl.* **5**, 51–55.

R. C. Rainey, F.R.S. (*Elmslea, Old Risborough Road, Stoke Mandeville, Bucks, U.K.*). During most years for a period of several months from about December to April, dense infestations are absent from the East African countries. Odiyo (in Rainey 1979) directed attention to an incursion of westerly winds at the African Rift Convergence Zone over Zambia in late 1971, which subsequently advanced into southern Tanzania at an appropriate time and area to have probably brought the parents of the first generation of the sequence of heavy infestations which later extended right up to Ethiopia and Yemen. In a number of years Kenya has recorded armyworm infestations coming in with incursions of westerlies.

In 1970, the wind-finding research aircraft recorded, close to the Nakuru light trap, the sharply defined leading edge of such an incursion of westerlies a few hours after the peak light trap catch of the season at this trap, and this catch proved to have sampled the parents of a further generation of armyworm in and around the Nakuru district (Haggis 1979).

It has been suggested that the windshift of such westerlies would be appropriate for experimental searches with the airborne radar for missing moth populations in the off-season.

Any further observations on the effects of westerly incursions on the armyworm could be highly relevant. Odiyo's investigation of 'expert systems' is particularly to be welcomed in the context of 20 years of successful armyworm forecasting.

References

Haggis, M. J. 1979 African armyworm. *Spodoptera exempta* (Walker) (Lepidoptera: Noctuidae) and wind convergence in the Kenya rift valley, May 1970. *E. Afr. agric. For. J.* **44**, 332–346.

Rainey, R. C. 1979 Control of the armyworm *Spodoptera exempta* in eastern Africa and southern Arabia: report of a mission to formulate an inter-regional project. *FAO-AGPP*: *MISC*/32.

Phil Trans. R. Soc. Lond. B **328**, 571 (1990)

Printed in Great Britain

PHYSICAL ELEMENTS

Introductory remarks

By K. A. Browning, F.R.S.

Meteorological Office, Bracknell RG12 2SZ, U.K.

To set the scene for this session on the role of physical factors, the chairman showed a number of slides, including figure 1 taken from Schaefer (1979). Both winds and rain are important elements in migrant pest problems, but this particular figure relating to the role of the windfield portrays as vividly as any the strong link between the weather and insect concentration, which is the theme of this meeting. The figure shows the density of moths accumulating at a sea breeze convergence line as observed by a downward pointing radar (Schaefer 1979) on an aircraft flying transverse to the sea breeze front. Other slides (not shown here) were presented from Simpson *et al.* (1977), which showed that the distribution of moths in figure 1 is consistent with them being carried by the gravity current circulation of cold air behind the sea breeze front. Relative to the frontal position, this flow is from right to left in figure 1, then rising in the kilometre or two behind the front before returning to the right above a height of 400 m. Accumulation of insects can be expected to occur in the rising part of such a circulation provided that they are capable of descending through the ascending air.

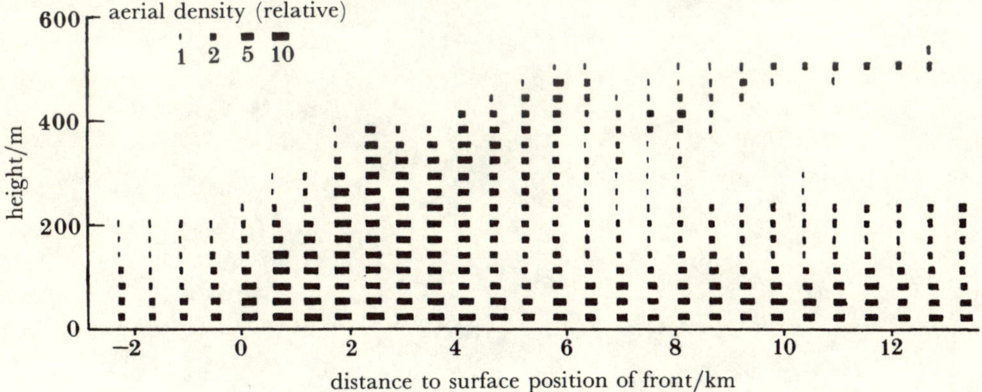

FIGURE 1. Density of moths accumulating at a sea breeze convergence line. New Brunswick, 22h37, 10 July 1976.

References

Schaefer, G. W. 1979 An airborne radar technique for the investigation and control of migrating pest insects. *Phil. Trans. R. Soc. Lond.* B **287**, 459–465.
Simpson, J. E., Mansfield, D. A. & Milford, J. R. 1977 Inland penetration of sea breeze fronts. *Q. Jl R. met. Soc.* **103**, 47–76.

PHYSICAL ELEMENTS

Introductory remarks

Phil. Trans. R. Soc. Lond. B **328**, 573–583 (1990)

Printed in Great Britain

Algerian case study and the need for permanent Desert Locust monitoring

By R. Kellou[1], N. Mahjoub[2], A. Benabdi[3] and M. S. Boulahya[4]

[1] *Ministry of Agriculture, Algiers*
[2] *CLCPANO Food and Agriculture Organization of the United Nations, Algiers*
[3] *National Institute for Plant Protection (I.N.P.V.), Algiers*
[4] *National Meteorological Office (O.N.M.), Algiers*

Through the experience gained by Algeria and the Maghreb Commission for Desert Locust control during the past recession and the present plague (1987, 1989), especially in the use of operational meteorological products of the World Weather Watch of the World Meteorological Organization for improved forecasts of swarm movement, an integrated acridometeorological watch system is suggested for the whole Saharan breeding area to avoid any surprise in the future. This permanent monitoring system should be built and operated jointly by the meteorological and the plant protection services at the national level. The regional and international coordination by the Food and Agriculture Organization of this Desert Locust monitoring could make use of the now experienced and integrated system of the World Weather Watch. The system proposed is not only useful for Desert Locust survey and control, but for the realistic use of the Saharan environment for a better life for nomads and the newly settled peasants.

1. Introduction

In October 1987, the Food and Agriculture Organization (FAO) stated in its Desert Locust Situation Summary and Forecast: 'Swarms produced south of the Sahara invaded Algeria and Morocco on a broad front in the second half of October... Summer breeding has probably terminated in western Sudan, Chad and Niger, but continues in Mali where it may become more widespread, and is probably starting in Mauritania'. The next FAO bulletin of November 1987 stressed that 'The scale of summer breeding south of the Sahara has been far greater than anticipated. Mali was invaded by swarms in early October, and Mauritania in late October. Breeding continues in both countries and is widespread in Mauritania...' In spite of these two warnings, no general mobilization was declared in the Maghreb and for good reason: the financially important decisions that had to be taken needed more precise field information than was available at that moment. In fact, the war against the Desert Locust would be declared only in February 1988, after the first big swarms coming from Mauritania and the Western Sahara had crossed the border of Algeria and Morocco, in association with a series of exceptional weather situations during 21–28 February 1988. These weather situations brought exceptional rain in the southwest of Algeria and Morocco and well-organized warm and strong winds, which provided a transport system of 10 days duration in Western Sahara.

In September 1988, all the national and international specialized authorities (FAO, Overseas Development Natural Resources Institute (ODNRI), National Desert Locust Services) were unanimous in predicting a severe Desert Locust invasion for the north African countries, starting from October 1988. The reason was the heavy rainfall in the Sudan and Sahel, which enhanced an unprecedentedly successful breeding season in the whole of the summer breeding area from the Red Sea to the Atlantic. Year one of the worst plague in this century was announced. In fact, Dr R. C. Rainey (personal communication), in preparing this discussion meeting, was urging the Algerian Desert Locust control team to follow closely this special situation and wrote: 'Algeria's forthcoming Desert Locust invasion of 1988/1989 could well be the most damaging the country has ever had, by reason not only of the scale of the parent generation in the Sahel, but also from the extent of agricultural development in Algeria, since the last comparable invasion of 1954–1958'.

Following these unanimous and pessimistic forecasts, Algeria, the Maghreb countries and the FAO mobilized huge and costly resources, but apart from Morocco, the scourge did not materialize because the majority of swarms had boarded the Atlantic cyclones. These cyclones, taking their origin from the heart of the Sahara and Sahel area, transported the Desert Locust swarms to the Caribbean Islands in a spectacular and apparently unpredictable way. The mechanism was explained, however, once we had the time to use the World Weather Watch (WWW) products prepared and disseminated by world forecasting centres such as European Centre for Medium-range Weather Forecasting (ECMWF) (Reading) or U.K. Meteorological Office (Bracknell).

In September 1989, the FAO Emergency Centre for Locust Operations (ECLO) Desert Locust Summary disseminated the following information: 'The continuing heavy rains in several parts of West Africa... have created excellent conditions for continued summer breeding. This could constitute the beginning of another upsurge, so every effort should be made to detect and destroy any infestations which may result.'

One month later, the FAO/ECLO Summary stated that: 'The general Desert Locust situation remained unexpectedly calm during the past month. Despite adequate rainfall and good ecological conditions in many areas, there is no evidence suggesting an upsurge in Western Africa as previously forecast'. In fact, in Algeria the Desert Locust control system had been fully mobilized during the whole 1989 season and no definite decision had been taken as of November 1989. This created an unbearable situation of 'no war, no peace', the main reason being the lack of adequate and reliable Desert Locust information in the vast breeding area in the Sahara.

Any decision-maker involved in the Desert Locust operations was therefore asking and being asked the same questions: 'Should we demobilize the whole system or should we maintain the current capability; or should we perhaps reorganize the resources of preventive control operations and set up a permanent Desert Locust Watch?'

As a matter of fact, the Desert Locust is feared not only for its destructive capacity of the food crops but also for the permanent threat that the 60 countries concerned have to bear. During these past years, we have experienced a true psychosis of the Desert Locust invasion in many African countries of the Sahara and Sahel regions. To protect the threatened countries from invasions that may not occur, the expenditures incurred are as heavy as those involved during a real Desert Locust plague. Nevertheless, the progress made in the scientific knowledge of this migrant pest should have changed this situation.

Some specialized authorities assure the public and the governments that it is now possible to answer positively the questions on *where*, *when* and *how* the desert plague develops and even *what are the means* and *methods* required to control rationally any scourge.

In our opinion, the solution to the Desert Locust problem is not only technical, but also organizational. It depends mainly on the willingness of all countries concerned, the donors and the interested international agencies, to unify their efforts to build an integrated system for a permanent World Locust Watch (WLW of FAO) complementary to the World Weather Watch system of the World Meteorological Organization (WMO). An invasion of the Desert Locust in its gregarious phase presents many similarities with meteorological phenomena: it knows no frontiers, concerns large areas, has great mobility, develops in areas difficult to reach and monitor, can take man by surprise, and is unpredictable in the long term.

After this review of the status of the actual Desert Locust information system and control organization (and before suggesting a realistic contribution to the strengthening of the organizational scheme and information network for a better plague forecast), let us look at the expenses incurred by Algeria and the lessons drawn from the recent Desert Locust campaign, which is not yet over because of the lack of information on the Desert Locust in its large breeding and development areas.

2. ALGERIAN CASE STUDY: THE 1987–1989 DESERT LOCUST CONTROL CAMPAIGN

Because of its importance and suddenness, the Desert Locust plague of 1987–1989, which occurred in West and northwest Africa, forced Algeria and Morocco into an unprecedented campaign covering more than 5 million ha† and requiring more than U.S. $100 million.

(a) *The invasion, as seen from Algeria*

In 1985, intensive breeding by the Desert Locust on the western edges of the Red Sea, though limited in space, started the formation of swarms that invaded Saudi Arabia in 1986 and left an important breeding potential in Sudan and Ethiopia. Insufficient control operations and favourable winds allowed the new swarms to reach the Sahel in July and August 1987. From September to October 1987 onwards the swarms settled in the northeast of Mauritania and in the Western Sahara where breeding was intense because of extremely favourable ecological conditions. This gave rise to winter-spring generations. These swarms invaded southwest Algeria during the first quarter of 1988 from February onwards. After that the monsoon rain, which fell in the Sahel from June to August 1988, led to the production of two locust generations that were expected to move to the north and the northwest from September 1988 in a massive invasion of the Maghreb countries. This scourge did not occur and many swarms ended in the Atlantic, with some reaching the Caribbean, forced by unusual westward winds. However, some swarms managed to reach western Algeria and southern Morocco.

This exceptional mass-cleaning phenomenon seems to have considerably reduced the locust potential in the west African and Sahel zone, enabling us to enjoy, since the end of 1988, a quiet period as regards acridian activity, despite the presence of a few populations in traditional breeding sites.

---

† 1 hectare = 10^4 m².

(b) The organization of the locust control system: efforts and means

Algeria has responded to the 1987–1989 locust invasion on more than 2 million ha representing more than 30 % of the treated area in the Maghreb (7 500 000 ha).

The dangerous evolution of the locust situation from October 1987 onwards had led the Algerian government to adopt exceptional procedures to face this plague so that minimum damage to agricultural production and the environment would result. The tasks of Desert Locust prevention and control are led by the National Plant Protection Institute (INPV), a specialized Public Service of the Ministry of Agriculture. During plague periods there is a strengthening of the preventive measures to ensure: (i) Government control; at the state's expense and conducted by the specialized public services in areas, such as in the Sahara and the Highlands, where swarms cannot be controlled by peasants. (ii) Collective control; to complement the public effort and locally conducted in the agricultural areas by peasants with coordination by professional associations.

The legal framework for this organization of locust control is provided by different laws, decrees and instructions:

(i) Decree No 67-177 dated August 8, 1967, appointing an interministerial committee responsible for the organization, coordination and evaluation of the Desert Locust control campaign.

(ii) Decree No 85-231 and 85-232 dated August 25 1985, related to the emergency organization and action at local level to prevent or face natural disasters, of which a Desert Locust plague is one.

(iii) Law No 87–17 dated August 1 1987, organizing the phytosanitary protection and prevention measures against pests and plant diseases, of which the Desert Locust is given top priority because of its potential damage to agriculture in both crops and pastures.

(iv) Instruction No 1 of July 13 1988, defining the tasks, organization and the creation of control groups during the invasion period.

(v) Instruction No 2 of October 16 1988, coordinating the collection and dissemination of Desert Locust monitoring and movement forecasts.

(vi) Instruction No 3 of March 1989, related to the protection of man and his environment.

Besides the legal form and organizational scheme at a high level, necessary to effectively control this plague over a vast territory, the following means have been progressively and rapidly deployed, beginning from February 1988: 50 light vehicles (4-wheel drive) for finding and monitoring the swarms; 60 heavy treatment vehicles fitted with exhaust nozzle sprayers; 200 heavy logistic transport vehicles; 1600 back sprayers (individual control in the oases); 60 airplanes (fixed wing and helicopters fitted with rotary atomizers); 5 million l of pesticides of which 3 million l have been sprayed; 500 plant protection technicians, and 1500 general workers (drivers, guides operators).

On the financial side, the control operations needed more than U.S. $36 million to meet the cost of pesticides and hired airplanes.

(c) The results and the weaknesses of the general mobilization

In tackling a national disaster such as a locust plague, it is difficult to quantify or even estimate all the positive results, therefore we concentrate on two main apparent effects. The first and most spectacular one was the successful protection of the major productive agricultural lands that cover the northern zone, and the range lands and main irrigated areas of the central and eastern Sahara. Some damage has been noticed in the natural pasture lands of the western Sahara, in irrigated agricultural land of the Imperial Valley in the Saoura region (southwest Algeria), and in the bee-keeping farms that have not been protected against the aerial spraying. The second and more indirect effect is the considerable decrease of the locust potential, which could have reestablished in the Sahelian summer breeding zones if the Maghreb countries, mainly Algeria, had not made a major effort particularly in hopper control. Indeed, in Algeria alone, more than 700000 ha of hopper bands were treated during June and July 1988. Thereby the life cycle of the gregarious Desert Locust has been disturbed and even cut.

Even though the actions were successful, the field operations against the Desert Locust have been hampered by the lack of bio-ecological real time information on the migrant pest and of real time and precise observations on the prevailing meteorological conditions over the vast source and target areas. This hindered the making of medium term forecasts of swarm movements, the strategic deployment of control equipment and the most efficient use of chemical spraying. In fact, the protection of the agricultural potential led the organization in charge of control of this plague to set up some techniques, means and procedures that were certainly efficient, but that left some negative effects on the environment by accumulating some pesticide residues in the water and food chain.

Although the efforts made by the countries of the Maghreb and of West Africa have been considerable and costly, better control of the locust phenomenon should in future be obtained through a strict coordination of the general prevention programmes, not only between these two subregions of Africa, but also between them and the central regions of East Africa and the Middle East. In this respect, the role of the international (FAO, ECLO, Organization for African Unity (OAU)) and regional organizations (CLCPANO, OCLALAV, DLCO-EA) appears predominant in locust information collection and its transmission and treatment. This is in addition to the training of specialists in control techniques and the coordination of the necessary field research.

3. THE COORDINATION AT THE REGIONAL AND BILATERAL LEVEL

To be able to take charge of the prevention and the control of Desert Locust plagues, the four Maghreb countries created in 1972, under the aegis of FAO, a regional specialized commission named CLCPANO (Desert Locust Control Commission for northwest Africa). In 1988 Mauritania joined the commission. Its members are the Agriculture Ministries in charge of Desert Locust prevention and control. An executive committee composed of experts from each country meets every year in recession periods and as often as necessary during invasion situations. The secretary of the commission installed in Algiers is in charge of the coordination with FAO, and the collection and regular dissemination of locust and meteorological information.

This commission has been very active in the training of locust specialists, the organization of workshops on locust control and on meteorological needs for better preventive control. It has been prominent in the conception of guides and manuals, and the conduct of locust prospection campaigns in summer breeding areas such as southern Algeria, northern Niger, Mali and Mauritania.

During the invasion period of 1987–1989, the Maghreb commission organized or contributed to meetings or seminars in:

(i) Tunis (Tunisia), March 1988; a common strategy for the control of Desert Locust.

(ii) Rabat (Morocco), April 1988; adoption of a strategy for the period May–September 1988.

(iii) Tamanrasset (Algeria), May 1988; coordination of the financial means for a common preventive strategy. During this meeting the creation of a Maghreb task force was agreed together with the strengthening of the meteorological observing network over all the summer breeding Saharan area with a focal point at Tamanrasset Regional Meteorological Centre.

(iv) Nouakchott (Mauritania), June 1988: a common Maghreb strategy was adopted at the ministerial level and presented by the Ministry of Agriculture of Mauritania to the Desert Locust Control Committee of FAO.

(v) Rabat (Morocco), September 1988.

These meetings have permitted the establishment of a Maghreb fund and a Maghreb prevention task force, which started working in association with neighbouring countries (Niger, Mali, Mauritania).

On the other hand, the CLCPANO commission has contributed to the conception of an inter-regional project for Desert Locust control, which aims to strengthen the plant protection services of eight countries from the Sahel and the Maghreb, which may be financed by FIDA (International Fund for Agricultural Development) and executed by FAO.

In addition to this regional system of coordination many bilateral agreements have been elaborated by Algeria and signed with Morocco, Tunisia, Libya, Niger, Mali and Mauritania. This was done, to coordinate the control effort, to normalize the observing system, to exchange locust forecasts and also to make rational use of mobile equipment, mainly in the border regions.

4. METEOROLOGICAL ASSISTANCE TO THE DESERT LOCUST CONTROL

In addition to a seminar on hydrological resources in the Sahara, a conference held in Tamanrasset in May 1980 on the need for meteorological data for Desert Locust Survey helped identify the necessity for reinforcement of the meteorological network in the border zones. A programme to install eight meteorological stations was made for data collection in central Sahara with the assistance of the National Plant Protection Institute.

Many seminars on acrido-meteorology have been organized and financed by the CLCPANO. These seminars have allowed preparations to progress for meteorological assistance in locust prospection and control.

Moreover, at the beginning of the first Desert Locust warning in August 1986, a new call for the reinforcement of the meteorological watch system in the summer breeding zones was made. Following an Algerian proposal, the African Regional Association of WMO adopted a resolution appointing a rapporteur for acrido-meteorological questions (Harare, December 1986) to suggest a pilot project to WMO/FAO to finalize these proposals for meteorological assistance to Desert Locust prevention and control.

This first alert has allowed the Algerian meteorological service to request and obtain a telecommunication line between Algiers, Dakar and Cairo (October 1986, December 1986). This is in addition to the existing links with Niamey (Niger) and Jeddah (Saudi Arabia).

Thus the Algerian meteorological service was linked to all centres of the regions concerned with Desert Locust problems. The first forecast of Desert Locust movement was made for the October 1987 invasion. On that occasion, the regional meteorological centres (Tamanrasset and the border stations of Tindouf. In Guezzam and Bordj Bordji Mokhtar), played a major role, but in spite of all the means involved, two main deficiencies were identified which are:

(i) The very low density of the meteorological network in the Saharan zone situated between the 15° and 25° north and between the Atlantic and the Red Sea. This is the main breeding and development area for the Desert Locust swarms before their invasion of the Maghreb region.

(ii) The lack of knowledge of the meteorological phenomena in the dry tropical zone and the difficulty to make a reliable medium range numerical weather forecasts (5–10 days) even by the most advanced meteorological centres.

From November 1987 onwards, information from the ECMWF was requested for humanitarian and emergency purposes, and this arrived three months later (February 1988). From the beginning of March and thanks to a link by telefax with ECMWF, the five-day forecasts of wind fields and rainfall at ground level and 850 hpa over the Maghreb area have been used for Desert Locust control. This was organized at the level of the central headquarters, bringing together all the services concerned (plant protection, aviation, meteorology, telecommunication, logistics). Later on, the importance of the plague on the regional side gave rise to a request for the numerical products of the U.K. Meteorological Office. These products covered the gregarious zone from India to the Atlantic via the Red Sea, which allowed a better understanding of the general situation and a better forecast for successive invasion five days in advance. These products, together with the experience gained in the field, have been used by members of the same team of meteorological forecasters from September 1988 onwards successively to reinforce the specialized services of FAO/ECLO. This action has been running now for 18 months and has been considered as advantageous by the Plant Protection Direction of FAO.

This experience, which has been going on intensively for two years, has proved that the meteorological products available at the local and especially at the European level, have contributed significantly to the organization of the Desert Locust prevention and control effort. The meteorological phenomena at the synoptic scale proved to be the major factors in the Desert Locust tragedy. The quick availability of these meteorological products by telefax transmission from the European meteorological centres to the national meteorological centre and then to the local control centres has been decisive in that a wind map is more expressive than a long telex. This technique of meteorological dissemination gave help in real time and enabled specification of some relations between the meteorological conditions and movement of the swarms.

However, some shortcomings have been felt and they concern mainly the meteorological conditions, which affect the locust reproduction in Sahara regions, and the effect of mesoscale meteorological phenomena on the aerial spraying of locust swarms. In view of those deficiencies, the need for a Desert Locust strategy was discussed at the regional seminar at Tamanrasset (May 1988), and later, at the international workshop on acrido-meteorology in Tunis (WMO/FAO, July 1988), it was decided to develop a permanent integrated meteorological and Desert Locust monitoring system to provide the necessary assistance for plant protection survey teams. This system, which will strengthen the World Weather Watch network, should be fully integrated into the actual monitoring network of each country and may be structured as:

(i) A regional meteorological centre for the Saharan area built on the strategic site of Tamanrasset which is in the heart of the Desert Locust breeding area common to Algeria, Niger and Mali and which hosts an old locust research base.

(ii) A national climatic network to complement the actual network in central Sahara, limited in the north by the Saharan Atlas, and which will help to improve the aerial assessment of the rainfall and of air corridors necessary for the development and progression of the Desert Locust.

(iii) A regional synoptic network consisting of 50 automatic weather stations fitted with data collection platforms for transmission via Meteosat and covering the whole area of Desert Locust activity from the north and central Sahara to the Atlantic and the Red Sea (the north of Chad, Niger, Mali, Mauritania, southern Algeria and Libya).

(iv) Five mobile meteorological 'centres', which will complement the basic network in areas where the access is extremely difficult, but which represent important sources during the invasion or remission periods.

This regional and national meteorological monitoring network will rely on and also strengthen the integrated system of the World Weather Watch. In parallel, the Desert Locust control plan should allow for the following actions:

(i) Establish a permanent locust watch (monitoring) system as a main component of the plan of action to strengthen and reorganize the preventive control of the Desert Locust plague at the national, regional and international scale. As a matter of fact, the Desert Locust research base of Tamanrasset could well present an important focal point because of its strategic position in the main Saharan breeding zone.

(ii) Maintain permanently the prospection teams in the field, to allow them, supported by the past, present and forecast meteorological information, to go through all the potential breeding areas where the ecological conditions might be favourable to Desert Locust development. When searching the large desert areas, contact could be made with the nomads or travellers, and even the newly settled peasants, to complete the survey and to cover wider areas.

(iii) Develop plans to train observers and scouts, plant protection inspectors, meteorologists, peasants and even local administrative authorities to build a unified team for locust control. Because of the wide development of agricultural production in the Sahara, all of these have a vested interest in such control.

(iv) Finally, the plant protection services of the different countries should develop a permanent system for the free exchange and dissemination in real time of locust observations and forecasts of their movement. Such exchange should be modelled after the World Weather Watch, which is coordinated by WMO and has already proven its efficacy as an example of international cooperation. This coordination, for instance between the national Maghreb services, should be done by the Regional Desert Locust Commission (CLCPANO) and may take advantage of the FAO regional Desert Locust control project, which is planned to cover eight neighbouring countries of the Maghreb and the Sahel. An exchange of basic locust information could be done directly between the regional centres of north Africa, the Sahel and the Red Sea, the international coordination being left to FAO/ECLO in Rome.

In conclusion, and having in mind the recent field experience of the 1987/1989 Desert

Locust plague, it is extremely urgent to consolidate and perpetuate an integrated monitoring system for locusts and meteorology (acrido-meteorology), knowing that the Desert Locust is a permanent calamity and a constant threat over a long time. This integrated system should allow a minimal use of pesticide, which used in large quantities represents another threat to the environment. Besides this important objective, the meteorological network covering those Saharan areas will also enable better use of the water resources, the range land and the development of agriculture for a better life for the local population.

Discussion

K. A. Browning, F.R.S. (*Meteorological Office, Bracknell, U.K.*). Most of the meteorological data that Mr Boulahya was using were synoptic data. There wasn't much said about mesoscale or local scale data, or about the smaller scale information obtainable from satellites. Is this kind of information used?

M. S. Boulahya. I started at the regional scale only because it is very important. On the local scale, satellite data are used to follow rainfall events. We were grateful in September 1988 to receive a satellite receiving system from the U.K. Government, which has been installed at Tamanrasset. There, it is being used to monitor probable rainfall distribution, with ground teams scouting to verify the information. Secondly, an automatic station transmitting through Meteosat (Data Collection Platform, DCP) was installed at In Guezzam, 400 km south of Tamanrasset at about 20° N, so ground proof of this rainfall information via satellite is also received. This local scale information is used mainly for rainfall distribution and for directing aircraft operations, forecasting low-level winds, etc., for where and when the planes should fly. But forecasting for general aviation was not new in Algeria, only its application to the anti-Desert Locust campaign.

D. E. Pedgley (*ODNRI, Chatham, U.K.*). We cannot fail to be impressed by the energy with which Mr Boulahya has tackled the severe and sudden problems faced by Algeria during the swarm invasions of 1988. He has made use of forecast wind fields, temperature and rainfall, derived from global numerical models, but I wonder to what extent he has found them to be realistic? I ask this because there is a pressing need to validate these forecasts in some parts of the world, not least over Africa.

K. A. Browning, F.R.S. As Mr Boulahya answers Mr Pedgley's point, could he also perhaps comment on the accuracy of the rainfall forecasts, which may be even more vulnerable to errors.

M. S. Boulahya. For winds, pressures and temperatures, the numerical forecasts were alright for our needs up to five days ahead, by using the products of Bracknell and Reading. The rainfall forecasts are not sensitive enough for the locust work, it is necessary to know where grass is going to grow and where the soil is wet enough for egg laying.

W. H. Lyne (*Meteorological Office, Bracknell, U.K.*). The Meteorological Office is a supplier of some of the products used by the Algerian Meteorological Service for locust and weather monitoring, and it was gratifying to learn of the good use to which they are being put by

Mr Boulahya and his colleagues. These products are sent only to Algeria at present, communications being a major problem in the supply of numerical weather prediction products to the African continent.

The Meteosat Meteorological Data Distribution (MDD) mission, now entering its demonstration phase, should provide a means of making such products more readily available to meteorological services within Africa and the Middle East. As well as the standard meteorological products, specialized products more directly related to the locust problem could be transmitted. Examples might include vertical velocity and convergence or divergence. The allied DCP mission should also enable more observational data from Africa both to reach the Numerical Weather Prediction (NWP) centres and to be redistributed back to the continent by MDD.

One of the more specialized products the Meteorological Office supplies is trajectory data, and these have recently been utilized by Mr Thomas of the Food and Agriculture Organization in an investigation into the arrival of locusts in the West Indies in the autumn of 1988. These trajectories were calculated from analysed data archived routinely at Bracknell from the numerical forecast model.

Numerical forecast models are under continuous development, with the present version to be replaced later in 1990. This will have a higher resolution, improved parameterization of physical processes, and should lead to improvements in the quality as well as the range of products.

M. S. BOULAHYA. We are working with Bracknell and the European Centre on the existing trajectory models for following air pollution and nuclear fallout, to refine them for use on the Desert Locust. They need to be more precise because the Desert Locust is an active element and not a passive one. We plan to do this work with colleagues from biological backgrounds. The existing model gives some idea, but not the right one, because it doesn't assume that the locusts stop flying at night, for instance, so we have to stop it at a certain time and then start it again at another time; we don't know how to do this within the present model.

The locust watch system started during the plague years and we hope to continue with it, as it is the best way to assist preventative control.

D. RIJKS (*World Meteorological Organization, Geneva, Switzerland*). On behalf of Professor Obasi, Secretary-General of WMO, I thank the organizers of the meeting for the invitation to participate in the discussions. Mr Boulahya has mentioned the role of weather information in the preparation of forecasts for the development and movement of Desert Locusts, and proposed arrangements to enhance this role. Such use of weather information is one of the public services that is being promoted under the Application of Meteorology Programme of WMO, that can show the economic and social benefits of the use of meteorological information. These economic benefits include not only the timely mobilization of resources for locust control, but also the timely demobilization. A demobilization coming too late can be quite costly. Meteorological information is used to increase the efficiency of the tactical movements of intervention forces. Spray pilots estimate that during the past campaign they spent about two thirds of their flying time in conveyance and one third on actual spraying. The total amount paid for flying time was about $100 million. Therefore, even a modest increase in efficiency of conveyance, through the use of trajectory estimates of expected locust movements, could mean a decrease of millions of dollars in spraying costs. Use of such trajectory estimates would also increase the number of hours of possible treatment of swarms, before flight. Other benefits of a social or environmental nature are more difficult to quantify in monetary terms, but they are recognized by national decision makers as important.

Although improvements in the forecasting of development and movement of Desert Locusts certainly need to be and will be made, the present capabilities have credibility. Therefore the step from a curative to a more permanent preventive locust observing and warning system seems feasible.

Mr Boulahya has stressed the need for a regional approach, which appears to be the only one practicable. To facilitate this, a coding of acrido-meteorological information on the meteorological communication system has been proposed, and will probably soon become operational. This would help to achieve a 'global Desert Locust watch' analogous to the world weather watch of WMO. In addition to this international cooperation, there is need for interdisciplinary cooperation (such as shown by the participants in the meeting in Tunis in July 1988, organized jointly by FAO and WMO), and increased cooperation between European and African scientists and technicians (as witnessed by Mr Boulahya's presence at this meeting). The results of actions on the recommendations of the Tunis workshop have been described in a circular letter from the Secretary-General of WMO, dated 13 October 1989, ref. WMO 36.229/M/AGDL.

The implementation of the 'Desert Locust Watch' does need the installation of about 50 automatic weather stations, transmitting via satellite links, to complement the existing synoptic network and to provide the ground-based information that enhances the value of remotely sensed information and vice versa. A start on the installation of this network has been made in cooperation with bilateral donor agencies. The payoff of such a network will not only be in Desert Locust control, but also in many other fields, including Climate System Monitoring and studies on climatic change.

WMO fully supports the training proposals made by Mr Boulahya. Perhaps, in addition, spray-pilots could be helped to understand better the physiology and habits of Desert Locusts, to help them achieve the highest possible efficiency in control operations. Such type of 'training' for agricultural aviation operators is generally provided and required for operation in the U.S.A.

The most important lesson from a locust control operation that cost about $300 million in the past few years, is that a permanent locust watch, capable of preventive control, at the cost of 1% or less of the curative operation, is a sensible proposition, to be implemented as rapidly as possible. It should, and can use the existing meteorological infrastructure, and provide for complementary infrastructure, if and where requested.

P. M. Symmons (*FAO, Rome, Italy*). FAO is obtaining a trajectory model; which will push swarms around by using the winds at varying heights. The work that will be done on our model will be primarily to explain past events. If the present location of the locusts is known, one can usually tell where they've come from, but that's very different from saying where they are going to go. Work on solitary locusts is even more problematic. We know very little about the controlling mechanisms that will cause non-swarming locusts to take off into the night sky. Predictions of the type Mr Boulahya has been talking about are certainly useful for the direct deployment of aircraft for swarm control, but then it is necessary to predict, to a scale of tens of kilometres in space. I would emphasize that one can go a long way if the reporting and reporting-back system for the control is working well.

Phil. Trans. R. Soc. Lond. B **328**, 585–606 (1990)

Printed in Great Britain

An airborne radar system for Desert Locust control

By the late R. C. Rainey, F.R.S.[1], and R. J. V. Joyce[2]†

[1] *Anti-Locust Research Centre*

[2] *Cranfield Institute of Technology, Cranfield, Bedfordshire MK43 0AL, U.K.*

All elements for a complete and self-contained airborne radar system have now been developed and extensively flight tested, separately and together, in seven countries. The system is capable, in principle, of systematically seeking, locating and maintaining contact with most airborne locust populations; and, under recession conditions, of undertaking immediate control. Against widespread and heavy infestations, the same system would be capable of detection and quantitative assessment of targets, and of quantitative assessment of results of control.

The first element is an airborne Doppler radar navigation system, with precision wind-finding facility, able to seek, locate and explore in detail the semi-permanent zones of wind convergence, towards which incontrovertible evidence collected over 30 years has shown airborne locusts move, and in which they accumulate. The second element is the Cranfield airborne insect-detecting radar for quantitative assessment. Finally, the targets assessed as appropriate would be attacked forthwith, in flight by air-to-air spraying methods refined from those already quantitatively tested and employed in large-scale control operations during the 1950s and 1960s.

1. Introduction

The recent upsurge of the Desert Locust *Schistocerca gregaria* (Försk.) appears to have established a case for radically new options both for monitoring and control. This paper is concerned with specific new options that have become available from research and development in recent decades with aircraft and radar in the context of 60 years of international cooperation in the study of this formidable migrant pest and its environment.

These recent findings have been provided by a series of coordinated programmes of field research, dating from what may have seemed at the time a far-fetched proposition tabled by the Desert Locust Control Organization for Eastern Africa (DLCO-EA) at a Food and Agriculture Organization (FAO) Working Group on Locust Control Methods at Rabat in 1963 (FAO 1963). This proposition envisaged that, by carrying appropriate wind-finding equipment, an aircraft would be able to seek and locate semi-permanent zones of wind-convergence, towards which airborne locusts necessarily move, travelling as they do, consistently downwind (Desert Locust Survey 1962*b*; Rainey 1963*c*), constrained, largely by temperature, to meteorologically low levels. Here they might be assessed by appropriate airborne radar, and attacked in flight by air-to-air spraying methods, which had already been developed, quantitatively tested and applied on a substantial scale during the 1950s and 1960s (Rainey 1958*a*, *b*; Desert Locust Survey 1962*a*). Those experiences provided evidence of greater effectiveness and less environmental contamination from such air-to-air spraying than from conventional methods. In 1963, however, appropriate potential airborne wind-finding equipment was only just becoming commercially available and no effective airborne insect-detecting radar system was yet in sight.

This paper outlines the evidence of progress in relation to the three elements of the proposed

† Present address: Maltfield, Berriew, near Welshpool, Powys SY21 8PG, U.K.

46-2

system, namely: reconnaissance using airborne wind-finding equipment, detection and assessment of flying insects by specially developed airborne radar, and air-to-air spraying of suitable concentrations of flying insects.

2. Airborne wind-finding for locating concentrations of flying insects

(a) History

Recognition of the consistently downwind displacement of swarms had an obvious value in seeking and following them. Already in 1952, special local early morning pilot-balloon observations were found particularly useful in determining the sector to be searched by aircraft for potential target swarms while they were still approaching (Rainey & Sayer 1953). Wind-finding directly from the searching aircraft was, however, found impracticable with the standard visual drift sight of those days, because of the degree of convective turbulence associated with the arid terrain and the flying heights (of the order of a hundred metres). Little was known of how gradual, or abrupt the wind transition might be in zones of wind convergence, and, indeed, doubt was expressed as to whether atmospheric discontinuities exist in the tropics, a view perhaps more relevant to 'oceanic' than continental conditions (Sawyer 1952). In 1953, however, H. J. Sayer had observed the zigzag tracks followed by swarms caught up in the ebb and flow of opposing winds over the Somali Horn, in the Inter-Tropical Convergence Zone (ITCZ) and the associated anabatic winds over the main escarpment. He was later able to show both the sharpness of this windshift, the opposing winds being shown by smoke generators dropped from aircraft at points less than 16 km apart, and the closeness of the association of this windshift with the position and the density of flying locust swarms, an association continuing, sometimes, for a month or more.

In the early 1960s Sayer drew attention to the commercial advent of the self-contained Doppler radar navigation with its facility of accurate wind-finding, and, in 1965, was able to arrange for a demonstration of this system in an Ambassador aircraft. This included an almost instantaneous response to yaw (provided by a touch to the rudder simulating a change in drift at a windshift), and the striking visual presentation of the decreasing windspeed during the final stages of landing and touch-down, manifested by decreasing groundspeed at steady airspeed.

Preliminary wind-finding trials of equipment fitted to a Pilatus Turbo-Porter were undertaken over East Anglia (Rainey 1972 b), the aircraft being supplied by CIBA-Pilatus Ltd of Switzerland as part of the programme of work of the CIBA Agricultural Aviation Research Unit located at the Cranfield Institute of Technology. Doppler radar equipment of similar specifications had recently been tested in a Canberra aircraft of the Meteorological Research Flight (MRF), and in comparing results in a discussion at the Royal Meteorological Society in 1970, D. N. Axford (1970) of the MRF agreed that, at the much lower airspeeds of the Porter, (around 100 knots† compared with 400 knots of the Canberra), r.m.s vector errors as low as about 3 knots were a reasonable expectation for the Porter aircraft, wind being the vector difference between course-and-airspeed and track-and-groundspeed vectors. The accuracy of this wind-finding, together with the manoeuvrability of the aircraft, were, indeed, found to make it possible to locate and explore in a detail not hitherto possible, a variety of wind features of significance to airborne pests (Rainey & Joyce 1972).

† 1 knot = 1.852 km h⁻¹.

[68]

(b) Observations in East Africa

The first field programme was undertaken in 1970 with the then East African Agricultural and Forestry Research Organization, where E. S. Brown's research on the African armyworm was begun, the larvae of this noctuid moth (*Spodoptera exempta* (Walker)) being responsible for devastating crop losses in 1960–61 throughout eastern Africa. Brown's work had shown a series of close analogies between armyworm outbreaks and locust migration (Brown *et al.* 1969) and these had led to weekly forecasts of the likelihood of armyworm outbreaks within districts in Kenya, Tanzania and Uganda. In mid-1970 this service was completing a successful first year (Betts *et al.* 1971; Betts 1976 and, with continuing support, has been maintained since (Odiyo 1979, and this symposium).

A long-standing problem of this species is the fact that in almost every year there is a period of several months when, in common with Desert Locust populations, all contact with the main population appears to be lost, commonly followed by a sudden re-appearance of moths in one or more of the light traps which Brown had established in a network covering much of eastern Africa.

A convergence zone that had already been found significant in relation to such armyworm moth concentrations (and, earlier, to important Desert Locust swarm movements (Rainey 1963c) occurs where westerly winds from the region of the Congo Basin meet the prevailing easterlies, and is known variously as the Congo or Zaire Air Boundary (ZAB), or the African Rift Convergence Zone. On at least two occasions, well established surges of the ZAB have been involved in the start of seasonal sequences of armyworm outbreaks from this area. One was in October 1971, when P. Odiyo (in Rainey 1979) noted that a marked incursion of westerly winds, observed at Lusaka in Zambia on 15 October, had already affected southwestern Tanzania at the time of the egg-laying inferred for the initial infestations of the 1971–72 armyworm season in eastern Africa. Quite independently, this same incursion of westerlies was selected to illustrate a typical surge for the ZAB (Bhalotra 1973). The second of these occasions was in December 1973. In this case, the synoptic analyses of the Malawi Meteorological Department for 21 and 22 December were later found (Rainey 1979) to have shown a temporary eastward surge of the ZAB, at the right time and in an appropriate area to have carried parent moths from known December infestations in Zambia into Mtwara district of southern Tanzania. Infestations there in January 1974 provided the putative progenitors of the sequence of infestations, from February to July, successively in Tanzania, Kenya, and particularly damaging in Ethiopia and Yemen (also in Nigeria at the same time). It would appear that the discovery and destruction of moth concentrations in zones of wind convergence on any of these occasions could have prevented the development of subsequent infestations.

In 1970, the Doppler wind-finding aircraft encountered the ZAB on 10 May as the sharply defined edge of a westerly incursion near Nakuru, Kenya, within 15 km of one of Brown's light traps and within a few hours of what was subsequently found to have been the peak armyworm moth catch of the season in that trap. This catch, moreover, proved to have sampled the parents of a further wave of armyworm attacks in the Nakuru district (Rainey 1972a; Haggis 1979).

(c) Observations in Sudan

The climate of the Sudan is dominated by the annual passage of the ITCZ, its movement northwards heralding the onset of the rains and its movement southward, the return of the dry season. In northern Sudan, between the latitudes of 14° and 19° N where the wind-finding work was conducted, the dry season is uninterrupted for seven months between November and June, and the advent of rains in July restores the opportunity for plant and animal life. The Inter-Tropical discontinuity (ITD), the boundary between the northerly trade and the southwesterly Monsoon winds, has long been recognized by meteorological services from Ethiopia westwards to the Atlantic as the major feature of surface synoptic analysis (World Meteorological Organization 1953; Clackson 1957; Tschirhart 1959, Walker 1958; Osman & Hastenrath 1969), and in northern Sudan its position is recorded on charts no less than eight times per day. These records show not only the seasonal progress and the regular daily movements of the ITD, but also irregular surges in which its position is displaced tens or hundreds of kilometres in a single 24-h period (Haggis 1982). The effect of these movements on the local climate is obvious, even without instrumentation, and insect activity is also clearly affected.

The Doppler equipped Porter located the ITD without difficulty at the first attempt and regularly afterwards, providing, between 29 September and 27 October 1970, a total of 31 traverses at latitudes between 19° and 14° N. Of these, 23 were made by day, between 06h00 and 11h00 (G.M.T. $+2$), mainly at 150 m above ground, and 8 by night, between 02h00 and 06h00 at 300 m. No adverse flying conditions were encountered by day or by night, associated, no doubt, with the fact that the rains of the ITCZ occur well to the south of the ITD, commonly several hundreds of kilometres away. On 26 of the 31 traverses, the main windshift occurred between two successive observations. On 16 of these 26 occasions, the two successive observations were separated by only 3 km and by 7 km on the remaining 10. These main windshifts were between predominantly northeasterly and southwesterly directions, respectively drier and more humid.

Sampling of airborne insects by suction traps from ground level to 15 m above ground and by specially designed nets fitted to the Porter (Spillman 1980 b) provided early Sudan evidence of increased numbers of small insects in the vicinity of the ITD (Bowden & Gibbs 1973; Russell-Smith in Joyce 1976; Rainey 1976 a). Particularly important subsequent observations using ground-based radar (Schaefer 1976) showed the ITD as a very sharply defined discontinuity with a line-echo some 200 m wide (mainly from airborne moths) passing overhead within two minutes of a surface windshift, the windshift and line-echo being in close agreement in position, alignment and speed of displacement with corresponding inferences from the three-hourly synoptic analyses from the Khartoum Meteorological Office. The sharpness of the windshift was indicated by marked differences in the track directions on the plan position indicator of the insect echoes ahead of the line-echo and of those behind it. Aircraft wind-finding similarly showed close agreement in position and alignment between windshift and line-echo at the ITD (Rainey 1976 a, figure 5.13).

By flying a box pattern extending some 25 km on either side of the ITD over the Sudan Gezira in September 1970 (Rainey 1976 a, figures 5.9 and 5.10), it was possible to give a quantitative estimate of wind convergence and demonstrate a net inflow of air into the box on the horizontal plane at a rate of $1.6 \times 10^{-3}\,\mathrm{s}^{-1}$. On the simplest assumptions, this degree of convergence could be expected to increase the area density of airborne insects at a rate of $0.25\,\mathrm{h}^{-1}$ with a standard deviation of $0.11\,\mathrm{h}^{-1}$.

An association of winter displacements of Desert Locust swarms along both coasts of the Red Sea with alternating surges of northwesterly and southwesterly winds, marking surges of another semi-permanent zone of convergence, has long been recognized. This Red Sea Convergence Zone (RSCZ) was investigated briefly in 1970 and, as with the ITD, its location and exploration with the instrumented Porter presented no problem (Rainey 1976a). Observations over the Sudan coast near Tokar showed an air-mass boundary between cooler northerlies and warmer easterlies in close agreement with early inferences on the RSCZ made during a detailed investigation by Pedgley (1966) at DLCO-EA. The windshifts encountered during the two aerial traverses over the sea (figure 5.19 in Rainey (1976a)) were noticeable for brief spells of turbulence in otherwise smooth air, and the wind sequences found on the two occasions were sufficiently similar to justify two-dimensional treatment. This enabled convergence in the main transition zone, which was 3–5 km wide, to be estimated directly from the wind components perpendicular to the orientation of the zone. The convergence so found was at rates of 5 and 6 h^{-1}, at 09h05 and 07h13, respectively, on 17 November.

(d) Observations in Burkina Faso

Evidence was secured during limited wind-finding observations for the World Health Organization in Burkina Faso, of effects on airborne insects of more vigorous, but less persistent wind systems, particularly storm outflows and line-squalls. These provided support for an earlier suggestion (Marr 1971) that such systems could be of importance in the distribution of *Simulium damnosum* Theobald, the blackfly vector of onchocerciasis, and incidentally provided initial circumstantial evidence suggesting involvement of line-squalls in the damaging armyworm invasions of Sierra Leone and neighbouring countries as in 1979 (Rainey *et al.* 1976; Rainey 1989, figure 75).

(e) Observations in Canada

Field work in Canada, undertaken on behalf of the Maritimes Forestry Research Centre of the Canadian Department of the Environment, in connection with the control of the spruce budworm, *Choristoneura fumiferana* (Clem.), similarly directed attention to the contrast between the vigorous convergence at systems with a short lifetime, particularly thunderstorms (Dickison *et al.* 1983, 1986), and convergence of relatively limited intensity persisting for twelve hours or more. The latter displayed little associated weather (though still manifested by appropriately detailed synoptic analysis), and a corresponding absence of adverse flying conditions (Dickison, this symposium). They included sea-breeze fronts and related marine flows (Neumann 1980), the persistence of which was found to result, on occasion, in striking concentration of the spruce budworm moth (Rainey 1976b, 1989; Greenbank *et al.* 1980).

An outstanding case was that of 16–17 July 1974, when well-developed sea breezes moved into New Brunswick across both southern and southeastern coasts and temporarily dammed up a major inflow of moths from the west (Rainey 1989). A line-echo developed over the Chipman ground-based radar at 19h40 A.D.T. (G.M.T.-3), simultaneously with the arrival of the sea breeze front from the east, and stalled just west of the site for 4h. Moth densities were augmented after 22h00 by moths brought into the area on the Fundy sea breeze, which reached Fredericton at 22h15. There, 1.5 h later, an aircraft trap catch provided evidence of an exceptionally high density of moths, averaging 4×10^{-3} m^{-3} from aircraft take-off to the initial cruising height of 460 m. This estimate was in close agreement with moth densities recorded at the time above Chipman, 50 km away, both by radar and by a second trapping aircraft. Still higher moth densities were recorded at 23h00 at Renous airstrip in the central uplands of the Province, 90 km N.N.W. of Chipman, with radar observations of a westward-moving line-echo

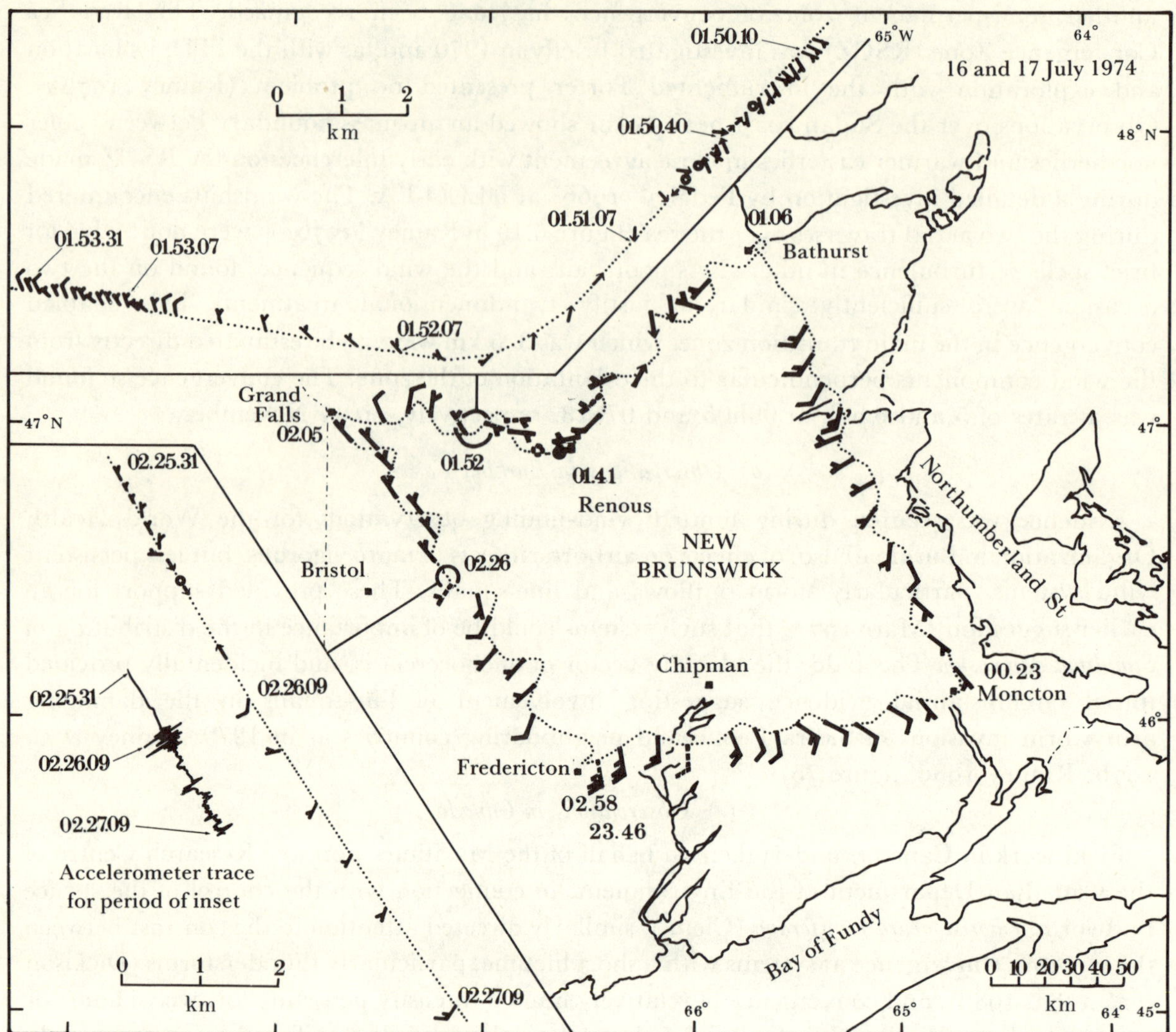

FIGURE 1. Winds measured by Doppler-equipped aircraft, New Brunswick, 16–17 July 1974, from Fredericton via
Moncton to Bathurst at 460 m above sea level (a.s.l.) and thence via Grand Falls to Fredericton at 910 m a.s.l.
Insets: details of winds measured along part of track indicated, and accelerometer trace for main area of
turbulence. Main figure, wind direction: solid arrows measured by multiple drift; dashed arrows by single drift.
Insets, only single-drift winds shown. Wind speed, 1 full barb = 10 knots; small circles denote clam. The 'dog
tooth' flight track was designed to enable multiple-drift wind measurements; times of observation A.D.T
(G.M.T-3) shown at main turning points on main figure, and in insets as h min s.

(suggested as the sea-breeze front from the east coast) with an exceptionally dense layer of
moths 100–250 m above ground at a marked inversion of temperature. This concentration was
confirmed by trapping from aircraft at a density of 10^{-2} m^{-3}. The wind-finding flight, passing
over Renous at 01h45 at 910 m a.s.l., was measured very light southeasterly winds, which were
abruptly replaced 35 km further west by 5–10 knots north to northwest winds (figure 1, top
inset). The Fundy front from the south was last traversed at 02h26 local time, at 150 km from
the coast, still moving inland, and marked by an abrupt onset of turbulence with an upward
gust of 0.43 g, suggesting an up-current of 3 m s^{-1} (figure 1, lower inset), coinciding with a
windshift from northwest to southwest 5–10 knots together with a 0.6° fall in temperature
within a distance of 1.5 km, and a reappearance of moths at an altitude of 910 m at a
temperature of only 12–14°. Very large numbers of moths were also reported on the ground
at Bristol, only 30 km away.   [72]

This case shows the very considerable accumulation of airborne insects that can occur at a front which is relatively inactive in respect of weather (and, in meteorological terms, very limited in vertical extent), in circumstances of a continued and substantial supply of insects arriving from up-wind, comparable, in fact, to the arrival of locusts, in swarms or individually at the ITD from distant sources.

3. Airborne insect-detecting radar (ABR)

The second new component of the Rabat proposition was airborne insect-detecting radar, at the time no more than a concept for the future. Development and production was eventually commissioned in 1974 from G. W. Schaefer and his colleagues at the Cranfield Institute of Technology by the Canadian Forestry Service (CFS), for use in the Service's research programme on spruce budworm moth flight in New Brunswick. Here, as reported by Schaefer (1979), the prototype set was successfully flight tested in 1975 in a Cessna 185 of the CFS, and two production sets were in use in extensive flying programmes the following year, in an Aztec aircraft as well as in a DC-3 in which the Doppler wind-finding system had already been in use during the previous three seasons.

The radar equipment incorporated the same low-cost Decca RM 925 marine radar transmitter/receiver as already used in extensive observations of insect flight from the ground in the Sudan (Greenbank *et al.* 1980.) The airborne installation used an aerial system with a downwardly directed parabolic reflector of 0.91 m diameter fed by a rotating dipole, giving a conical beam of half-power width 2.4°. With a wavelength of 3.18 cm, a peak pulse power of 25 kW, a pulse-length of 0.1 μs, and a pulse repetition frequency of 1760 Hz, individual budworm moths (of body length about 1 cm and weight about 35 mg) could be detected down to the forest canopy from an aircraft height of 750 m, and moths at commonly recorded densities of 10^{-3} m^{-3} detected from a height of 1500 m. Correspondingly, Desert Locusts with a body length of 5 cm and weight 2.5 g could be detected by such a radar system at proportionately greater ranges.

The detected echoes were fed to a range-gating and processor system which sampled echo intensity at 32 levels below the aircraft, starting at 50 m down and ending at about ground level. On the aircraft, the density profile was displayed, pulse by pulse, on an A-scope; and the profile was integrated to display the area density of moths per hectare on a strip-chart. Complete intensity profiles were stored on tape at a rate of 128 s^{-1}. Each 32-level height distribution was recorded in 8 ms; 16 successive measurements were then averaged at each level, to produce smoothed profiles of airborne moth density (and of the predominant moth alignment) for 32 levels at a rate of eight profiles per second, i.e. for every 8 m of aircraft track.

The performance of the equipment was well demonstrated by a unique series of observations of the concentration of spruce budworm moths into a New Brunswick sea-breeze front (Schaefer 1979; Greenbank *et al.* 1980).

Again on 15 July 1976, detailed re-analysis of current local synoptic data directed attention at the pre-flight briefing to an inconspicuous feature over western New Brunswick in which budworm moths were observed by ABR in flight at increased densities (Dickison, this symposium). It was later suggested by C. Miller of the Maritimes Forestry Research Centre, Fredericton, that these moths had been involved in significant immigration of egg-laying moths into the Upsalquich area (Dickison, this symposium, figure 3).

Contrasting conditions were shown on the following night (16–17th), when the ABR equipment recorded moths in numbers 70 km downwind from budworm infestations of unprecedented severity in the Cape Breton area of Nova Scotia, among thunderstorms out to sea over the Cabot Strait and halfway to Newfoundland. These Cabot Strait observations helped to direct attention to the manner in which moth invasion from distances of hundreds of kilometres is likely to have been involved in the unprecedented severity of subsequent infestations in Newfoundland.

4. Air-to-air spraying as control technique of choice

The third element of the Rabat proposition was envisaged as direct attack from aircraft on flying locusts in the concentrations located and assessed by the airborne radar system, by using the special spraying techniques and materials which had been developed in the course of more than a decade of field and laboratory research and development (Kennedy *et al* 1948; Gunn *et al.* 1948; Rainey & Sayer 1953; Joyce 1962, Rainey 1963a), and applied on a substantial scale. These were found to provide rates of insect kill, in numbers of dead locusts per unit of insecticide applied, which have not been approached by any other pesticide application method. This conclusion may be shown by the fully assessed kill of 179 million locusts (about 350 tonnes), with a standard deviation of 24 million, which followed the application of 270 l of 85% Diazinon (half the load of a Beaver aircraft), from a height of 30–60 m among moderately dense flying locusts in a large settling swarm near Hargeisa at 18h05 on 8 September 1957 (Rainey 1958b); visibly contaminated dead locusts were found up to 20 km from the sprayed site. This result would correspond to a complete kill of a swarm covering 4 km² at typical density of 50 million km⁻², namely a swarm larger than most recorded recession swarms.

The technique of air-to-air spraying of locust swarms was placed on a sound theoretical basis by the work of the Porton Laboratory in the 1940s and 1950s, and the mathematical model constructed by Sawyer (1950) allowed the performance of an individual spray-line, of known physical and insecticidal properties, applied within swarms of given density and orientation, to be calculated. This was based on the dispersal of droplets of the spray in accordance with their fall velocity, and the collection of droplets by individual locusts flying in the spray cloud. The resulting graphical treatment proved invaluable in the interpretation of the results of individual field operations against flying swarms (Rainey & Sayer 1953; Rainey 1958b), and in radically improving earlier spray tactics.

More recent studies in Cranfield, particularly by Spillman (1976), have emphasized the potential importance of very small droplets whose numbers in the droplet spectrum could not have been measured in the earlier work. Spillman pointed out that, while catch efficiency increases with droplet diameter and with increase in the speed of the insect relative to the air, it decreases with increase in the diameter of the part of the insect in the droplet's path. Taking the catch efficiency of a droplet of 50 μm diameter as 1.00, Spillman showed that the catch efficiency associated with the heads of flying insects only falls significantly when the droplet diameter is less than 40 μm, whereas for the smaller parts, such as the legs, antennae, and sensory hairs, the catch efficiencies drop by more than 10% only when droplets have diameters smaller than 30 μm. However, the probability of a droplet being caught depends on the product of its numbers and lifetime in the vicinity of the targets, that is, within the swarm. Contact with the ground is the most probable fate of large droplets in a swarm of typical density of 0.01 m⁻³, and this is to be avoided. In still air, the catch probability increases greatly with the increase in impaction parameter for the small droplets. In particular, a 10 μm droplet has a 7.5 times greater probability of being caught by a locust head than does one of 50 μm diameter. This ratio rises to 18.5:1 when considering the catch probability of the smaller parts of the insect.

The Sawyer theory of air-to-air spraying, later elaborated by MacCuaig & Yeates (1972), visualized the release of a curtain of droplets above a swarm, the size being selected to ensure penetration through the depth of the swarm. Because of the low probability of direct hits on locusts by large droplets falling through a swarm, Sayer, in the 1960s, visualized suspending a cloud of small droplets within swarms and, since individuals comprising a swarm are in continual motion, both horizontally and vertically, and the life of the cloud was effectively commensurate with that of the swarm, all individuals would eventually pass through the cloud and collect a lethal dose. This concept receives vindication from the recent Cranfield studies. Moreover, there are strong statistical reasons for using small droplets (Spillman 1980a). Suppose that droplets, each containing one LD_{50} dose, are released in just a sufficient number that, on average, each target receives one contact. Then, assuming each droplet has an equal probability of contacting a target, it can be shown that about 37 % of the targets are likely to collect one droplet, 37 % none and 26 % more than one. If the diameter of the droplets is halved and the same volume of liquid applied, the number of droplets is increased eight times (but each only one eighth as lethal) so that the average number of droplets per target is increased to eight. Now, less than 1 % of the targets will be missed completely and nearly 70 % collect sufficient droplets to provide a dose between 0.75 and 1.25 of the LD_{50}, and less than 30 % are dosed to a greater level. The LD_{50} of 85 % Diazinon is about 1.5 μg g^{-1} of locust, and the object of spraying is to ensure that the vast majority of the individuals in a swarm collect at least two LD_{50} doses, and as few possible more than two. A single LD_{50} dose is contained in a droplet of approximately 240 μm diameter. Application of a spray curtain comprised exclusively of this droplet size, would, at best, result in a large fraction of the swarm collecting no insecticide. On the other hand, by reducing the droplet size to, say, 30 μm diameter, the number of droplets is increased 500-fold and, although 1000 hits are required for an individual to collect two LD_{50} doses, the probabilities are that nearly 100 % of the targets will collect between 0.75 and 1.25 of two LD_{50} doses and virtually none less or more. Unlike the 1960s, atomizers designed for aircraft are now available which will break down a spray liquid to give a narrow droplet spectrum in which the mode of droplet diameter is about 30 μm.

The model constructed by Sawyer took little account of turbulence, which was assumed to affect the locusts equally as it affected the droplets. However, the fall speed of the droplets now considered is so different from that of the locusts, that account must be taken of air turbulence and diffusion around the loci of their path within the swarm. Spillman (1980a) showed that small droplets used in conditions of low turbulence have a relatively high catch probability, particularly on the smaller parts of the insect.

It has long been recognized that the height to which locusts fly is limited by temperature, so that the topmost individuals in a swarm are at a level where the vertical upward velocity of the air has the velocity of the falling speed of a locust, about 1 m s^{-1}. If high flying 'cumuliform' swarms are to be attacked, great care must be exercised in choosing the height at which spray droplets are injected into the swarm. It is likely that this should be near to ground level, outside the swarm, so that droplets entrained in the same air that brings the locusts together impact on the flying insects as the entrained air rises within the swarm.

All air-to-air spraying involves problems of potential hazards of locust impact on the aircraft, for which appropriate airframe and engine modifications were incorporated from 1951 onwards to meet requirements of airworthiness authorities and aviation insurance (Rainey 1958*b*, 1963*a*). With these modifications, more than 3000 h of flying, involving varying degrees of physical contact with flying locusts, have been undertaken without accident attributable to locust impact.

5. Retrospective evidence for possible applications of the ABR system

(*a*) *In recession conditions*

The original Rabat proposals were tabled towards the end of a year of relatively trivial Desert Locust infestations, in striking contrast to the widespread and heavy attacks of the previous 14 years. The ABR was accordingly envisaged, in the first instance, for recession conditions (which, in fact, continued for five years). Moreover, this was the first period of locust recession to be centrally monitored with the aid of current daily synoptic weather charts and analyses by a Locust Forecasting Service, supported and itself monitored by all countries concerned. This made possible, in 1967, a systematic study, in unprecedented, though still not comprehensive detail, of the records of those five years of recession, by staff who had been responsible for the day-to-day monitoring, forecasting and warning service over this period. This study (Rainey & Betts 1979, figure 2) provided a new synthesis demonstrating a quite unexpected and significant degree of continuity of recorded locust populations, by reason of their mobility, and despite the inevitable spatial and temporal gaps in information. These results now make possible preliminary retrospective considerations of where and when the proposed airborne radar system might have been deployed during such a period, and how the operation of the system might have expected to have influenced subsequent locust developments.

In 1967, an important role in a developing upsurge was played by extensive populations of flying locusts produced in May on the Red Sea coast near the Egyptian–Sudan border and with which all contact had been lost during June–October 1967. To these missing populations can be attributed the important subsequent reported breeding in areas of the Atbara valley in the Sudan and the Mourdi depression at the same latitude in Chad, both of which would have come under the influence of the ITCZ during these months. In 1967 (as also in 1950), the ITCZ and its rains appeared to have extended further north than usual in Sudan and Chad. As in 1950, when breeding which took place in the normally arid areas of the extreme north Sudan (such as Wadi Howar) generated swarms which invaded Upper Egypt, the missing locusts of June–October 1967 likewise probably spent enough of these months in the vicinity of the ITD to have made this appropriate for search with a wind-finding aircraft in the same area and at the same time of the year as the first Porter ITD traverse 80 km north of Atbara on 29 September 1970.

From such a base as Dongola, in northern Sudan, a hypothetical 'search and strike' aircraft, with the same radar equipment and a performance comparable with that of a DC-3 (as used in Canada), could have maintained a daily patrol along perhaps 1000 km of the ITD, while also carrying sufficient insecticide (say, 500 l of 85 % Diazinon) to deal in a single sortie with a whole swarm of a few km², the kind of size which appears to be characteristic of recession and early upsurge situations.

On the other side of Africa, early stages of the recent upsurge have suggested a closely analogous involvement of the ITD with missing locust populations and a further retrospective opportunity for the airborne radar system. After a whole year (1984) with no reports of Desert Locust swarms anywhere, the only slender clue to suggest that the species might still be present in numbers somewhere in west Africa was a report of four locusts showing the pinkish colouration which is manifest only by locusts from swarming populations, which were seen in April 1985 near Aioun-el-Atrouss in southeastern Mauritania. Further reports were lacking, and by June it was suggested that Desert Locust numbers might be at their lowest ebb since systematic collection of data started, some 60 years earlier. In late September, however, three ships reported scattered flying locusts at sea off the coasts of Mauritania and Sénégal, extending N.N.E.–S.S.W. over a distance of some 500 km. In early October many hundreds of square kilometers of advanced-instar hoppers (including hopper bands) and pink adults were discovered in southwestern Mauritania, necessitating control on a substantial scale. By the end of the month, there was the first report of a young swarm of the next generation. It was clear that there had been undiscovered breeding on a large scale, on the heavy rains which had been recorded at a number of points in the area in July and early August, and that the locusts seen at sea would have represented some of the last parent population.

Evidence of the close association between the ITD and the position and movements of a swarm in this area had been provided by the ground observations of G. B. Popov at Tamchakett in July, 1959. In view also of the evidence on the ITCZ for the summer of 1985 in west Africa provided by satellite and rainfall data (FAO Desert Locust Summaries 1985), there would appear to be little doubt that the ITD over Mauritania would have been an appropriate feature for ABR to search for the missing west African Desert Locusts in 1985 also.

Sea breeze systems have long been known to have significant effects on Desert Locust swarms. On the north coast of the Somali peninsula, for example, an afternoon sea breeze from a northerly quarter, reinforced by anabatic flow up the neighbouring escarpment, develops during the northern summer, and meets the opposing prevailing S.W. monsoon. During 1943–47, the additional meteorological observations of wartime helped to show that swarms landing on this coast in summer were probably mainly of local origin, swept out to sea by the powerful morning monsoon and subsequently brought in again by the sea breeze front (Brooks & Durst 1935; Rainey & Waloff 1948). Analogous effects of opposing monsoon and sea breeze have been recorded in the coastal areas of Gujarat in northwestern India, with swarms reported almost daily in the vicinity of the front between a regular southwesterly sea breeze and the north–northeasterlies above, clearly isolated from the main infested areas further inland for a two month period from mid-December 1954, and again for a three-week period of November–December 1962 (Rainey 1963 c).

The significance of the RSCZ was again emphasized in 1985, when sufficient locusts to involve control operations had been found on the Ethiopian coast in August and in southwestern Arabia in September. Following sighting of Desert Locusts in numbers on a fishing trawler 40 km off Gizan on 10 November 1985, substantial infestations were discovered in Saudi Arabia from December onwards, initially in coastal areas. By early January, the total area infested between Lith and Qunfidhah was estimated to be 2500 km²; 200 hopper bands had already been controlled since 10 December and a further 200 bands with fledglings and three swarmlets were controlled in mid-January. It may be that the missing locusts of the Red Sea area in October 1985 were in the vicinity of the RSCZ, where they might profitably have been sought.

[77]

During the 43 months of the 1964–67 study period, there were gaps of information, as in 1967, totalling 23 months, with each of these gaps similarly beginning with a disappearance of recorded locust populations and followed by a reappearance of such populations within areas and periods under the influence of the ITD or the RSCZ, accordingly suggesting, retrospectively, appropriate opportunities for ABR search for missing locust populations. In each of these individual cases, the evidence for this inferred continuity is admittedly circumstantial, but the cumulative significance of such an amount of independent circumstantial evidence must be near indisputable. Moreover, from 1935 to 1971, there was, in fact, no period of more than 4 months without swarms somewhere, and, on occasion, swarms were found to show adult life-times of more than 6 months (Rainey 1989). As evidence of the degree of continuity of recession populations began to accumulate in the early 1960s, attention was directed to the potentially cumulative effects of control against sequences of such populations. Possible evidence of such cumulative effects was indeed suggested by the west African sequence of September 1965–March 1966 and that in Pakistan and India of April 1964–September 1965 (Rainey & Betts 1979, figure 2).

With the increased effectiveness to be expected of both the 'search' and 'strike' elements of the proposed ABR system, a corresponding enhancement of the potentially cumulative effects of recession control could be expected.

6. ABR IN PERIODS OF UPSURGES AND HEAVY INFESTATIONS

No study comparable with that of 1963–68 is yet available for a recent period of heavy infestations, but loss of contact with significant locust populations is a problem not confined to periods of recession. In the course of the exceptional and heavy infestations of 1988, contact appears to have been lost for nearly a month with substantial populations of young swarms moving out of northwest Africa (and perhaps also the western Sahel) in late June, until they reappeared as an unexpectedly heavy invasion of Chad and western Sudan in late July, with corresponding difficulties in determining the full extent and distribution of the resulting oviposition and nymphal infestations. In this region and time of year, problems of repeated alternating northerly and southerly movements of locusts with the ITD, night flight and long-range displacement at low densities, may well all have been involved, so that effective reconnaissance would probably have been practicable only with the full facilities of the proposed ABR system.

The ABR system, moreover, provides an opportunity to study in quantitative terms the natural dynamics of locust populations which, at times, exceed 10^{11} individuals and probably rarely total as few as 10^8 (Rainey *et al.* 1979). These individuals are also in almost continuous movement over a total area three times the size of Europe at speeds up to several thousand kilometres per month. If they were evenly distributed over the whole of the 30 million km² of the invasion area, they would constitute no economic problem. However, from time to time, a very large fraction of the total world population is concentrated in a very limited area. For example, the Northern Region of the Somali Republic was invaded between July and August 1960 by nearly 2000 km² of swarms bred in an area of over 500 000 km² in Somalia, Ethiopia and southern Arabia during the previous months. Corpse counts after aerial spraying indicated that this invasion was of the order of 10^{10} individuals and represented the entire Desert Locust population of eastern Africa accumulated and confined by the wind systems already described within an area of some 5000 km² (Desert Locust Survey 1962*a*). Again, analysis of radar photographs made of swarms over New Delhi by the Cloud Physics Research Centre demonstrated the presence of flying locusts within 100 km of Delhi over a total of about 900 km^2 on 27 July and 1400 km^2 on 28 July 1962. Estimated volume densities gave an average value of between 0.07 and 0.13 m^{-3}, with numbers at heights up to 1500 m. These measurements provided an estimate of a total of 10^{11} locusts involved in this invasion. This, moreover, was at a time when swarms were present in a number of other places in India up to 600 km south of Delhi (Rainey *et al.* 1979).

10^{11} locusts, each weighing 2.5 g, represent 250000 tonnes of insects, each eating its own weight of food each day. It is concentrations such as these which create the unique catastrophe of locusts as crop pests. It is such concentrations that can overwhelm local control organizations, but their destruction could have profound effects on the future development of an upsurge. It is at such concentrations that control must be primarily directed and the ABR concept provides the necessary mobility to striking power.

7. FURTHER CONSIDERATIONS

In the 'search and strike' system envisaged against locusts, the role of the airborne insect-detecting radar would be the detection and assessment of locusts in flight, their densities being displayed as smoothed profiles recorded for 32 levels, and giving one such profile for about every 10 m of aircraft track. Downward visual observation from an aircraft above a flying swarm is difficult or impossible by reason of lack of contrast with the usually variegated background (Rainey 1963 *b*). With the radar, satisfactory profiles could be expected throughout the vertical extent of the highest swarm likely to be encountered (say 2000 m), by reason of the relatively modest area densities even of high-flying swarms. The corresponding volume densities would be of particular value in target selection. Where necessary, target identification might be confirmed (as with the spruce budworm) by netting with the specially designed trap (Spillman 1980 *b*). These facilities are of particular potential value in relation to a locust problem so far intractable, namely the assessment of the scale of escaping fledglings moving out of an area of hopper development, commonly at very low density, before assembling into swarms, at times at distances of a hundred kilometres or more. A complication in searching for airborne locusts in the vicinity of the ITD in its northern seasonal position on the fringe of the Sahara, is the marked reduction in flight activity by day to be expected by reason of high temperature, and the probably associated substantial flight activity by night instead. Successful search for locusts under these conditions would be readily possible for the ABR system.

While, with the ABR system in no more than two or three aircraft, no larger than a DC-3 or equivalent, very substantial improvement in overall coverage of Desert Locust infestations could be reasonably expected, this would still be far short of complete and continuous coverage of the entire area in which Desert Locusts may at times occur. The reports of the desert nomads, particularly of the Sahara, have long been recognized as a unique source of information on the desert rains (Dubief 1953). The special degree of intimacy of the desert nomads with the weather, the vegetation and the fauna of their continuously changing environment (Abdallahi 1979; Abdallahi *et al.* 1979) means that fuller use of their information would provide an essential complement to the data provided by the ABR and other remote sensing, of which the satellite infrared imagery described, for example, by Hielkema (this symposium) would be of especial value.

Moreover, in conclusion, perhaps the most important of all considerations on the ABR system would be the outstanding opportunities it could provide for nationals of the developing countries concerned, already with relevant professional training and expertise, for example, from national airlines, to participate and, in due course, accept responsibility for bringing to bear appropriate contemporary technology on what could prove to be a key element in a long-term solution of a major international problem.

REFERENCES

Abdallahi, O. M. S. 1979 Concluding discussion. *Phil. Trans. R. Soc. Lond.* B **287**, 475–476.

Abdallahi, O. M. S., Skaf, R., Castel, J. M. & Ndiaye, A. 1979 OCLALAV and its environment: a regional international organization for the control of migrant pests. *Phil. Trans. R. Soc. Lond.* B **287**, 269–276.

Axford, D. N. 1970 Divergence measurements from an aircraft. Discussion. *Q. Jl R. met. Soc.* **96**, 765.

Betts, E. 1976 Forecasting infestations of tropical migrant pests: the Desert Locust and the African Armyworm. *Symp. R. ent. Soc. Lond.* **7**, 113–134.

Bhalotra, Y. P. R. 1973 Disturbances of the summer season affecting Zambia. (16 pages.) Lusaka: Department of Meteorology.

Bowden, J. & Gibbs, D. G. 1973 Light-trap and suction-trap catches of insects in the northern Gezira, Sudan, in the season of southward movement of the Inter-Tropical Front. *Bull. ent. Res.* **62**, 571–596.

Brooks, C. E. P. & Durst, C. S. 1935 The circulation of air by day and night during the south-west monsoon near Berbera, Somaliland. *Q. Jl R. met. Soc.* **61**, 167–177.

Betts, E., Rainey, R. C., Brown, E. S., Mohamed, A. K. A. & Odiyo, P. 1971 Armyworm forecasting service. *Rec. Res. E. Afr. agric. For. Res. Org. 1970*, pp. 103–105.

Brown, E. S., Betts, E. & Rainey, R. C. 1969 Seasonal changes in distribution of the African armyworm *Spodoptera exempta* (Walk.) (Lep. Noctuidae), with special reference to eastern Africa. *Bull. ent. Res.* **58**, 661–728.

Clackson, J. R., 1957 *Tech. Notes, Br. W. Afr. Met. Serv. Lagos* no. 5.

Desert Locust Survey 1962*a* *Report of the Desert Locust Survey for June 1955 to May 1961*, Nairobi: East African Common Services Organization.

Desert Locust Survey 1962*b* *Final Report of the Desert Locust Survey 1961–1962*. Nairobi: East African Common Services Organization.

Dickison R. B. B., Haggis, M. J. & Rainey, R. C. 1983 Spruce budworm moth flight and storms: case study of a cold front system. *J. Clim. appl. Met.* **22**, 278–286.

Dickison, R. B. B., Haggis, M. J., Rainey, R. C. & Burns, L. M. D. 1986 Spruce budworm moth flight and storms: further studies using aircraft and radar. *J. Clim. appl. Met.* **25**, 1600–1608.

Dubief, J. 1953 *Essai sur l'hydrologie superficielle au Sahara.* (457 pages). Government of Algeria.

FAO 1963 Report on the 2nd symposium on Insecticidal Control of the Desert Locust: Aerial control, Rabat, Morocco. FAO-UNSF/DL/M/3. Rome: FAO.

Greenbank, D. O., Schaefer, G. W. & Rainey, R. C. 1980 Spruce budworm (Lepidoptera: Tortricidae) moth flight and dispersal: new understanding from canopy observations, radar and aircraft. *Mem. ent. Soc. Can.* **110**.

Gunn, D. L., Graham, J. F., Jaques, E. C., Perry, F. C., Seymour, W. G., Telford, T. M., Ward, J., Wright, E. N., & Yeo, D. 1948 Aircraft spraying against the Desert Locust in Kenya, 1945. *Anti-Locust Bull.* **4**.

Haggis, M. J. 1979 African armyworm *Spodoptera exempta* (Walker)(Lepidoptera: Noctuidae) and wind convergence in the Kenya Rift Valley, May, 1970. *E. Afr. agric. For. J.* **44**, 332–346.

Haggis, M. J. 1982 Distribution of *Heliothis armigera* eggs on cotton in the Sudan Gezira: spatial and temporal changes and their possible relation to weather. *Proceedings of the International Workshop on Heliothis management: 15–20 Nov. 1981.* India: ICRISAT, Patancheru, A. P.

Joyce, R. J. V. 1962 *Report of the Desert Locust Survey for June 1955 to May 1961.* Nairobi: East African Common Services Organization.

Joyce, R. J. V. 1976 Insect flight in relation to problems of pest control, *Symp. R. ent. Soc. Lond.* **7**, 135–155.

Kennedy, J. S., Ainsworth, M. & Toms, B. A. 1948 Laboratory studies on the spraying of locusts at rest and in flight. *Anti-Locust Bull* **2**.

MacCuaig, R. D. & Yeates, M. N. D. B. 1972 Theoretical studies on the efficiency of insecticidal sprays for the control of flying locust swarms. *Anti-Locust Bull.* **49**.

Marr, J. D. M. 1971 Observations on resting *Simulium damnosum* at a dam site in northern Ghana. *WHO/ONCHO/*71, 85.

Neumann, H. H. 1980 Prediction of New Brunswick sea breezes. *Am. met. Soc./Can. met. Ocean. Soc.* Second Conference. March 1980. Fredericton: Meteorology of Northern New England and the Maritimes.

Odiyo, P. O. 1979 Forecasting infestations of a migrant pest: the African armyworm *Spodoptera exempta* (Walk.). *Phil. Trans. R. Soc. Lond.* B **287**, 403–413.

Osman, O. E. and Hastenrath, S. L. 1969 On the synoptic climatology of summer rainfall over Central Sudan. *Arch. Met. Geophys. Bioklim.* B **17**, 297–324.

Pedgley, D. E. 1966 The Red Sea Convergence Zone. Part I: The horizontal pattern of winds. Part II: Vertical structure. *Weather, Lond.* **21**, 350–358, 394–406.

Rainey, R. C. 1958*a*. The comparative assessment of locust control methods. *Proc. Int. Congr. Ent., Montreal 1956* **10**, 263–268.

Rainey, R. C. 1958*b* The use of insecticides against the Desert Locust. *J. Sci. Fd. Agric. Lond.* **9**, 677–692.

Rainey, R. C. 1963*a* Tactics and strategy of the use of aircraft in locust control. *2nd Internat. Agric. Aviation Congress*, 219–227.

Rainey R. C. 1963*b* Aircraft reconnaissance and assessment of locust populations. *2nd Internat. Agric. Aviation Congress*, 228–233.

Rainey, R. C. 1963*c*. Meteorology and the migration of Desert Locusts: applications of synoptic meteorology in locust control. *Tech. Notes. Wld Met. Org.* **54**.

Rainey, R. C. 1972*a* Wind and the distribution of the Desert locust. *Proc. Int. Conf. Acridology, London 1970*, pp. 229–237.

Rainey, R. C. 1972*b* Flying insects as potential targets: initial feasibility studies with airborne Doppler equipment in East Africa. *Aeronaut. J.* **76**, 501–506.

Rainey, R. C. 1976*a* Flight behaviour and features of the atmospheric environment. *Symp. R. ent. Soc. Lond.*, **7**, 75–112.

Rainey, R. C. 1976*b* New prospects for the use of aircraft in the control of flying insects and in the development of semi-arid regions. *5th Internat. Agric. Aviation Congress*, pp. 229–233.

Rainey, R. C. 1979 Control of the armyworm *Spodoptera exempta* in eastern Africa and Southern Arabia: report of a mission to formulate an inter-regional project. AGPP: MISC/32. Rome: FAO.

Rainey, R. C. 1989 *Migration and Meteorology.* (308 pages.) *Oxford University Press.*

Rainey, R. C. & Betts, E. 1979 Continuity in major populations of migrant pests: the Desert Locust and the African armyworm. *Phil. Trans. R. Soc. Lond.* B **287**, 359–374.

Rainey, R. C., Betts, E. & Lumley, A. 1979 The decline of the Desert Locust plague in the 1960s: control operations or natural causes? *Phil Trans. R. Soc. Lond.* B **287**, 315–344.

Rainey, R. C., Haggis, M. J. & Coles, R. 1976 An exploration of the wind-systems of the main areas of reinvasion using aircraft. *WHO/OCP/SAP/76/WP6.*

Rainey, R. C. & Joyce, R. J. V. 1972 The use of airborne Doppler equipment in monitoring wind-fields for airborne insects: some recent results. *7th Internat. Aerospace Instrumentation Symp. Cranfield*, pp. 8.1–8.4.

Rainey, R. C. & Sayer, H. J. 1953 Some recent developments in the use of aircraft against flying locust swarms. *Nature, Lond.* **172**, 224–228.

Rainey, R. C. & Waloff, Z. 1948 Desert Locust migrations and synoptic meteorology in the Gulf of Aden area. *J. Anim. Ecol.* **17**, 101–112.

Sawyer, J. S. 1952 Memorandum on the Inter-Tropical Front. *Met. Rep. Lond.* **2**.

Sawyer, K. F. 1950 Aerial curtain spraying for locust control: a theoretical treatment of some of the factors involved. *Bull. ent. Res.* **41** 439–457.

Schaefer, G. W. 1976 Radar observation of insect flight. *Symp. R. ent. Soc. Lond.* **7**, 157–197.

Schaefer, G. W. 1979 An airborne radar technique for the investigation and control of migrating pest insects. *Phil. Trans. R. Soc. Lond.* **287**, 459–465.

Spillman, J. J. 1976 Optimum droplet sizes for spraying against flying targets. *Agric. Aviation.* **17**, 28–32.

Spillman, J. J. 1980*a* The efficiency of aerial spraying. *Aeronaut. J.* Feb. 1980, 60–69.

Spillman, J. J. 1980*b* The design of an aircraft-mounted net for catching airborne insects. Trends in airborne equipment for agriculture and other areas. *Proc. Seminar UN/ECE, Warsaw*, 169–180.

Tschirhart, G. 1959 *Monog. Mét. Nat., Paris.* no. **13**.

Walker, H. O. 1958 The Monsoon in West Africa. *Symposium on Monsoons of the World. New Delhi, 1958*, 35–42.

World Meteorological Organization 1953 The Inter-Tropical Convergence Zone. *Final report on the First Session of the Regional Association for Africa* (pp. 173–198). Antananarivo, Madagascar.

Discussion

M. J. Haggis (*ODNRI, Chatham, U.K.*). I should like to enlarge on the practicality of using an aircraft instrumented for wind-finding and with the various other equipment Professor Joyce has just described. The object of the exercises in which we engaged was, in Dr Rainey's words, 'to locate and explore zones of wind convergence likely to concentrate flying insects'. The most likely place to find such conditions is at a windshift, though more subtle situations occur (as explained by David Pedgley (this symposium)). Having located our potential target, we would measure the extent to which inflow of air exceeded outflow and the relative volume-densities of insects within and around the target area. Thus we hoped ultimately to identify and attack good air-to-air spray targets.

The fine detail included in published results (see, for example, Rainey 1976) was reached after applying numerous corrections to the readings. Such rigorous treatment is not necessary to gain a very fair picture of the windfield immediately, in-flight, even with the limited facilities of the small aeroplane from which so many of our observations were made.

First, how did we measure the wind from a moving aircraft? The wind is the vectorial difference between the direction and speed of movement of an airborne object relative to the air and its simultaneous movement relative to the ground beneath. By quantifying each of these, we could work out the wind by basic trigonometry.

The inertial navigation systems available in the late 1960s measured the aircraft's position relative to fixed beacons, possibly many thousands of kilometres away. For our purposes, using a small, low-flying and slow-moving aircraft, they did not provide a fine enough resolution, you can't draw an accurate triangle of velocities with one side 2 km long and the other two sides 2000 km, possibly even 12000 km long. We therefore used the Doppler system, which is self-contained within the aircraft, not dependent on far-distant beacons, and compact enough to be installed in a single-engined aircraft. In this system, three or four slightly divergent radar beams are simultaneously transmitted from beneath the aircraft, and the back-scatter of the return echo received from the ground is measured to compute the aircraft's displacement, registered as drift and groundspeed. Unlike current models, the version we used did not provide any read-out of windspeed or direction.

The Doppler display includes a roller map which moves under a pen to record the aircraft's position. The map, on which topography and other ground features are shown, is prepared in advance against a predetermined heading which, with map scale, is logged into the navigation computer. The equipment was guaranteed to an accuracy of 5 %; our experience confirmed this and, by manually annotating the trace in-flight, we could retrospectively isolate the major areas of error and identify to within 2 km the ground over which a computed wind observation had been made. In-flight corrections were made by re-setting the Doppler chart above a reference point such as the field base.

Standard aircraft instruments give compass heading, indicated airspeed (IAS) and pressure altitude. From IAS and pressure altitude we obtained true airspeed (TAS) by applying standard corrections, read off a Jeppesen CSG circular slide-rule. TAS and compass heading (with appropriate correction for magnetic variation) give aircraft movement relative to the air; groundspeed and track (true heading plus port drift or minus starboard drift), from the compass and Doppler meter, give aircraft movement relative to the ground. Initially, these vectors were drawn on the Jeppesen computer to obtain the wind vector, and indeed this method proved as quick and reliable as the programmable battery calculator that we also used in later field work.

In addition, we recorded height above ground from the radar altimeter, wet-bulb and dry-bulb temperatures, and the time of observation, a total of nine readings per set.

To minimize instrument errors, in particular the airspeed indicator (ASI), we regularly began field trials by flying, in fairly steady winds, a pattern that permitted multiple-drift wind-finding. The simplest such pattern is reciprocal headings on zero drift, but usually we flew a sexagonal track with each side just long enough to make 3–4 readings. The vectorial mean wind from all readings combined was almost always within 5 knots accuracy and often within 3 knots (Rainey 1972), and provided a reliable standard against which to compare the individual readings for systematic errors. If any were found, e.g. over- or under-reading of the ASI, we could then use a routine 'correction' to the single-drift wind computations.

For the flight trials in U.K. and all the African field work, we used a single-engined Pilatus Turbo-Porter aircraft, in which space was limited to the pilot and one observer in the cockpit, and a net operator with perhaps a second observer in the rear, behind the spray tank. An insect collecting net designed at the Cranfield Institute of Technology to soft-land the catch (Spillman 1980) was mounted under the port wing; the end-cap could be changed in-flight.

In the Porter all recording was done manually, usually at one-minute intervals, though for one observer (M.J.H.) half-minute readings were possible for limited periods of up to half an hour. As much of our flying was done at night, the recording sheets were attached to an aviation knee-pad fitted with a shielded lamp, so as not to interfere with the pilot's night vision, and careful use of a torch was needed for reading the Mk Vb strut psychrometer. This robust wartime instrument was chosen for measuring dry-bulb and wet-bulb temperatures, being suitable for the conditions we expected to encounter, as it was fully calibrated for dynamic heating and lag and the detailed corrections are published (Meteorological Office 1945). These corrections were only applied in retrospective analysis, whereas the others I have mentioned had always to be applied.

In Canada, we were provided with a DC-3 aircraft, which regularly carried a flight crew of two pilots, professional navigator and technician, and 1–4 scientists. All recording was automated: the instrument panel, comprising ASI, Doppler meter, compass, pressure altimeter, radar altimeter and clock was photographed usually at 2- or 12-second intervals, the frequency altered in-flight as required. Wet-bulb and dry-bulb temperatures were recorded as continuous traces from thermistor probes already calibrated for pressure correction. In addition, there was continuous trace output from the accelerometer, recording the fine detail of vertical turbulence and, after the first year, standard weather radar, recorded photographically. These facilities enabled in-flight draft mapping of winds. In 1976 the DC-3 also carried an insect detecting radar, which Mr Wolf operated. Like the Porter, the DC-3 was fitted with a Cranfield insect collecting net.

Coming now to how we used this equipment: it is important to emphasize that all flight plans were finalized and filed only after consultation with the forecasters at the local meteorological office and, in Canada, with additional guidance from Professor Dickison. This meant that we already knew the readily identified synoptic or subsynoptic windfield in which we were to fly and approximately where to be most alert to the details of its features.

For example, before our first interception of the Inter-Tropical Discontinuity (ITD) in Sudan on 29 September 1970 (figure 5.8 in Rainey (1976)) we were informed that the ITD was near Atbara; on arrival there we were told it had moved north about an hour earlier. We found it some 80 km to the north, perceived in-flight by the marked increase in port drift, decrease in groundspeed and change in humidity, and confirmed by working out the winds on the ground at the Station 10 desert airstrip before our return flight.

To quantify the rate of convergence across the ITD, we aimed to fly a 'box' pattern straddling the windshift, as mentioned by Professor Joyce. Although the pattern was planned in advance, its position could only be determined when and where we actually located the ITD: with in-flight wind computing by the second observer, our first 'box' pattern (Rainey 1976, figure 5.9) was begun 12 min after the initial interception of the windshift. A later refinement was to locate the ITD on the outbound flight so that on the return we could change insect nets before the anticipated traverses and again after completing the 'box', thus exposing nets entirely to the south of the ITD, across it, and entirely to the north (Rainey 1976, figure 5.11); a significantly higher insect density was found across the ITD than in either single airmass. On these and other occasions, the actual rate of convergence (e.g. Rainey 1976, figure 5.10) was computed retrospectively.

The next stage was to fly multiple closed patterns across a windshift line, a technique we developed in 1971 for the sea breeze front of southern England, which was being investigated by the Reading Department of Geophysics. Aided by their briefing on the advance of the front, we located it and began the zigzag pattern (Rainey 1976, figure 5.20) about 5 min later; by marking on the Doppler map the areas of turbulence encountered, we were able to reduce the size of zigzags on the return leg of the flight. The same technique was later used on the ITD in Sudan on an occasion of marked turbulence and very sharp windshift, particularly at the western edge of the pattern in figure 5.14 (Rainey 1976) where the groundspeed dropped by 6 knots in about 10 s, precluding any meaningful reading of instruments; however, although we tried to follow the line of turbulence marked on the Doppler map, the 'pattern' was spoiled by the abrupt northward displacement of the ITD by some more than 100 km in 3 h, as shown by subsequent synoptic analyses. This same technique was used in Canada to explore a minor but well-defined front (figure 13 in Rainey (1978)) at which significant convergence was measured (figure 14 in Greenbank et al. (1980)).

A further refinement to the closed pattern for measuring the rate of convergence across the ITD was at the same time to fly a very slow climb and descent, at a rate of about 60 m min^{-1} to obtain a vertical profile through it.

In conclusion, given an aircraft with the equipment described by Dr Rainey and Professor Joyce, and aided by no more than the standard information available at the meteorological office at the local civil airport, locating wind convergence zones, particularly the ITD, is fully practicable. The technology and techniques were available and flight tested 20 years ago, indeed our first flight trials were in December 1969. The techniques will be simpler with present day technology that includes read-out of ready computed winds. Searching for locust swarms might still be partly visual, but the day-to-day area of search could be narrowed down to a belt of, say, 300 km. Scattered locusts, flying even in low numbers, would be picked up on the airborne insect-detecting radar, their volume-density measured and variations in density with height, such as layers, identified. Having located and delimited a sprayable target, the aircraft could then fly a sampling run at the level of the highest insect density to check the identity of the insect echoes and measure the wind within them for optimum direction of spray run, before putting down a spray line. The exact location of each activity would be recorded on the Doppler map, eliminating the need for environmental contamination by repeated spraying over the same ground. The location of visual observations of insects or of other information like green areas (potential breeding areas), of relevance to both air and ground surveys, would also be precisely recorded on the Doppler map.

References

Greenbank, D. O., Schaefer, G. W. & Rainey, R. C. 1980 Spruce budworm (Lepidoptera: Tortricidae) moth flight and dispersal: new understanding from canopy observations, radar and aircraft. *Mem. ent. Soc. Can.* **110**.
Meteorological Office 1945 *Meteorological air observer's handbook*. (73 pages). London; HMSO.
Rainey, R. C. 1972 Flying insects as potential targets: initial feasibility studies with airborne doppler equipment in East Africa. *Aeronaut. J.* **76**, 501–506.
Rainey, R. C. 1976 Flight behaviour and features of the atmospheric environment. *Symp. R. ent. Soc.* **7**, 75–112.

Rainey, R. C. 1978 The evolution and ecology of flight: the 'oceanographic' approach. In *Evolution of insect migration and diapause* (ed. H. Dingle), pp. 34–48. New York: Springer–Verlag.
Spillman, J. J. 1980 The design of an aircraft-mounted net for catching airborne insects. In *Trends in airborne equipment for agriculture and other areas*, pp. 169–180. *Aero-Agro 1978. Proceedings: Seminar UN/ECE, Warsaw.* Oxford: Pergamon Press.

J. R. RILEY (*ODNRI, Malvern, U.K.*). I suggest that the satellite-based Global Positioning System (GPS) should now provide a more accurate and much less expensive method of wind-finding from aircraft, than the Doppler systems described by Professor Joyce and Miss Haggis. I think also, that a forward looking, modified meteorological (precipitation) radar would be more effective in finding swarms than a downward-looking system, especially in the case of high-flying, cumuliform swarms. A downward-looking radar would be useful to make quantitative measurements of the density distribution within the swarms, once they had been found.

D. E. PEDGLEY (*ODNRI, Chatham, U.K.*). There is little doubt that an aircraft can be equipped to seek out windshift lines and sharply defined convergence zones, but Professor Joyce's strategy is based on the hypothesis, put forward by Dr Rainey in the 1940s, that flying locust swarms should tend to accumulate in convergence zones. How often does that happen? Examples can be quoted easily, but how typical are they? I ask because, as far as I am aware, convergence zones did not have a substantial role to play in slowing the movement of swarms that were examined for the WMO *Technical Note* (Rainey 1963) or for the 55 case studies prepared for the *Desert Locust forecasting manual* (Pedgley 1981). The majority of swarms are reported nowhere near convergence zones, either recognizable on synoptic weather maps or inferable from known mechanisms of formation. They are often highly mobile, and when they do slow down is it not usually from low temperatures or onset of breeding? I accept that convergence zones may go unrecognized, especially near coasts and mountains. It is in such limited parts of the Desert Locust area that the feasibility of a search-and-strike aircraft might best be tested for the control of swarms trapped in convergence zones. Two such areas are around the southern Red Sea and near the coast of northwest Africa, both of which are known for the accumulation of swarms at certain times of the years.

What are the chances of finding a worthwhile target in, say, 10 h flying time, and hence the costs in relation to the likely savings resulting from control?

References

Pedgley, D. E. (ed.) 1981 *Desert Locust forecasting manual.* vol. 1, (268 pages); vol. 2, (142 pages). London: Centre for Overseas Pest Research.
Rainey, R. C. 1963 Meteorology and the migration of desert locusts: applications of synoptic meteorology in locust control. *Tech. Notes Wld met. Org.* **54**.

R. J. V. JOYCE. First, regarding persistent zones of wind convergence, I agree these exist only in a few special localities. A notable one occurs in the northern Region of the Somali Republic between July and September, occupying day after day some 5000 km². This zone acts as a temporary sink into which drain the Desert Locust swarms which have bred over some 500 000 km² on the Short Rains in the Somali Republic, the adjacent parts of Ethiopia and in southern Arabia. There have been occasions, as in 1960, when a major fraction of the world population of Desert Locusts has accumulated in this zone, providing a unique opportunity for overall regulation of Desert Locust numbers. A similar persistent convergence zone appears to trap locusts in the Souss Valley of Morocco.

The convergence zone in the Northern Region of the Somali Republic has a diurnal movement; swarms settling in it in the early evening are disrupted when they take off the following morning in the strong southwesterly winds which, during the night, have replaced the convergence zone, and the swarms reform as the zone again moves south during the afternoon. The cohesion and density of the swarms is determined by the intensity of the convergence, opposing winds being separated by only a few kilometres.

Mr Pedgley states, that convergence is largely restricted to mountainous areas and it is here alone that 'Search and strike' would be useful, and that most swarms are in transit, not in zones of wind convergence, so that the contribution of the method to control is likely to be small. I find it difficult to accept this for the following reasons.

Dr Rainey has pointed out that, whilst the convergence hypothesis derived from largely circumstantial evidence based on studies on synoptic scales of 10^2 to 10^3 km^2, it is now supported by evidence on the structure of convergence on much smaller scales involving the flight of individual insects (and not always gregarious ones). These began with the radar studies of Professor Schaefer who observed not only locusts and grasshoppers concentrating in and moving with zones of wind convergence, but also moths of several species. Dr Riley, who continues these radar studies, observed individual armyworm moths moving from both sides into a front caused by the cold outflow from a rain-storm in Kenya at closing speeds of 25 km h^{-1} from points only one kilometre apart.

Again the rate of convergence of 0.25 h^{-1} across the ITD, measured by Rainey and Haggis, is of course the average over the whole 500 km^2 of the 'box' pattern; at the discontinuity itself the rate would be very much higher. The ITCZ, traversing Africa from east to west, is a particularly important regular seasonal system for accumulating swarms in transit during at least five months of the year, and search for locust concentrations along this discontinuity is, I believe the key to locust control. The ABR system makes such surveys possible and is, indeed, the only method which can provide quantitative information on the locusts in transit.

Mr Popov stresses the importance of source areas where breeding has taken place on a scale sufficient to produce swarming populations and the value of satellite imagery in providing a broad indication of areas of current rainfall or supporting vegetation where such breeding may be taking place. These areas can be vast, not less than 5 000 000 ha and frequently 10–100 times more, and must be searched in weeks rather than months. Many such areas are inaccessible to ground parties, either for security reasons or through terrain. They can be searched by airborne observers for gregarious hopper bands because, when basking, even first instars can be readily seen for at least one kilometer from the aircraft track. A properly calculated system of sequential sampling, requiring systematic random search, can be achieved only by aircraft fitted with accurate track-guidance. By this means, areas supporting an unacceptably high population could be identified, subjected to detailed search and, if necessary, control, and estimates could also be made of the numbers of locusts which could be expected to escape from those areas where the survey had indicated populations were within the acceptable range.

In this opening remarks, Dr Rainey drew attention to the two conflicting views on the origin of Desert Locust upsurges, but to me, the requirements of both point to the need to employ a specially instrumented aircraft, which we now connote the ABR system.

The actual cost of the Insect Detecting Radar is quoted by the Cranfield Institute of

Technology at U.S.\$75000 and to this capital expenditure must be added the cost of suitable spray gear and a few other essential items for which a further U.S.\$25000 is allowed. Installation of specialized equipment into the aircraft is estimated at U.S.\$230000, giving a total capital expenditure of some U.S.\$330000.

Operational costs include the wet charter of a suitable aircraft with an endurance of 6 h, interest on capital outlay and a provision of 15 % of the capital for maintenance and up-dating of equipment. For a nine months charter and 600 h flying with a crew consisting of a pilot, two navigators and one operator of the specialized equipment, I estimate the cost at U.S.\$900 per flying hour (table 1). Thus the annual recurring budget for the 'Search and strike' operation is estimated at U.S.\$640000 in which estimate is a provision of U.S.\$100000 for insecticide.

TABLE 1. SEARCH AND STRIKE FOR DESERT LOCUST CONTROL USING AIRBORNE RADAR SYSTEM

	U.S. \$	
Capital costs		
airborne insect detecting radar system as specified by CIT		
including 4-channel strip chart recorder	50000	
data analysis		
microcomputer for data analysis & storage		
printer and plotter		
software for data collection		
consumables (paper, magnetic tapes etc)	25000	
Camp & domestic equipment	25000	
		100000
Preparation		
fitting of spray gear		
(Cranfield Venturi system)	50000	
navigation system	55000	
insect detecting radar	90000	
certification	35000	230000
total costs		330000
Operational costs		
interest on capital costs (at 8%)	26400	
maintenance & up-dating capital equipment	15000	
wet lease for 9 months		
(aircraft, crew, maintenance, insurance)	362000	
flying costs, say, 600 h at \$219 h^{-1}	131400	
Total for 9 months operations		534800
Cost per flying hour	900	
Contingency costs		
subsistence: 4 crew at \$40 per day	50000	
insecticide	50000	100000
Total recurrent budget provision (9 months)		634800
Total recurrent budget provision (in £ sterling)		400000

R. B. B. DICKISON (*Department of Forest Resources, University of New Brunswick, Canada*). Controlling eastern spruce budworm populations by spraying against migrating moths is obviously appealing to pest managers in Canada, but there are practical limitations. First, the feeding period of the insect (the larval period) is past, and the only accomplishment would be to limit the transport of viable eggs to new areas to establish or augment a larval feeding population in the following year. This is partly mitigated because 50 % or more of the eggs are deposited before the female will emigrate. Secondly, environmental concerns have made it impossible, from a socio-political point of view, to spray under conditions where the ground surface impact area cannot be determined. Current regulations in New Brunswick prohibit spraying within one mile of human habitation, even for air-to-ground applications. It may be that the environmental impact is more perceived than real, but those who make political decisions are very reluctant to make such distinctions. This, of course, applies only to use of chemical insecticides, but, although pheromone control is under investigation, there is little encouraging evidence of an alternative to chemical insecticides against moths. Indeed, pheromones cannot be used against migrating spruce budworm moths, since mating takes place several days before emigration. Such environmental concerns are overwhelming for many perceived strategies of spraying airborne insect targets. Wherever human population densities are high, and environmental consciousness is keenly honed, I expect the strategy is impractical.

P. R. JONAS (*Institute of Science and Technology, University of Manchester, Manchester, U.K.*). In the paper, attention is drawn to the sizes of the droplets needed to optimise the effectiveness of spraying into an insect cloud. The results apply only to non-volatile sprays. If more environmentally friendly sprays are developed with an aqueous or other volatile base, then evaporation may significantly influence the effectiveness of spraying. Evaporation is particularly rapid for the smaller drops which, on the basis of the calculations, appear to be the most effective.

The calculations of spray effectiveness rely on the ratio of the mean time for droplets to be captured by insects to the residence time of the spray droplets in the cloud. Laboratory experiments on droplet capture by droplets suggest that moderate levels of turbulence in the droplet cloud may significantly increase the capture efficiency for droplets around 10 μm radius. It is possible therefore that your calculations of spray effectiveness may be rather pessimistic. It would be of interest to compare the results of the calculations with detailed observations of the capture by insects, possibly by spraying an inactive tracer.

In weather modification experiments, where the objective is to place material into particular regions within an orographically forced cloud, ground based aerosol generation is often used, with the air motion carrying the material into the cloud. Is the use of ground based sprayers feasible in this application, especially where pests are concentrated by stationary convergence induced by topographic features which might be predicted some time in advance?

R. J. V. JOYCE. With regard to the formulation employed, we have always insisted that it must have a high degree of non-volatility, so the droplets which are produced at the nozzle remain that size until they are collected by the target. I don't think I need enlarge very much on capture efficiency, except to say that although it is greatest for droplets of diameter of about 50–60 μm, they will occupy the same air as the flying locusts for such a short time, that the decreased efficiency with which droplets of, say, 10 or 20 μm are captured is fully compensated by the fact that they are in the same air for so much longer.

As for injecting spray from the ground, I don't think that is at all practicable in the Sahel of Africa, either in terms of locating the convergence zone, or for that matter, in logistics.

Phil. Trans. R. Soc. Lond. B **328**, 607–617 (1990)

Printed in Great Britain

Detection of mesoscale synoptic features associated with dispersal of spruce budworm moths in eastern Canada

By R. B. B. Dickison

Department of Forest Resources, University of New Brunswick, Bag Service no. 44555, Fredericton, New Brunswick, Canada E3B 6C2

Radar studies in eastern Canada of spruce budworm moth distribution patterns and associated windfields frequently revealed mesoscale synoptic features induced by the strong thermal contrast between the heated land surface and the surrounding coastal waters that resulted in significant redistribution of the airborne moths. Experience gained during a four-year study in New Brunswick enabled meteorologists to identify the synoptic situations favouring the development of these mesoscale features. This paper examines the details of a particular case, on 15–16 July 1976, when insect detection teams were alerted in advance to the existence and location of a significant wind convergence zone.

1. Introduction

A collaborative study of spruce budworm moth dispersal in eastern Canada in the mid-1970s (Greenbank *et al.* 1980) involved the use of ground-based and airborne radar for detecting insects, and the use of a Doppler radar equipped aircraft for determining windfields. The project was coordinated by the Canadian Forest Service, but the radar and atmospheric components of the study were mainly carried out by researchers from the Cranfield Institute of Technology and the (then) Centre for Overseas Pest Research. The author provided meteorological forecast interpretation during the study.

This paper deals with operations in 1976, when two aircraft were used: a DC-3, primarily for windfield determination but also equipped with insect-detecting radar, and a Piper Aztec equipped for insect detection. The equipment and operating procedures are described in detail by Schaefer (1979) and Greenbank *et al.* (1980).

(a) *The eastern spruce budworm and its dispersal*

The eastern spruce budworm *Choristoneura fumiferana* (Clem.) spends only about one week of its one-year cycle in the moth stage. In eastern Canada, phenological spacing spreads this period throughout the month of July. Female moths emigrate from the forest canopy in vast numbers each evening, and remain aloft in dispersal flights of a few hours duration (Greenbank *et al.* 1980), at altitudes mostly 100–300 m above ground. While in flight the moths are often subject to significant redistribution within the general windfield, either due to migratory meteorological systems (Dickison *et al.* 1983, 1986) or systems, such as sea breezes, which are features of the local geography.

(b) *Significance of convergence zones to dispersal*

Wind convergence zones are not only of concern with respect to their effect on any particular occasion, but collectively, they may indeed be the key factor in establishing new outbreaks and thus sustaining the species. It was once thought that outbreaks resulted when conditions locally favoured a rapid increase in an endemic population – the 'epicentre' concept (Hardy *et al.* 1983). Later, Royama (1984) presented analytical evidence against the epicentre concept. Attempts at modelling the spread of epidemics suggested that dispersal in the moth stage was an absolutely crucial factor and more complex than originally postulated (Clark 1979); it was this requirement in the population model which led to the intensive study of moth dispersal. Even after incorporating this new knowledge into the model, Clark (1979) concluded that extensive dispersal alone cannot account for new outbreaks, unless a population threshold of some critical level is exceeded. Rainey & Haggis (1988) then offered the 'new hypothesis' that, although general uniform moth dispersal may not be a critical factor, immigration in situations with marked convergence zones embedded in the general windfield would lead to a new outbreak 'without necessarily involving any prior build-up or even existence of local low density populations.' T. Royama (personal communication) maintains that this still does not fully explain the mechanisms for population outbreaks.

Early in the study it became apparent that sea breeze effects, especially those induced by the very cool Bay of Fundy to the south, played a significant role in the distribution and movement of the moths. Inland air temperature in July is markedly warmer than that of the sea which borders New Brunswick to the south and east. This thermal contrast results in changes in the synoptic pattern, which is frequently characterized by mesoscale features induced by regional geography. Neumann & Mukammal (1979) examined critical conditions for the establishment of marine air intrusions into the province and Burns *et al.* (1980) examined the particular case of easterly flows. If the windfield features (particularly zones of convergence) can be detected and forecast, and their occurrence and movement considered in relation to insect development and behaviour (period of moth activity, time of exodus flights, temperature and light controls, duration of flights), an opportunity exists for better monitoring and management of the movement of insect populations and better control of epidemic outbreaks.

2. Identification and forecasting of wind convergence zones

The analytical and forecast problems related to passive insect dispersal are mesometeorological in scale – specifically the meso-β scale as defined by Orlanski (1975), with time-scales 2–24 h and space-scales 20–200 km. As such, they are barely sub-synoptic in most regions (and certainly sub-synoptic in New Brunswick, where, at the time of the exercise, there were only five synoptic meterological observing stations in an area of nearly 75000 km^2). Most mesoscale concerns relate to severe weather phenomena, for which weather radar and (to a lesser extent) satellite imagery can provide the observational base. Phenomena significant to insect dispersal, however, are likely to be too benign to show up on radar or satellite imagery.

A workshop on the problem of forecasting mesoscale features, sponsored by the Canadian Meteorological and Oceanographic Society (1983), stated that 'forecasting of the probability of the occurrence of mesoscale events is often possible through deductions made from synoptic scale information,' and that 'It is also possible to infer the presence of mesoscale structures from the time sequence of events at the stations of a conventional synoptic scale network.' However, many of the features important to insect dispersal are not only difficult to detect by radar or satellite imagery, but are so obscure and generally irrelevant as to escape the notice of an operational forecaster. On the other hand, an alert analyst, if conditioned by local experience and solely dedicated to the insect problem, can often identify and forecast the location and behaviour of significant features sufficiently well to provide valuable guidance to the pest manager.

This paper reviews the conditions for the development of non-migratory mesoscale structures, detectable from synoptic data and therefore forecastable, focusing mainly on a particular case: the evening of 15 July 1976, during one of the insect dispersal/windfield detection exercises conducted in New Brunswick. In this case, and practically all of those which received special attention during those exercises, the features were apparently coastal in origin, induced by the thermal contrast between land and sea.

3. Marine flow convergence zones

Features of coastal origin have been called 'sea breezes' and 'marine flows,' but the distinction may be largely semantic. Neumann (1980), examining sea breezes during the spruce budworm moth dispersal study in New Brunswick, used a method of classification based on the lake breeze work of Biggs & Graves (1962) and Lyons (1972). This method was successful in predicting many of those sea-breeze cases in New Brunswick which resulted in line concentrations of airborne moths detected by radar (Schaefer 1979). The deep inland penetration of the sea breeze (Schaefer 1979) resembled cases described by Simpson *et al.* (1977) in southern England.

Burns *et al.* (1980) described synoptic situations that could be recognized subjectively by an analyst as favouring development of marine flow convergence zones. They reviewed cases from the moth dispersal study to show the synoptic conditions that favour these developments – essentially, whenever a flat pressure field tracks into the region at the time of day when the

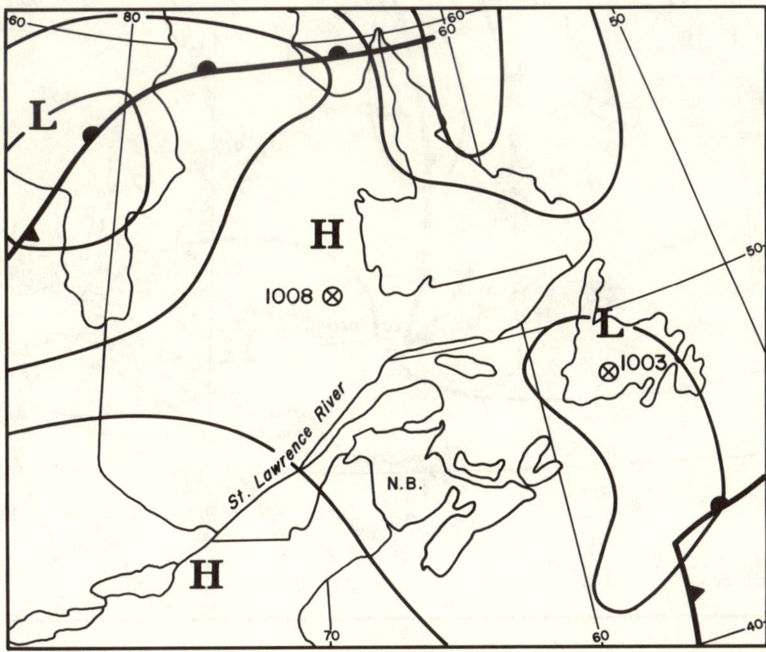

FIGURE 1. Canadian Meteorological Centre surface analysis for 15h00 A.D.T. (18h00 G.M.T.) 15 July 1976. Isobar interval = 4 mbar.

thermal contrast is greatest. Although the evening of 15 July 1976 was not among those cases described by Burns *et al.* (1980), it is nevertheless an exemplary case, and its detection allowed researchers to locate and document significant moth concentrations.

(a) Synoptic situation on 15 July 1976

At 18h00 G.M.T. (15h00 A.D.T.) the Canadian Meteorological Centre showed New Brunswick in a region of slack pressure gradient (figure 1). A weak ridge of high pressure lay just north of the St Lawrence River, with a very weak northwesterly gradient flow across New Brunswick behind a weak low pressure centre over Newfoundland. The diabatic effect of daytime heating over the land led to slight pressure decreases, and a marine flow developed. At the time of the pre-flight briefing, 18h00 A.D.T., an easterly flow was evident over most of the province and a north–south convergence zone was located over western New Brunswick. Initially it was positioned near the Maine–New Brunswick border, but post-analysis (figure 2) suggests a location at 18h00 A.D.T. about 50 km further east. Four to six hours later, the insect detection flight showed concentrations of airborne moths at about the position shown for the front at 18h00 A.D.T. There is no concrete evidence on which to judge the motion of the front, but a westward movement of 50 km in that time seems reasonable.

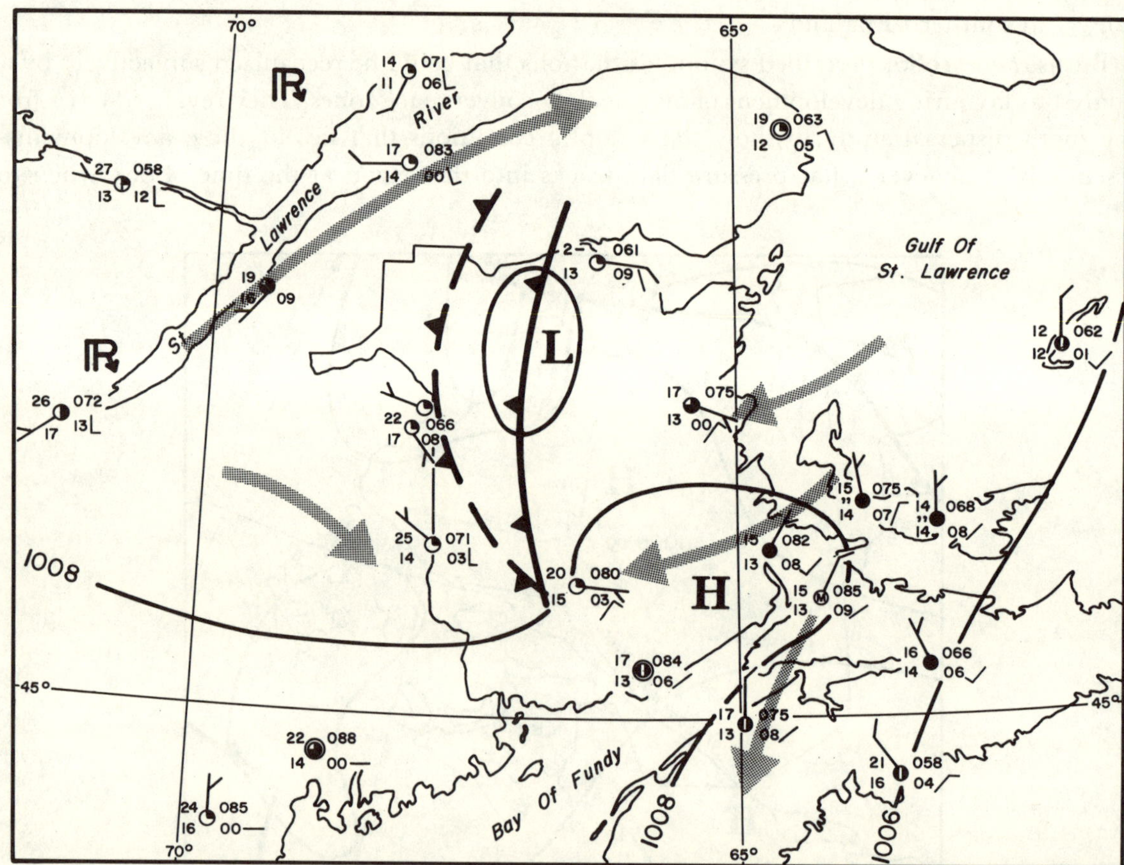

FIGURE 2. Detailed surface analysis for 18h00 A.D.T. (21h00 G.M.T.) 15 July 1976. Isobar interval = 2 mbar. Broken line, analysed position of sea-breeze front shown at pre-flight briefing; solid line, position determined in final analysis.

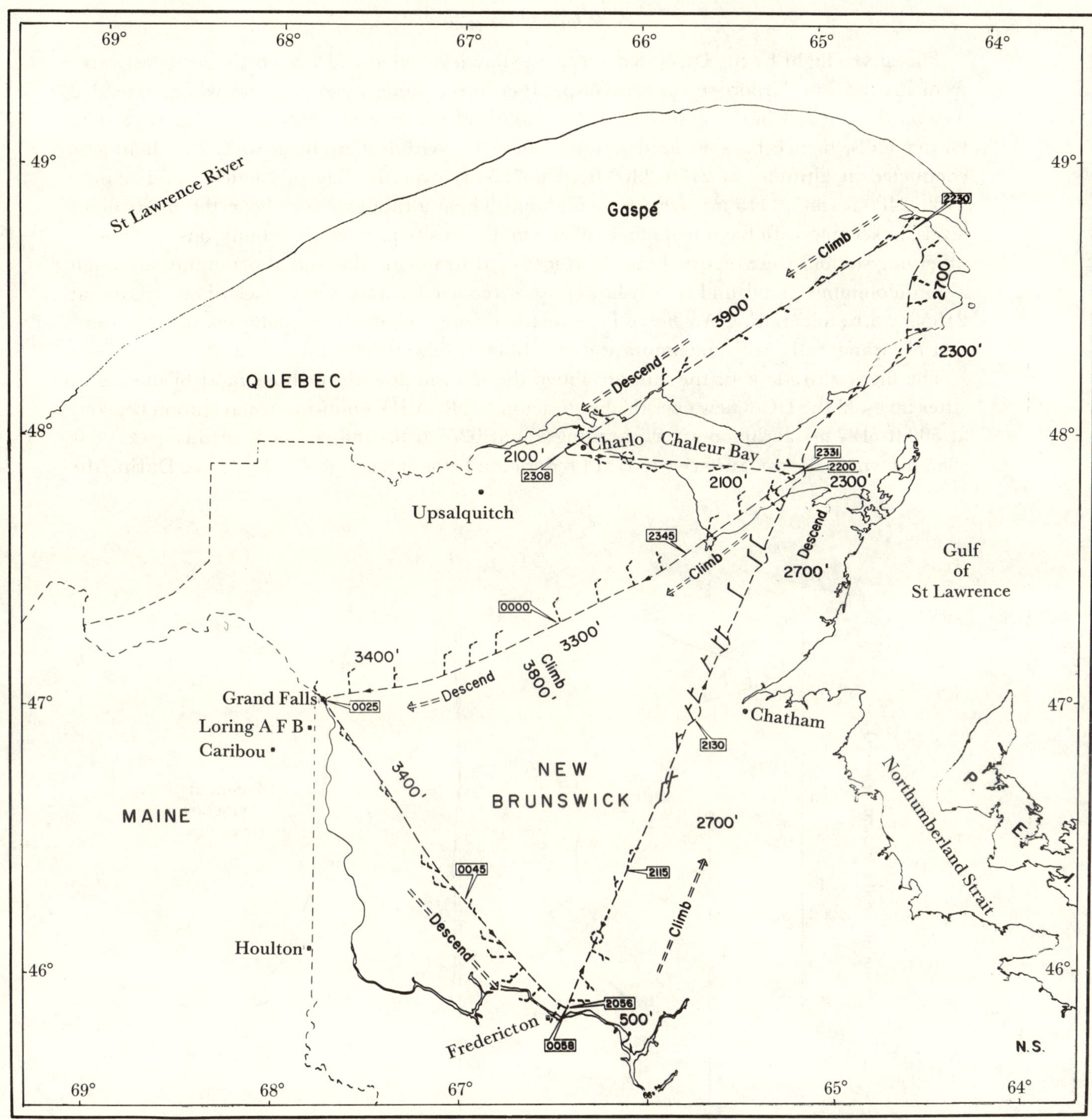

FIGURE 3. Flight track of DC-3, and Doppler drift wind measurements over New Brunswick and Gaspé, 20h56–00h58 A.D.T. 15–16 July 1976. Aircraft winds: ⌐——, multiple-drift winds: ⌐_ _ _, single-drift winds; 1 full barb ≡ 10 kt (5 m s⁻¹). ◯, calm; →----, aircraft track (flight altitudes shown in feet above mean sea level, alongside track); times of observations are shown boxed. (Data supplied by R. C. Rainey and M. J. Haggis.)

(b) Windfield detection flight

The survey flight by the DC-3 to describe the low-level windfield was north–northeastwards from Fredericton Airport to eastern Gaspé, thence returning by zigzag legs which extended westward to near Charlo, eastward to the mouth of the Bay of Chaleur, southwestward to Grand Falls, thence back to Fredericton (figure 3). Within New Brunswick, the flight was conducted at altitudes of 2100–2300 ft (690–755 m) over the Bay of Chaleur, and mainly 2700–3400 ft (885–1115 m) over the mainland. These altitudes were above the moth flight levels, in keeping with the intent of measuring moth density profiles. Sky conditions at synoptic observing stations were reported mostly as scattered stratocumulus and altocumulus, although a cumulonimbus cloud and a rainshower were reported within sight of Caribou, Maine at 21h53 A.D.T., and Loring Air Force Base weather radar registered a small area of echoes just east of Grand Falls, with maximum tops at 16000 ft (about 5250 m).

The flight altitude kept the aircraft above the marine flow throughout most of the flight. After take-off, the DC-3 flew over the Fredericton VOR (VHF omni-range navigation beacon) at 500 ft (197 m) at 20h56 A.D.T., heading about 025° true, and climbed steadily to 2700 ft (885 m) within 65 km. Surface winds at Fredericton before takeoff were 120°/5 kts. During the

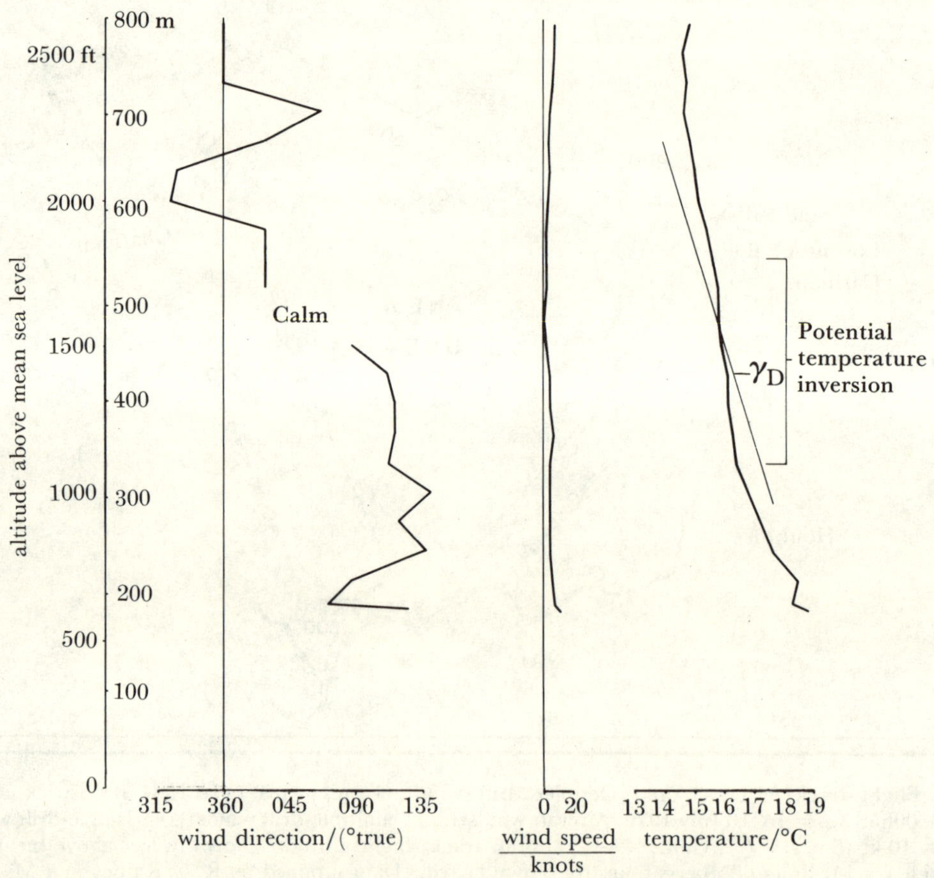

FIGURE 4. Vertical profiles of wind direction and speed, and air temperature, measured by DC-3 climbing northward from Fredericton, 20h57–21h15 A.D.T. 15 July 1976 (data supplied by R. C. Rainey and M. J. Haggis). γ_D indicates neutral (dry adiabatic) lapse rate.

climb, winds remained southeasterly to about 1500 ft (460 m), then shifted to northerly (figure 4); this windshift, and the corresponding potential temperature inversion through the layer 1100–1800 ft (330–550 m), placed the marine frontal surface at about 500 m. Beyond Chatham, winds were generally northwesterly and the aircraft was clearly above the marine air.

The intrusion of marine air, as evidenced by surface observations, extended to well west of Fredericton. Reports from Houlton, Maine showed a wind shift to southeasterly around 21h00–22h00 A.D.T., accompanied by a temperature drop of 5° C over two hours. Measurements from the aircraft, however, showed that the marine air was confined to a fairly low layer, as it was encountered on descent northwest of Fredericton, probably just above 2000 ft (near 600 m). Single-drift wind calculations above 2000 ft indicated backing of about 35°/100 m – from 297° 08 kt at 2500 ft (670 m) to 223° 13 kt at 1900 ft (580 m) – during descent within a marked potential temperature inversion of about 1° C per 100 m.

(c) *Insect detection flight*

The Piper Aztec was dispatched to northwestern New Brunswick to document moth populations within the convergence zone identified at the pre-flight briefing (figure 2). Figure 5 shows the track of the aircraft during five 24-min flight segments of radar sampling, over the period 22h01–00h17 A.D.T. The aircraft was flown at heights of 500–540 m above ground.

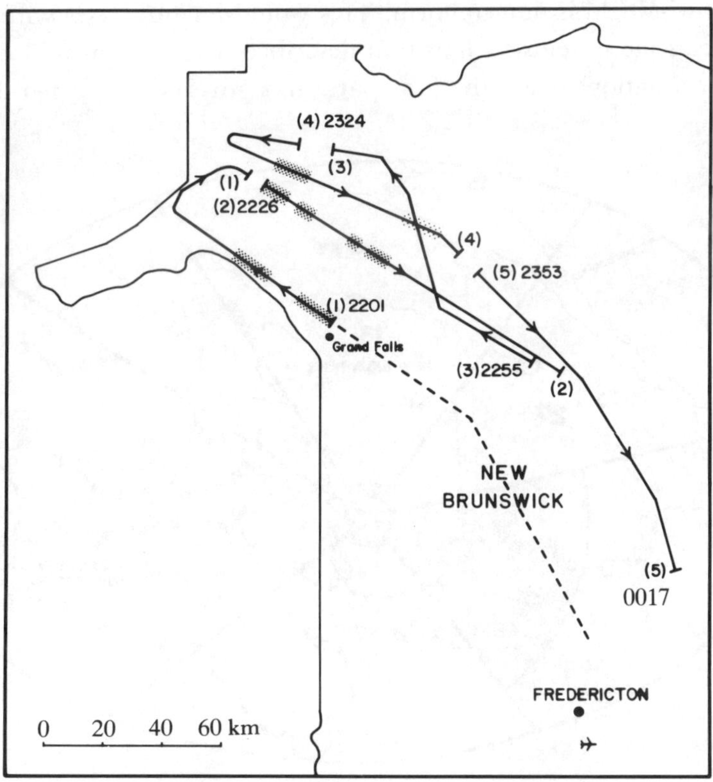

FIGURE 5. Track of Piper Aztec during radar insect-detection flight over northwestern New Brunswick, 15–16 July 1976, 22h01–00h17 A.D.T. (data supplied by K. Allsopp, Cranfield Institute of Technology). Stippled segments show zones of dense concentration of airborne moths; data in lightly stippled zone uncertain.

Areas of airborne moth concentration were observed on flight segments 1, 2 and 4 (figure 5), generally in the western part of the area sampled. The available information gives relative integrated echoes of moth populations in two layers, 75–315 m below the aircraft, and from thence to ground level (rather than detailed profiles as given by Schaefer (1979) for a case five days earlier). The moth density data cannot, then, be used to determine the structure of the meteorological feature; however, the presence of the concentrations is consistent with the existence of a zone of wind convergence in the area identified in advance by analysis of the surface meteorological observations.

4. APPLICABILITY OF THE RESEARCH EXPERIENCE TO PEST MANAGEMENT

The population dynamics of the eastern spruce budworm are related closely to the availability of extensive areas of mature forest. Management intervention in recent years has focused on keeping this forest in a marketable condition during an epidemic (Clark 1979). If it were feasible to control populations by spraying airborne moths – which, for sociopolitical reasons, is considered unacceptable – meteorological service support would be essential. But, during an epidemic, management strategy should at least include a programme of monitoring moth dispersal, which would include the use of radar equipped aircraft for insect detection and windfield determination (Rainey & Haggis 1988) with strong meteorological interpretation support.

During the moth dispersal phase intensive monitoring could be restricted to those evenings when the regional windfield is non-uniform. This would include cases with transient systems, and those with mesoscale systems such as that described here. One would need to distinguish, therefore, between situations when the flow pattern is governed by general synoptic patterns

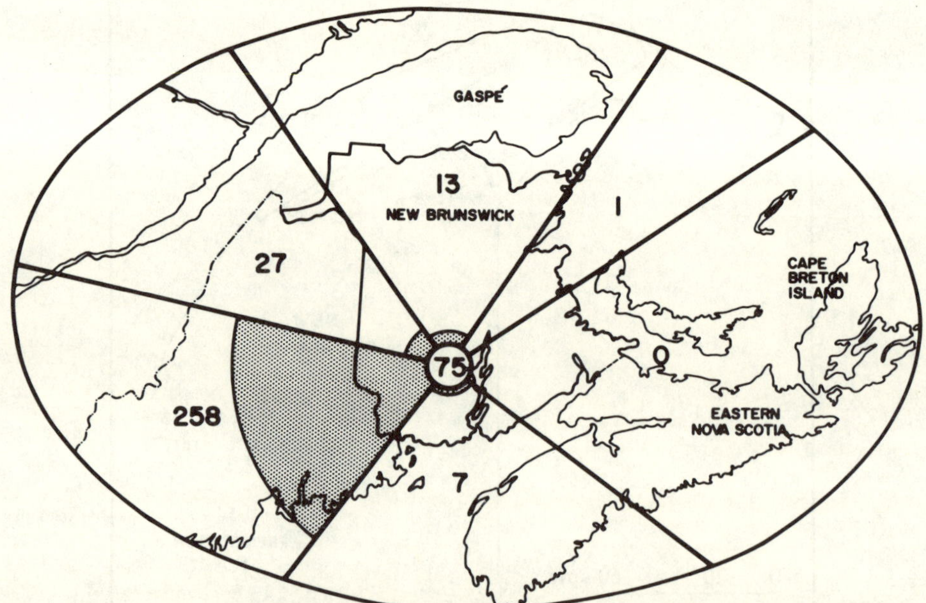

FIGURE 6. Frequency of occurrence of gradient-level winds flows within specified sectors over south/central New Brunswick during 586 July evenings with temperatures favourable for spruce budworm moth exodus, 1958–1976. The value in the centre of the circle is the frequency of light/variable wind conditions. On the remaining 205 evenings, temperatures were too cool for exodus flights. After Dickison (1989).

and situations when sub-synoptic features may develop. The latter become of consequence only when the gradient of the general pressure pattern weakens. In an examination of synoptic maps of 586 July evenings (00h00 G.M.T., or 21h00 A.D.T.) during a 19-year period (Dickison 1989), 205 evenings were too cool for moths to undertake exodus flights from the canopy. For the remaining cases pressure gradient patterns were examined for the area centred on Fredericton. It was found that gradient-level flows were predominantly from westerly sectors (figure 6). Easterly flows at the synoptic scale were found to be extremely rare. This does not, however, represent an appraisal of the likelihood of easterly sea breezes, because these develop when general pressure gradients are too weak to be classified with respect to direction. Realistically, the 75 cases (12.8 % of evenings) with light/variable wind conditions represent evenings when sub-synoptic features could be expected to develop.

I thank Dr Rainey and Miss Haggis for their support and encouragement in preparing this and other manuscripts, and for the extraordinary stimulation which our associations have provided. We are grateful for the support of the late Dr Glen Schaefer and Keith Allsopp, Cranfield Institute of Technology, and to Dr Frank Webb and David Greenbank of the Canadian Forest Service.

REFERENCES

Biggs, W. G. & Graves, M. E. 1962 A lake breeze index. *J. appl. Met.* **1**, 474–480.

Burns, L. M. D., Dickison, R. B. B. & Neumann, H. H. 1980 Easterly flows in New Brunswick during July. *Am. met. Soc./Can. met. Ocean. Soc. Second Conference. March 1980.* Fredericton: Meteorology of Northern New England and the Maritimes.

Canadian Meteorological and Oceanographic Society. 1983 *Mesoscale Meteorology, Research Planning Workshop. January 24–26. 1983.* Toronto: C.M.O.S. Scientific Committee.

Clark, W. C. 1979 Spatial structure and population dynamics in an insect epidemic system. Ph.D. thesis, University of British Columbia, Vancouver.

Dickison, R. B. B. 1989 Forest insects: recent research on development, larval behavior and adult dispersal, and efficacy of control methods. In *Proceedings Symposium/Workshop Climate Applications in Forest Renewal and Forest Production. November 1986*, pp. 165–170. Orillia, Canada: *Can. For. Serv./Ont. Min. Nat. Res./Can. Atmos. Environ. Serv.*

Dickison, R. B. B., Haggis, M. J. & Rainey, R. C. 1983 Spruce budworm moth flight and storms: case study of a cold front system. *J. Clim. appl. Met.* **22**, 278–286.

Dickison, R. B. B., Haggis, M. J. Rainey, R. C. & Burns, L. M. D. 1986 Spruce budworm moth flight and storms: further studies using aircraft and radar. *J. Clim. appl. Met.* **25**, 1600–1608.

Greenbank, D. O., Schaefer, G. W. & Rainey, R. C. 1980 Spruce budworm moth flight and dispersal: new understanding from canopy observations, radar, and aircraft. *Mem. ent. Soc. Can.* **110**.

Hardy, Y., Lafond, A. & Hamel, A. 1983 Epidemiology of the current spruce budworm outbreak in Quebec. *For. Sci.* **29**, 715–725.

Lyons, W. A. 1972 The climatology and prediction of the Chicago lake breeze. *J. appl. Met.* **11**, 1259–1270.

Neumann, H. H. 1980 Prediction of New Brunswick sea breezes. *Am. met. Soc./Can. met. Ocean. Soc. Second Conference. March 1980.* Fredericton: Meteorology of Northern New England and the Maritimes.

Neumann, H. H. & Mukammal, E. I. 1979 Incidence of mesoscale convergence lines as an input to spruce budworm control strategies. *Eighth International Congress in Biometeorology September 1979.* (Shefayim, Israel.)

Orlanski, I. 1975 A rational subdivision of scales for atmospheric processes. *Bull. Am. Met. Soc.* **56**, 527–530.

Rainey, R. C. & Haggis, M. J. 1988 Spruce budworm moth flight: possible implications of field research findings for population dynamics and management. *Proceedings of Symposium. Aerial Application of Pesticides in Forestry.* October 1987, pp. 275–279. Ottawa: *National Research Council/Can. For. Serv.*

Royama, T. 1984 Population dynamics of the spruce budworm. *Ecol. Monogr.* **54**, 429–462.

Schaefer, G. W. 1979 An airborne radar technique for the investigation and control of migrating pest insects. *Phil. Trans. R. Soc. Lond.* B **287**, 447–455.

Simpson, J. E., Mansfield, D. A. & Milford, J. R. 1977 Inland penetration of sea-breeze fronts. *Q. Jl R. met. S.* **103**, 47–76.

Discussion

P. J. MASON (*Meteorological Office, Bracknell, U.K.*). To assist in understanding insect movements and for use in any control actions, it would be worth considering the application of mesoscale models, which have been shown to capture the dynamics of sea breeze circulations very well. They provide both details of flow for scientific studies and forecasts of expected sea breeze positions.

K. A. BROWNING, F.R.S. (*Meteorological Office, Bracknell, U.K.*). Presumably, there are some clouds in these systems; if there are clouds moving in different directions on either side of the sea breeze front, then with satellites, one would be able to observe the convergence directly.

R. W. LUNNON (*Meteorological Office, Bracknell, U.K.*). I think one has to be a little careful in using satellites for cloud track winds in this sort of context, because it is quite likely that the steering level, to which the cloud motion pertains, might be subtly different on opposing sides of a sea breeze front. Quantitative estimates of divergence from such measurements might then be misleading. However, I think qualitatively one could use the regular 30 min geostationary satellite images in this way to get a very good indication of the geographical position of a zone of strong convergence.

D. E. PEDGLEY (*ODNRI, Chatham, U.K.*). What non-standard techniques of analysis, such as time sections, were used to get the maximum value out of the sparse, routine observations? And is there a case for installing a temporary meso-net during the short migration season? Concerning outbreaks, what proportion is derived from the re-concentration of moths streaming away from earlier and distant outbreaks rather than from local, low-density populations? I ask this, because there is growing evidence from studies of the African armyworm moths that some caterpillar outbreaks are not derived from earlier ones, particularly those at the beginning of the season that may lead to widespread infestations downwind in subsequent generations (Pedgley *et al.* 1989).

Reference

Pedgley, D. E., Page, W. W., Mushi, A., Odiyo, P., Amisi, J., Dewhurst, C. F., Dunstan, W. R., Fishpool, L. D. C., Harvey, A. W., Megenasa, T. & Rose, D. J. W. 1989 Onset and spread of an African armyworm upsurge. *Ecol. Ent.* **14**, 311–333.

R. B. B. DICKISON. First, I did not do any time diagrams on this particular occasion, as the feature had apparently already tracked past the most interesting location and I anticipated that it would stall as it moved farther inland. However, it was a technique that I employed on other occasions.

Secondly, in the past one or two years, there has been a meso-network operating, but only for documentation and not for forecasting. This was before the days of the present data-logging and communications equipment, and we were not able to interrogate that network in real time. A network is soon to be established, which would enable that sort of analysis in the future, if there were any interest in it for insect dispersal. Incidentally, 1976 was the peak of the epidemic and since then there has been no appreciable build-up of budworm populations in the province. This is another reason why a lot of the financial support has been withdrawn.

J. R. RILEY (*ODNRI, Malvern, U.K.*). Firstly, is there any evidence from ground observations that moths have appeared where mesoscale convergence zones were predicted? Secondly, in the light of political constraints on air-to-air spraying in Canada, has any practical use been made of the new knowledge about the entrainment of spruce budworm in convergence zones?

R. B. B. DICKISON. Spruce budworm eggs are laid in July and the consequent feeding of the next generation does not begin until May, 10 months later. The traditional approach has been to sample for egg masses and then design the next year's air-to-ground larval insecticide control programme on that distribution of eggs. The suggestion was made by Rainey & Haggis (1988) that information about sea breezes could be used to carry out the sampling more efficiently. But there has been no meteorologist dedicated to the programme to capitalize on this.

R. J. V. JOYCE (*Cranfield Institute of Technology, Bedfordshire, U.K.*). Professor Dickison's case study from which he draws the conclusions: 'Wind convergence zones...collectively may indeed be the key factor in establishing new outbreaks...', supporting the new hypothesis put forward by Rainey and Haggis (1988) that: 'moth immigration in situations with marked convergence zones embedded in the general windfield would lead to new outbreaks without necessarily involving any prior build-up or even existence of local low density populations' is of immense, and possibly revolutionary importance to concepts of pest outbreaks.

A pest outbreak occurs when the pest density in a crop reaches the economic injury level, defined as the level at which damage is roughly equal to the cost of control. It is customary to regard the 'economic threshold' as the pest density at which control measures should be applied to prevent an increasing pest population from reaching the economic injury level. This includes an element of short-term prediction. However, with almost every insect species studied it has been found that the flight activity has been grossly underestimated or wrongly assumed that, when it occurs, immigration is roughly compensated by emigration. Insect dispersal would be such a random process if the air that transports the insects were not structured. Wind fields, however, represent air structured in the horizontal and the vertical planes, so that insects entering this air leave eventually at very different densities. This can, on occasion, give rise to massive redistribution of populations so as to deplete numbers in one area and overwhelm natural controlling agents in another, thus giving rise to an outbreak. This is most clearly manifest in the case of insects that are obviously migratory, such as locusts, but I have had clear evidence of the same mechanism in the case of outbreaks of cotton pests such as jassids, whitefly and bollworm. It must be accepted, I believe, that the term 'economic threshold' can have no meaning unless it contains a statement of scale, and this scale is a function of the flight activity of the species which, rarely, is confined to field or farm boundaries. Insect pest control must therefore be regarded as lying in the domain of ecology and not agronomy.

Professor Dickison has performed a great service in establishing the dramatic effect that meteorological features 'so obscure and generally irrelevant that they are likely to escape notice' can have on one very important pest species.

Phil. Trans. R. Soc. Lond. B **328**, 619–630 (1990)

Printed in Great Britain

Recent airborne radar observations of migrant pests in the United States

By W. W. Wolf[1], J. K. Westbrook[1], J. Raulston[2], S. D. Pair[1] and S. E. Hobbs[3]

[1] *Insect Biology & Population Management Research Laboratory, USDA ARS, Tifton, Georgia 31793-0748, U.S.A.*

[2] *USDA ARS, Weslaco, Texas 78596, U.S.A.*

[3] *Cranfield Institute of Technology, Cranfield, Bedfordshire MK43 0AL, U.K.*

An airborne radar was used to detect insets dispersing from an area of about 200 000 hectares of maize infested with *Heliothis zea* (Boddie) and *Spodoptera frugiperda* (J. E. Smith), in the lower Rio Grande River Valley (LRGRV) of northeast Mexico. During the night of 20 June 1989, a ground-based radar, located at the downwind edge of the maize production area, detected early evening take-off and departure of flying insects. Simultaneously, an airborne radar detected low numbers of airborne insects 50 km downwind. Large numbers of flying insects were detected 200–700 m above ground as the aircraft approached the source. The edge of the emigrating cloud of insects was clearly marked by a rapid change in numbers of airborne insects; the cloud was subsequently intercepted at various locations downwind of the source area. The downwind edge of the insect cloud appeared to travel northward at least 400 km in 7.7 h. Maximum *H. zea* moth emergence occurred on 18 June from maize in the LRGRV. Three days later, maximum oviposition occurred in cotton near Uvalde, 370 km from the LRGRV and downwind on the night of the mass migration observed.

1. Introduction

Zea mays (L.) is an important cereal crop in many countries and often serves as a host to lepidopteran insect pests that ultimately disperse to other high value crops (Sparks *et al.* 1986). Maize may produce large populations of insects imagoes, if the economic value of the maize crop does not warrant chemical control of larval stages. Raulston *et al.* (1986 a) described a maize-growing area in the Lower Rio Grande River Valley (LRGRV) of northern Mexico that acts as an insectary capable of producing billions of corn earworm, *Heliothis zea* (Boddie) moths. Circumstantial evidence indicates that *H. zea* moths can migrate long distances (Hartstack *et al.* 1982). An understanding of insect movement from source areas is needed to improve pest management in high value crops that are vulnerable to pest migration.

A ground-based radar (GBR), previously operated at the northwestern edge of the LRGRV maize-growing area, documented the emigration of a large number of moths from the Mexican side of the LRGRV (Wolf *et al.* 1986). The quantity of moths leaving the maize measured by the GBR agreed with the estimated number of imagoes produced on the maize. Earlier estimation of the dispersal of these emigrating moths relied on estimated dispersion parameters and extrapolated meterological data (Wolf *et al.* 1986). However, incorrect dispersion parameters may result in large errors in estimated dispersion. Also, interpolations of National Weather Service (NWS) data often differ from local nocturnal field measurements. Therefore, on-site measurements of flight duration, dispersion rates and meteorological parameters are needed to improve and validate insect dispersion models.

Schaefer (1979) reported an airborne radar (ABR) technique for investigating the vertical distribution of insects along a flight path. He obtained height-density and height-orientation profiles for various locations while studying moths of spruce budworm *Choristoneura fumiferana* (Clem.) in New Brunswick, Canada. We used an improved airborne radar system to detect insects flying downwind of the LRGRV source area during June of 1989.

2. Methods

(a) Site description

The LRGRV in northern Tamaulipas, Mexico and southern Texas, U.S.A. lies in a region described by Correll & Johnston (1970) as the Tamaulipan brush lands. The climate is semi-arid subtropical. About 200000 ha† of maize (85% of the maize in the LRGRV) is grown annually on irrigated land south of the Rio Grande River (centred at 25° 30′ N 98° 00′ W. Cotton, sorghum and sugar cane are the major irrigated crops north of the river. Insignificant numbers of host plants for *H. zea* occur on non-irrigated land during June. Land use outside the valley is mostly dryland sorghum production or cattle ranching (Raulston & Houghtaling 1986). Earlier ecological investigations of insect and crop phenology show that the LRGRV is a major source area of dispersing insects (Raulston *et al.* 1986*b*).

A site near Uvalde, Texas (370 km northwest of the LRGRV), was selected for sampling emergence of local moths and monitoring potential fallout of immigrating moths. The area around Uvalde, Texas, is also Tamaulipan brush land, with the major crops being irrigated cotton and maize (about 16200 ha of maize).

(b) Entomology

Soil samples were dug from 100 maize fields in the LRGRV and 40 maize fields near Uvalde. The sampling period spanned the corn earworm pupation cycle in fruiting corn during June 1989. Two soil samples (1 m² × 10 cm deep per sample) per field were screened for pre-pupae and pupae. Species composition, sex, and mortality factors were determined. Each live field-collected pupa was re-buried inside a small emergence cage with a plastic tube leading to a screened collection cage on the soil surface. Re-burial sites had soil and vegetation similar to the respective collection site. Cages were examined daily for emergence to determine emergence profiles for the two areas.

(c) Airborne radar

Hobbs & Wolf (1989) describe the ABR principles and signal processor used in these studies. Our ABR transceiver had a 25 kW peak power, a 3.2 cm wavelength, a 100 ns pulse length and a 2560 Hz pulse repetition rate. The transceiver could be electronically switched between a dummy load or circular parabolic reflector antenna. The dummy load position of the waveguide switch prevented any target echoes from reaching the receiver so that radar-noise levels could be recorded. A 0.61 m diameter antenna was mounted inside the aircraft's camera port so a radome was not used. The resulting conical beam pointed downward below the aircraft and had a 2.5° half-power width. A rotating feed caused the beam polarization to rotate at 10 Hz around the beam centreline.

The ABR system was installed in a twin-engine Aero Commander‡ airplane. This airplane

---

† 1 hectare = 10⁴ m².

‡ Mention of company names or products does not imply endorsement by the U.S. Department of Agriculture.

has a true airspeed of 268 km h^{-1} at normal power settings. The crew consisted of a pilot, copilot and radar operator. Position coordinates were obtained from a Morrow Inc. Model 604 Loran receiver. A Model 100 Radio Shack computer automatically recorded position coordinates and plotted a map of aircraft track on the computer screen each minute. Position information was additionally obtained from the aircraft's direction and distance measuring receivers. Indicated airspeed, altitude, and heading were manually noted in the radar log.

A signal processor obtained target samples from 48 ranges (distances) spaced at 15 m intervals starting 146 m below the aircraft. These samples were multiplexed onto three channels of a frequency-modulated analogue tape recorder. Sample altitude was the difference between airplane altitude and radar range. The fourth channel of the tape recorder contained synchronizing signals for decoding the data.

During analysis of data, voltages from two channels of the tape recorder were passed through an analogue adder, a low-pass filter and recorded on a strip chart recorder. These voltages represented the sum of insects (area density) flying above 250 m; so, the vertical axis of each strip chart recording was scaled and labelled insects per hectare. The scale factor was empirically determined during high insect density and when the airborne radar was within two miles of the GBR.

Tape recorder voltages were digitally sampled at times corresponding to strip chart maxima, minima, and inflection points. Each digital sample consisted of nine seconds of data from each of 48 radar ranges. The mean voltage and standard deviation for each range were calculated. The difference between mean voltage and mean radar-noise was equal to mean target amplitude for each of the 48 ranges. A graph of mean target amplitude (representing insect numbers) versus altitude showed the vertical distribution of insects (insect profile).

(d) Ground-based radar

The GBR was located within 100 m of the Rio Grande River, and within 20 km of the northwest corner of the source area. Data from the GBR were used to calculate profiles of insect density versus altitude. Profiles were measured at 15 min. intervals during the first two hours of insect flight and thereafter at 30 min. intervals until the overhead density of flying insects decreased by two orders of magnitude from the peak. Operating and analysis procedures were similar to those described earlier by Wolf *et al.* (1986). Insect density profiles were integrated to obtain area density values.

(e) Meteorology

Meteorological observations at the GBR site included visual cloud cover assessment, upper-air soundings, and automated temperature, humidity, wind vector, and pressure measurements at the surface. An instrument package (radiosonde) attached to an ascending helium-filled balloon transmitted wet and dry bulb temperature and barometric pressure to a ground-based radio receiver. The balloon azimuth and elevation angles were measured with an optical balloon-tracking theodolite. An Apple II computer automatically logged theodolite and radiosonde values at five-second intervals. Uninstrumented helium-filled balloons were tracked hourly to provide additional upper wind information.

(f) ABR missions

Four ABR missions were flown during the night of 20 June 1989. Intervals between missions allowed for insect displacement, refuelling, and planning for subsequent missions. The objectives of these missions were: (*a*) to measure insect density near the GBR while the most dense portion of the insect cloud was passing the GBR; (*b*) to determine positions of the downwind and crosswind edges of the insect cloud; (*c*) to measure density and vertical distribution of insects ahead of, and within, the dispersing insect cloud; (*d*) to determine duration of insect fight. A fifth mission (while returning to Weslaco, Texas after sunrise) provided data during low insect density.

3. Results

Peak emergence of adult *H. zea* from maize in the LRGRV occurred 15 days before peak emergence at Uvalde (figure 1). At Uvalde, maximum oviposition by *H. zea* on cotton preceded maximum emergence by 12 days. An estimate of the total number of reproductive females produced on maize and the potential number of hectares that they could infest (table 1) was based on: (*a*) 10% survival of pupae to reproductive status; (*b*) 1000 viable eggs per female; (*c*) an economic infestation level of 25000 eggs ha^{-1}; (*d*) a uniform distribution of

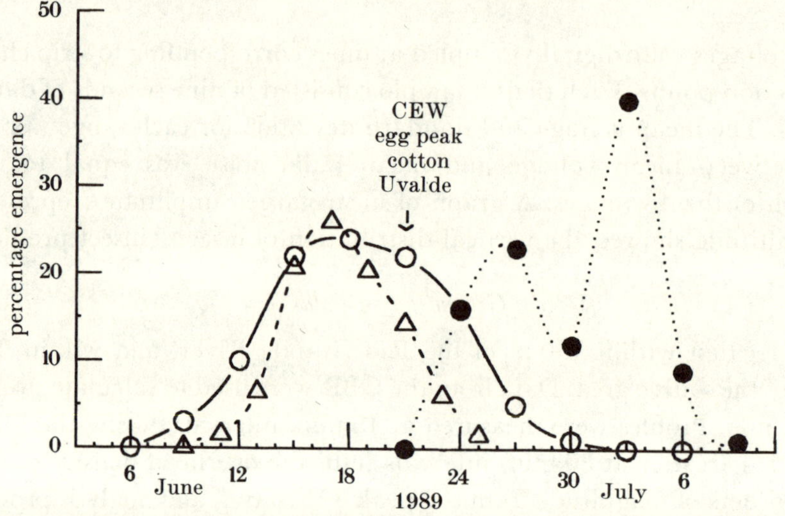

FIGURE 1. Emergence of fall armyworm (FAW) and corn earworm (CEW) from maize fields in the Lower Rio Grande River Valley and near Uvalde. (Open triangles, LRGRV FAW; open circles, LRGRV CEW; closed circles, Uvalde CEW.)

TABLE 1. TOTAL *HELIOTHIS ZEA* PUPAE PRODUCED IN THE LOWER RIO GRANDE RIVER VALLEY AND UVALDE AREAS, AND ESTIMATED AREA OF POTENTIAL INFESTATION BY REPRODUCTIVE FEMALES DURING JUNE, 1989

location	maize area	number of soil samples	number of pupae produced[a]	number of adult females[b]	area of potential infestation[c]
	(ha)		($\times 10^9$)	($\times 10^8$)	(ha)
LRGRV	200000	200	6.70	3.35	13400000
Uvalde	16200	80	0.72	0.36	1440000

[a] Based on pupae recovered per square meter per soil sample.
[b] Assumed 10% survival between pupa and reproductive stage.
[c] Assumed each reproductive female lays 1000 eggs, and an economic infestation level of 25000 eggs per ha.

females. An estimated 7.9×10^9 fall armyworm, *Spodoptera frugiperda* (J. E. Smith), emerged from maize fields in the LRGRV with peak emergence on 17 June 1989.

Meteorological soundings during the night of 20–21 June 1989 showed a low-level jet wind from 142° at 21h30 Central Daylight Savings Time and 161° at (GMT-5) 00h40 (figure 2). Upper-air temperature profiles showed the bottom of the atmospheric inversion layer to be near the top of the jet wind.

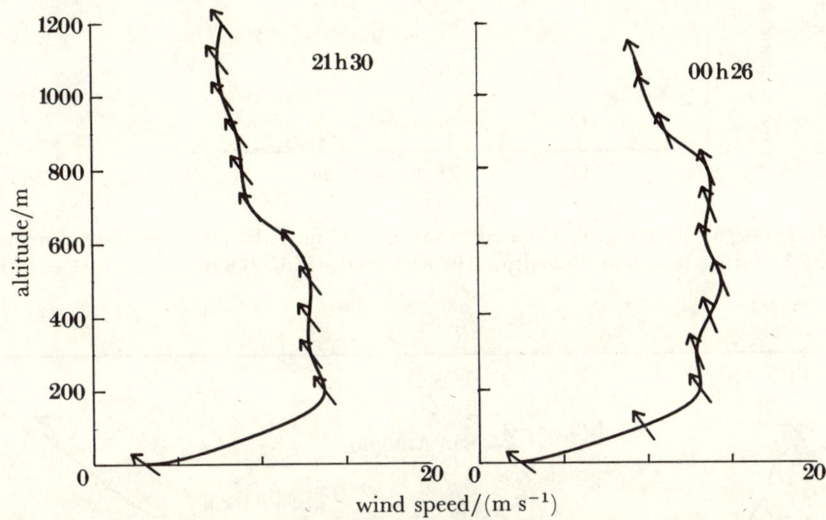

FIGURE 2. Wind speed profiles measured at the ground based radar during night of 20–21 June 1989.

NWS upper wind data (19h00 and 07h00) were spatially and temporally interpolated to estimate wind vectors. These interpolated wind vectors were then used to calculate the path or track of passive particles at 600 m altitude. Three tracks during the night of 20–21 June 1989 had north-northwestward displacements (350°).

The GBR detected numerous insects passing overhead from 21h00 to 23h00 during the night of 20 June 1989 (figure 3). Insect density peaked at 21h33. Species composition of insects detected by the GBR and ABR was unknown; however, peak radar voltages from individual targets were similar to voltages from corn earworm or fall armyworm moths. From the GBR data, the edge of the insect cloud passing overhead was defined as the occurrence of 50 % of the night's maximum density (i.e. at 21h16 and 22h04 for the leading and trailing edges, respectively, figure 3). Insect cloud edge no. 0 (table 2) was located at the GBR at the time 21h16. From the ABR data, cloud edge was defined as the location where insect density changed by 50 %.

The first ABR mission started at 21h30 when insect density above the GBR was near maximum. The aircraft flew downwind beyond the insect cloud (21h39) and then turned upwind toward the McAllen, Texas airport (figure 4). ABR missions could not be flown into Mexico so the aircraft turned eastward at McAllen and passed the GBR at 21h53. At this time, the ABR display indicated numerous insects as high as 750 m above ground. After passing the GBR, the aircraft descended briefly to 600 m altitude for visual monitoring of insect flight activity. One insect per second could be seen in the aircraft's landing lights. The time at this altitude was about one minute. Five minutes later the aircraft landed at the McAllen airport; the remains of twelve insects were on the windscreen and wings and thorax of two *H. zea* moths were recovered from the engine cooling fins.

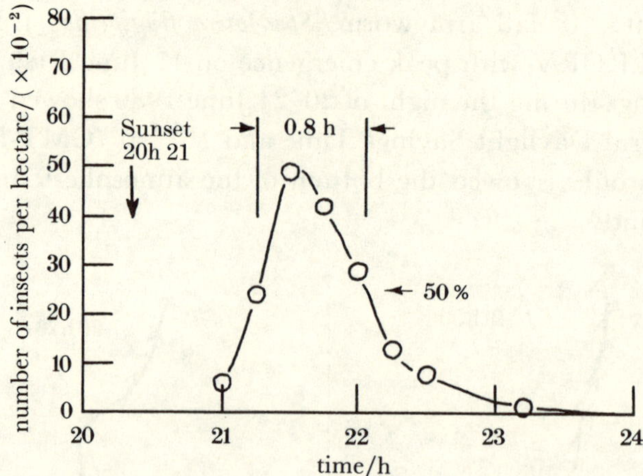

FIGURE 3. Density of insects passing the ground-based radar located near the northern edge of maize production area in the LRGRV during the night of 20 June 1989. (Cloud width $(0.8\ \mathrm{h} \times 54\ \mathrm{km\ h^{-1}}) = 43$ km.)

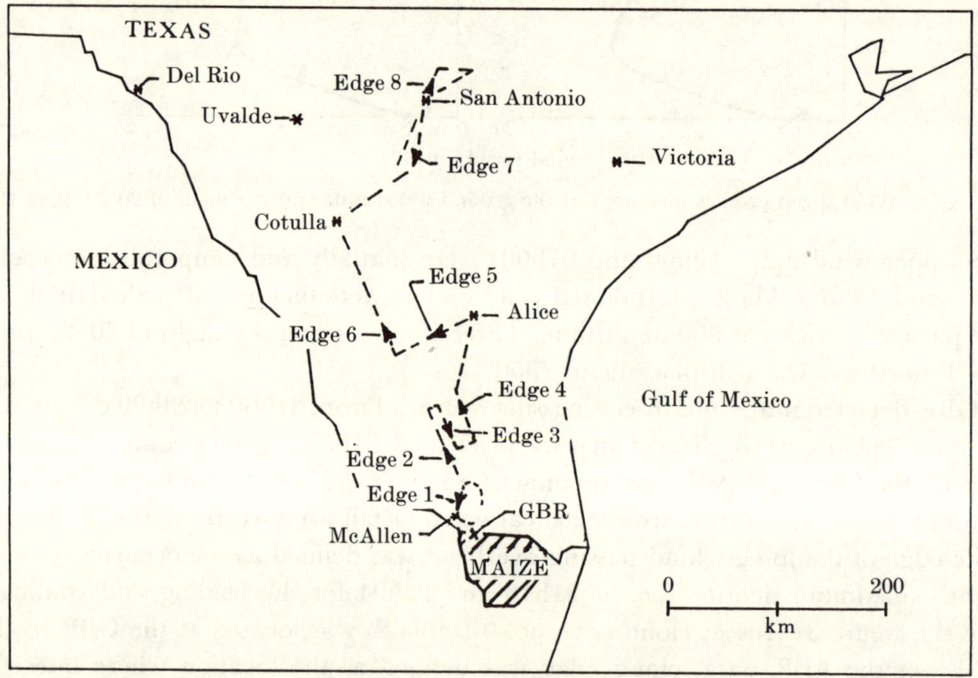

FIGURE 4. Track (dashed lines) of aircraft during four airborne radar missions during the night of 20–21 June 1989. A solid arrow indicates direction of aircraft displacement and location of insect cloud edge (50 % change in insect density along aircraft track). GBR is location of ground-based radar.

The strip chart recording of insect density† showed that the edge of the insect cloud was encountered at 21h46 while flying upwind toward McAllen at location number 1,26 km from the GBR (figures 4 and 5b). The leading edge of the insect cloud was thus moving at 13.9 m s⁻¹ (table 2). During three successive missions the insect cloud was detected seven times (figures 4, 6, 7, 8 and 9). Edge locations 2, 3, 6, 7 and 8 were interpreted as sequential locations of some

† The conversion of radar mean volts to insect density was an approximation so strip chart recordings showed relative insect density fluctuations.

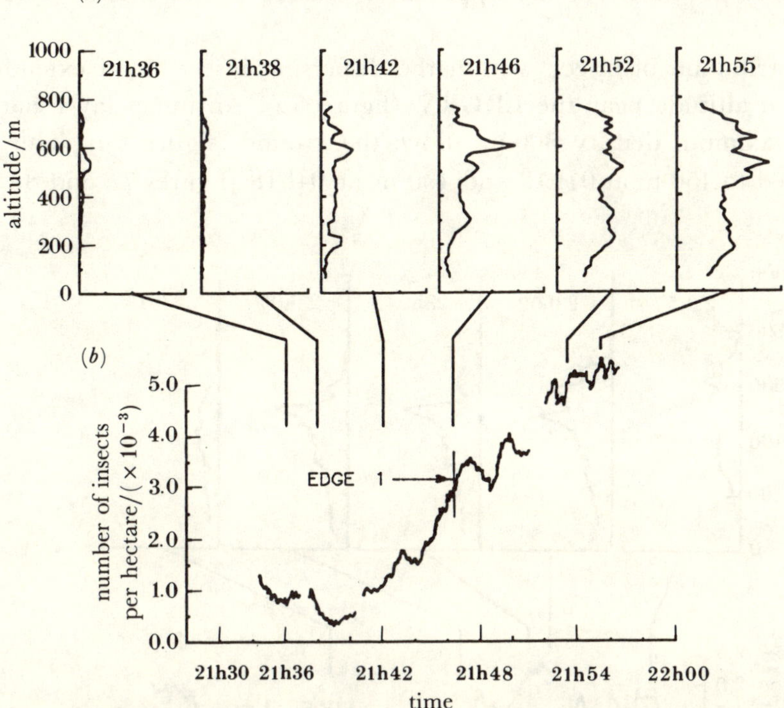

FIGURE 5. (a) Insect profiles. (b) Numbers of insects per hectare near the source area during mission 1 (between Weslaco and McAllen) 20 June 1989. A 50% change in density was designated as a cloud edge.

TABLE 2. GROUND SPEED OF LEADING EDGE[a] OF INSECT CLOUD DETECTED BY GROUND AND AIRBORNE RADARS DURING NIGHT OF 20–21 JUNE 1989

cloud edge location number	time of night	distance from source	cloud edge displacement		
			distance	duration	speed
	(h)	(km)	(km)	(h)	(m s⁻¹)
0[b]	21h16	0	—	—	—
1	21h47	26	26	0.52	13.9
2	23h08	85	59	1.36	12.1
6	01h21	204	119	2.21	14.9
7	04h04	343	139	2.72	14.2
8	04h56	421	78	0.87	24.9

[a] Cloud edge defined as location where insect density changed by 50%.
[b] Location no. 0 was at edge of maize production area in LRGRV.

portion of the northern edge of the cloud. These interpretations were based on a clockwise wind shift of 40° during the night, for locations downwind from the LRGRV, derived from wind measurements at the GBR, interpolated NWS data, and differences between aircraft heading and displacement vectors. Edge locations 4 and 5 (figures 4 and 7) were interpreted as the eastern or northeastern of the cloud.

The last evidence of an insect cloud occurred at 04h56 at location 8 which was about 420 km from the source area (figure 9). Insect density fluctuations and low average densities prevented an accurate measurement of this cloud edge location. The error in measuring the location of edge no. 8 probably accounts for the apparent increased displacement velocity listed in table 2.

The vertical distribution of insects within the dispersing insect cloud extended from the surface to *ca.* 800 m altitude near the LRGRV (figure 5*a*). An insect layer had formed by 23h00 with the maximum density 500 m above the ground (figure 6*a*). This insect layer gradually descended to 450 m at 01h12 and 400 m at 04h18 (figures 7*a* and 8*a*). The insect

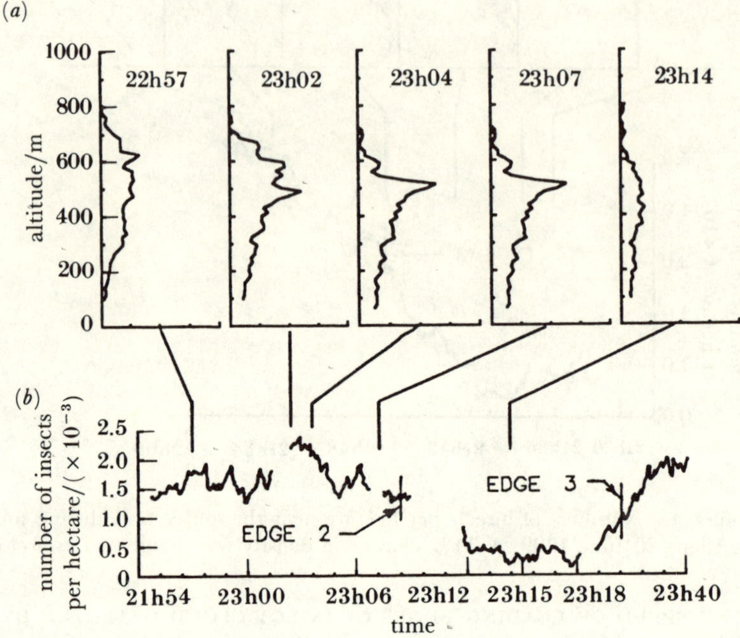

FIGURE 6. (*a*) Insect profiles. (*b*) Numbers of insects per hectare for first portion of mission 2 (between McAllen and Alice, 20 June 1989). A 50 % change in density was designated as a cloud edge.

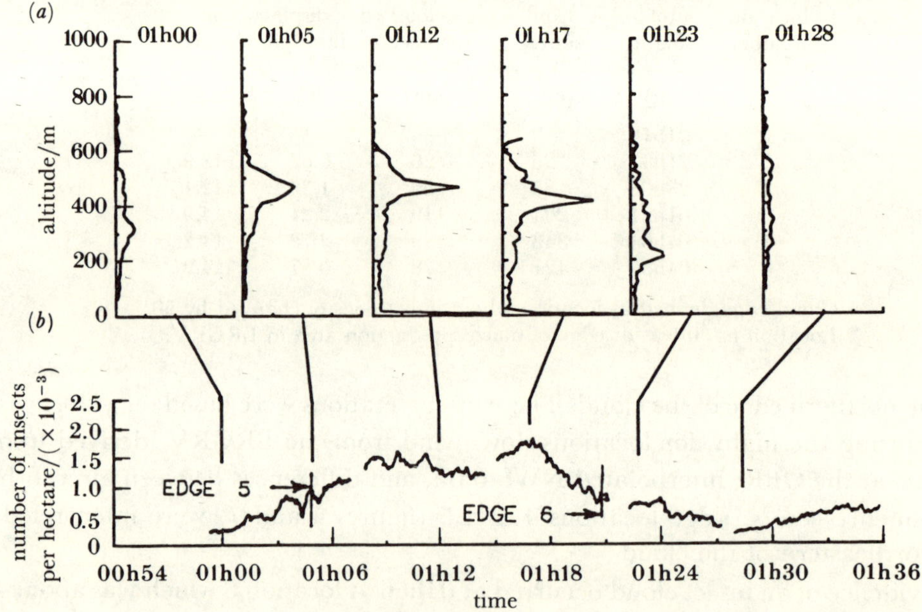

FIGURE 7. (*a*) Insect profiles. (*b*) Numbers of insects per hectare for mission 3 (between Alice and Cotulla, 21 June 1989). A 50 % change in density was designated as a cloud edge.

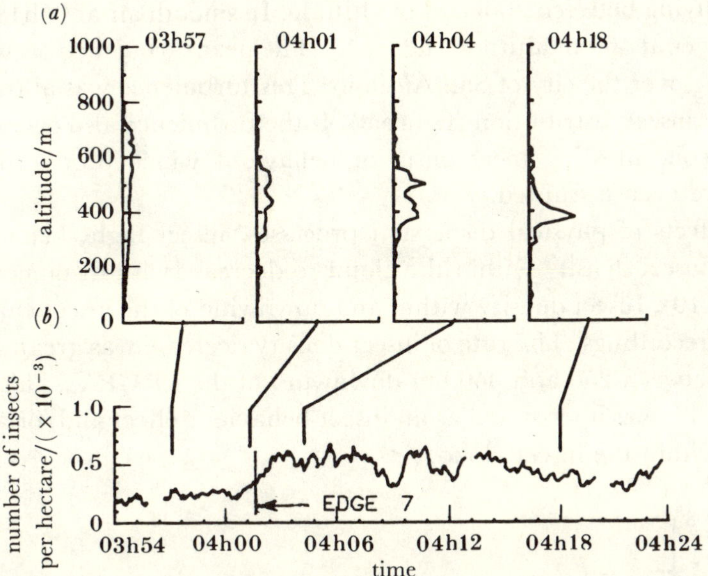

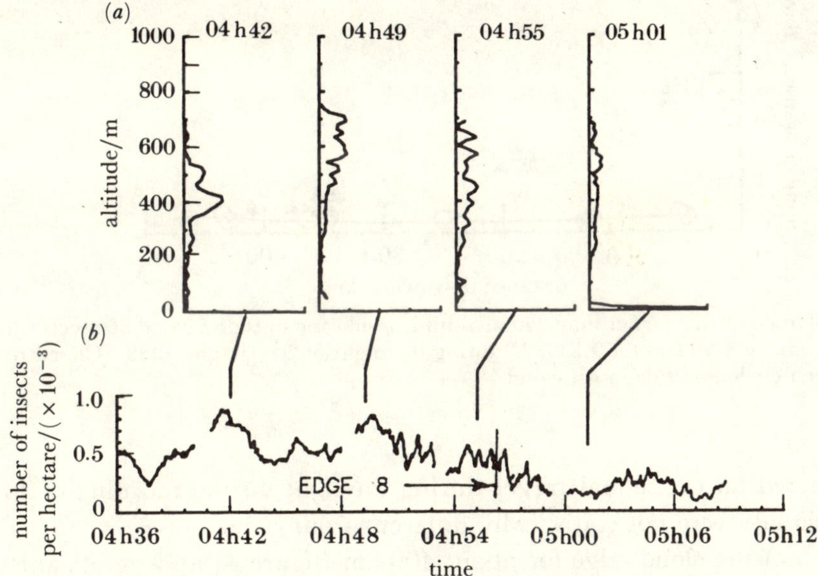

FIGURE 8. (a) Insect profiles. (b) Numbers of insects per hectare for first part of mission 4 (between San Antonio and Cotulla, 21 June 1989). A 50% change in density was designated as a cloud edge.

FIGURE 9. (a) Insect profiles. (b) numbers of insects per hectare for last part of mission 4 (between Cotulla and San Antonio). A 50% change in density was designated as a cloud edge.

layer extended from 400 m to 700 m altitude (04h49) and from 300 to 600 m altitude (04h55) as the aircraft passed over the northern and southern limits of the city of San Antonio, respectively (figures 4 and 9a). We interpreted the increase of insect altitude over the city to be an urban thermal effect.

The shape of the insect profiles changed when the aircraft encountered mild turbulence. Between San Antonio and Cotulla, Texas mild turbulence occurred from 03h54 to 04h09 (previous flights had been smooth even during ascent and descent). During turbulence at

04h04, insects were flying between 300–600 m altitude. In smooth air at 04h18 the insects were concentrated in a layer at 400 m altitude (figure 8). The next turbulence occurred from 04h42 to 04h49 while flying over the city of San Antonio. This turbulence was also accompanied by an increased vertical insect distribution (figure 9). If the turbulence also occurred at the insect flight altitude, the concentrating mechanism or behaviour which produced a thin layer in smooth air may have been disrupted.

The combined effects of physical dispersion processes, insect flight behaviour, and flight termination caused insect density within the cloud to decrease with distance downwind from the LRGRV (figure 10). Insect density within and downwind of the insect cloud was obtained from the strip chart recordings. The rate of insect density decrease was greatest between 0 and 100 km and again between 250 and 350 km downwind of the LRGRV. This bimodal rate of change in insect density was interpreted as an insect behaviour effect and/or an artefact of the aircraft's locations within the insect cloud.

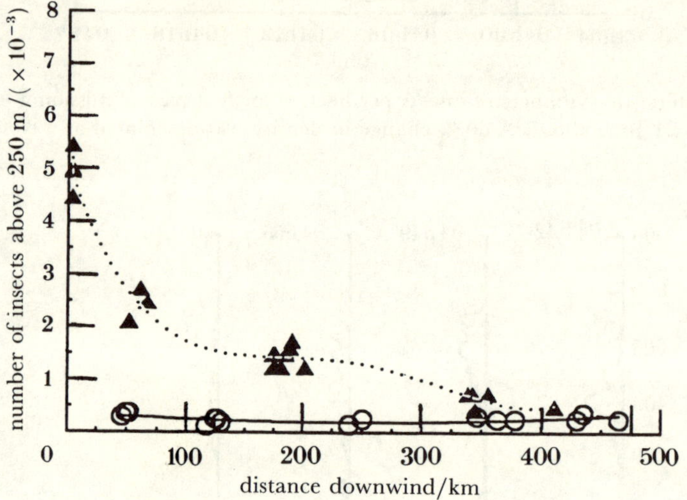

FIGURE 10. Density of insects flying higher than 250 m altitude inside and outside a cloud of insects which emigrated from a maize production area in the LRGRV during the night of 20–21 June 1989. (Closed triangles, inside insect cloud; open circles, outside insect cloud.)

The ABR detected an occasional target during daylight on the morning of 21 June, but averaged radar signals were not statistically different from radar noise.

The progression of the cloud edge for about 400 km (figure 4) and significant numbers of insects within the cloud for 7 h 40 min (figure 10) are evidence for long distance dispersal during a single night. Several nights of such long distance dispersal could have resulted in maximum corn earworm oviposition at Uvalde two weeks prior to local emergence (figure 1).

REFERENCES

Correll, D. S. & Johnston, M. C. 1970 *Manual of the Vascular Plants of Texas*. Renner, Texas: Texas Research Foundation.

Hartstack, A. W., Lopez, J. D., Muller, R. A., Sterling, W. L., King, E. G., Witz, J. A. & Eversull, A. C. 1982 Evidence of long range migration of *Heliothis zea* (Boddie) into Texas and Arkansas. *Southwestern Entomol.* 7, 188–201.

Hobbs, S. E. & Wolf, W. W. 1989 An airborne radar for studying insect flight. *Bull. ent. Res.* 79, 693–704.

Raulston, J. R. & Houghtaling, J. E. 1986 Circumstantial ecological evidence for *Heliothis virescens* migration into the lower Rio Grande valley of Texas from northeastern Mexico. In *Long-Range Migration of Moths of Agronomic Importance to the United States and Canada: Specific Examples of Occurrence and Synoptic Weather Patterns Conducive to Migration* (ed. A. N. Sparks), 34–37. *USDA/ARS-43.*

Raulston, J. R., Pair, S. D., Sparks, A. N., Westbrook, J. K., Loera, J., Summy, K. R. & Rummel, D. R. 1986a Production of *Heliothis zea* on corn in northeastern Mexico and the Lower Rio Grande Valley of Texas: a potential source for corn and cotton infestation on the high plains of Texas. *Proc. Beltwide Cotton Conf.* pp. 222–225. Memphis: National Cotton Council of America.

Raulston, J. R., Pair, S. D., Martinez, F. A. P., Westbrook, J. K., Sparks, A. N. & Valdez, V. M. S. 1986b Ecological studies indicating the migration of *Heliothis zea*, *Spodoptera frugiperda*, and *Heliothis virescens* from Northeastern Mexico and Texas. In *Insect flight dispersal and migration* (ed. W. Danthanarayana), pp. 204–220. Berlin: Springer–Verlag.

Schaefer, G. W. 1979 An airborne radar technique for the investigation and control of migrating pest insects. *Phil. Trans. R. Soc. Lond.* B **287**, 459–465.

Sparks, A. N., Raulston, J. R., Westbrook, J. K., Wolf, W. W. & Pair, S. D. 1986 The potential importance of late spring-early summer bollworm moth migration to cotton production. *Proc. Beltwide Cotton Conf.* pp. 147–149. Memphis: National Cotton Council of America.

Wolf, W. W., Westbrook, J. K. Sparks, A. N. 1986 Relationship between radar entomological measurements and atmospheric structure in south Texas during March and April 1982. In *Long-Range Migration of Moths of Agronomic Importance to the United States and Canada: Specific Examples of Occurrence and Synoptic Weather Patterns Conducive to Migration* (ed. A. N. Sparks), pp. 98–104. *USDA/ARS-43.*

Discussion

J. R. Riley (*ODNRI, Malvern, U.K.*). Mr Wolf was kind enough to acknowledge the ways in which he has benefitted from radar entomology in the U.K., and I would like to say in return, that his own dedicated and enthusiastic contributions to the subject have also been of considerable benefit here. The airborne radar study of long-range moth migration described in Mr Wolf's paper is unique in that he was able to maintain contact with the moths for at least 400 km. Many of the data have still to be analysed, but the results presented here demonstrate again the great power of airborne radar systems when used in migration studies. I would like to ask if any evidence has been found for the maintainance of orientation during the flight, and also if there is any suggestion that moths take off in winds from particular directions?

W. W. Wolf. Regarding orientation, Dr Riley was there operating my ground-based radar on a previous occasion and saw crosswind orientation during the night, very frequently. On the particular night I've described here the ground-based radar did not show much orientation, except at very low altitudes. I have not yet extracted the orientation information from the full data set, but I hope to do so and I think the results will be very interesting. If the airborne radar shows consistent crosswind orientation downwind it will definitely affect dispersion parameters.

On the subject of take-off, we've always seen take-off at the same time of night, about 30–45 minutes after sunset. The only indication that I have of preferred winds comes from when we were operating the radar in northwest Texas during the fall. When the wind was from the south it seemed that the flight was at a much lower altitude than when the wind was from the north.

P. J. Mason (*Meteorological Office, Bracknell, U.K.*). It is of some interest to compare Mr Wolf's observations with our knowledge of the dispersion of passive materials under similar nocturnal conditions. Under nocturnal conditions, dispersion due to small-scale three dimensional eddies is limited and the main effects are movements in various directions at separate heights. It is common to see an initially well-mixed region of material split in different layers, each moving in the direction appropriate to its height.

R. J. V. JOYCE (*Cranfield Institute of Technology, Bedfordshire, U.K.*). Mr Wolf, with characteristic generosity, has paid tribute to the pioneer work done in radar entomology by the late Dr Glen Schaefer and by Dr J. Riley. I simply wish to add that my own programme on the flight activity of the noctuid moth *Heliothis armigera* (Hübner) in Sudan, in which Dr. Schaefer conducted the investigations which inspired Mr Wolf, was undertaken because of the evidence collected by him, A. N. Sparks and others, of long distance flights of *Heliothis spp.* over the Gulf of Mexico. These included catches of moths in specially designed black-light traps mounted over 300 m above ground and on oil-platforms 100 miles offshore.

In the Sudan Gezira, *H. armigera* was disappointing as a long-distance flyer, only a small fraction of the population climbed much above the crop boundary layer and the majority of moths engaged in intermittent flight in the lower 10 m of air-space. Displacement was invariably downwind, and massive redistribution of populations (and subsequent oviposition on cotton) occurred when strong but short-lived cold-fronts, resulting from out-flows from the convective storms which can sweep across the Gezira characteristically at the time of the maximum flight activity of *H. armigera* moths, were found by Schaefer (1976) to concentrate airborne insects 60-fold in a single hour.

Reference

Schaefer, G. W. 1976 Radar observations of insect flight. *Symp. R. ent. Soc. Lond.* **7**, 157–197.

J. R. RILEY (*ODNRI, Malvern, U.K.*). Regarding Professor Joyce's comment, we found exactly the same effect with *H. armigera* in central India. The moths persisted in flying below radar cover, but by using an infra-red video system we could look at the flight above the crop. It was quite clear that the reason we were not finding them on the radar was that they were flying low, as Professor Joyce had previously found.

Phil. Trans. R. Soc. Lond. B **328**, 631–653 (1990)

Printed in Great Britain

Concentration of flying insects by the wind

By D. E. Pedgley

Overseas Development Natural Resources Institute, Central Avenue, Chatham, Kent ME4 4TB, U.K.

The role of horizontal wind convergence in concentrating flying insects is examined. Because of the inverse relation between intensity and persistence of convergence zones, a tenfold increase in volume density is unlikely to be exceeded in a single atmospheric disturbance, and then only if the insects are not taken through the convergence zone. Radar has shown that the commonest disturbances associated with insect concentrations are windshift lines: modifications of the broad-scale wind flow due to buoyancy and blocking effects. Even in the absence of concentration, a tenfold increase in area density can be brought about by vertical circulations at windshift lines. Insect concentrations are likely to be most frequent and persistent at night, and over and near mountains, where searches are most difficult.

1. Introduction

If windborne insects accumulate in particular kinds of atmospheric motion patterns, in numbers and at volume densities to make air-to-air spraying effective, then the possibility arises of improving control by seeking out such patterns. This paper reviews several of the mechanisms that might concentrate flying insects, and discusses their application to the operational monitoring of insect populations. For an ecological discussion of the significance of aerial concentration on colonization by windborne insects, see Drake & Farrow (1989).

There is considerable evidence that many migrant insect species are windborne; for recent reviews see Pedgley (1982, 1983), Drake & Farrow (1988), and McManus (1988). The evidence continues to grow: it refers to not only small, weak fliers such as vectors of organisms causing serious diseases in crops (Kisimoto 1984: Liu 1984; Rosenberg & Magor 1987; Wada *et al.* 1987; Irwin & Thresh 1988), and man (Baker *et al.*, this symposium), but also larger, stronger fliers such as grasshoppers (Reynolds & Riley 1988), locusts (Symmons 1986; Casimir 1987) and moths. Among moths, there has been emphasis on various crop pests: armyworm moths in China and Japan (Chen & Bao 1987; Miyahara 1987; Hirai 1988; Chen *et al.* 1989), Africa (Rose *et al.* 1987; Pedgley *et al.* 1989) and North America (McNeil 1987; Johnson 1987; Pair *et al.* 1987), as well as bollworm moths in Australia (Wilson 1983), and green clover worm moths (Wolf *et al.* 1987) and celery looper moths (Peterson *et al.* 1988) in the U.S.A. Presumably all these windborne migrant species are open to concentration by the air motion.

2. Dispersion and concentration of flying insects

Insects moving like a cloud carried downwind may respond to each other as a coherent swarm. This happens with locusts and bees but it has not been demonstrated in other species, although large numbers of independently flying individuals may give the impression of a swarm. Instead, a cloud of insects might be expected to thin out as the mean separation between individuals increases under the influence of dispersion, like smoke or by individual

[113]

flight behaviour. Field evidence in support of such thinning comes from radar studies of insect volume density decreasing downwind from a source (Riley *et al.* 1983; Riley & Reynolds 1983) and of observed variability in heading within a cloud of insects, particularly where they are likely to be largely of one species (Schaefer 1979; Greenbank *et al.* 1980; Riley & Reynolds 1983; 1986; Riley *et al.* 1981, 1983, 1987; Drake 1983, 1985*b*; Mueller & Larkin 1985; Drake & Farrow 1985; Chen *et al.* 1989). Mark-and-capture studies of distances flown by insects from a common source provide additional evidence, the dispersion then being dominated by day-to-day and hour-to-hour changes in the wind field (Li *et al.* 1964; Hendricks *et al.* 1980; Rose *et al.* 1985; Showers *et al.* 1989). There appears to be no theoretical study of dispersion of a cloud of insects having a variety of headings in a turbulent wind. Models developed for pollution studies might be adapted, as they have been for clouds of young caterpillars of the gypsy north, suspended on silken threads (Fosberg & Peterson 1986), but they would be even more complex than for particles, because of the need to incorporate insect flight behaviour.

The reverse of dispersion, concentration, might be expected if there is some mechanism to bring flying insects together. Concentration of insects *onto the ground* can be brought about by preferential settling, e.g. in areas sheltered from the wind. This is well known to occur in the lee of windbreaks (Pasek 1988), but it also occurs, on a larger scale, in the lee of hills. It may account, for example, for the association of African armyworm outbreaks with topography (M. J. Haggis, personal communication) due to concentration of the night-flying moths, similar to the concentration of day-flying sod webworm moths in New Zealand (Cowley 1987).

There is certainly evidence for the occurrence of dense insect clouds *in the air*: the sudden appearance of large numbers in flight, and the presence of innumerable radar echoes in the form of bands one or two kilometres wide and tens of kilometres long, moving at up to 10 m s^{-1}. Volume densities of moths and grasshoppers in these echo bands are typically 10^{-3}–10^{-4} insects m^{-3} (see, for example, Schaefer 1976, 1979; Riley *et al.* 1981; Pedgley *et al.* 1982; Riley & Reynolds 1983; Reynolds & Riley 1988). This contrasts with the much greater densities of Desert locusts in swarms: typically 10–10^{-3} insects m^{-3}, measured by radar (Ramana Murty *et al.* 1964) and by vertical photography (Waloff 1972). With typical depths 0.5–1.0 km, and therefore cross-sections of about 1.0 km^2, each kilometre length of band contains 10^5–10^7 insects. Some bands have been followed by radar for distances varying from 7 km (in 25 min, Pedgley *et al.* 1982) to 38 km (in 99 min, Schaefer 1976).

Increased volume density of flying insects has been associated with windshift lines, where there is a more or less well-defined boundary between two airstreams with different directions or speeds, or both. Such an association may be due to a change in the source of the insects, a triggering of take-off, a vertical concentration of individuals near the ground with the onset of rain, or a horizontal concentration of individuals by convergent winds. The first three are considered briefly; emphasis is put on the fourth.

A change in volume density with a change in wind direction can be identified by means of an hourly recording trap or by radar. An example of the use of radar comes from Muguga, near Nairobi, on the night of 17–18 April 1980, when a light trap caught many more armyworm moths than on the previous and following nights. Differences in trap catch could not be attributed to differences in trap efficiency (due to moonlight or wind speed), nor was there any significant rain. After the dusk take-off there were few echoes until just after midnight, when the wind changed fitfully from northeast to west and echoes moving from the west increased to a maximum at 01h45. There was no echo band and no windshift line (D. R. Reynolds, personal communication.).

There do not seem to have been any field studies of take-off being stimulated by a windshift line during the short time that it takes to pass overhead, although Bergh (1988) found increased take-off frequency in the laboratory with passage of cold fronts.

Large numbers of night-flying moths seen at lights or caught in traps are well known in rainy weather (Brown *et al.* 1969; Dickison *et al.* 1983, 1986; Tucker 1983; Tucker & Pedgley 1983). There is some radar evidence (Greenbank *et al.* 1980; Riley *et al.* 1983) to suggest that these large numbers can be due to descent to near the ground, or even landing, at onset of rain.

3. Wind convergence and concentration of flying insects

Rainey and co-workers have for long associated the concentration of flying insects with horizontal convergence of winds (Rainey 1951, 1963, 1972, 1974, 1976, 1980; Sayer 1962; Brown *et al.* 1969; Joyce 1973, 1983; Dickison *et al.* 1983). Horizontal wind convergence occurs where there is a net inflow into a given marked volume of air. Conversely, net outflow is divergence. Convergence and divergence are potentially mappable, as are other properties of the wind, such as speed and direction. A dense network of surface observing stations may be used, or Doppler radar where rain drops are tracers of the wind (Browning & Wexler 1968; Rabin & Zrnic 1980).

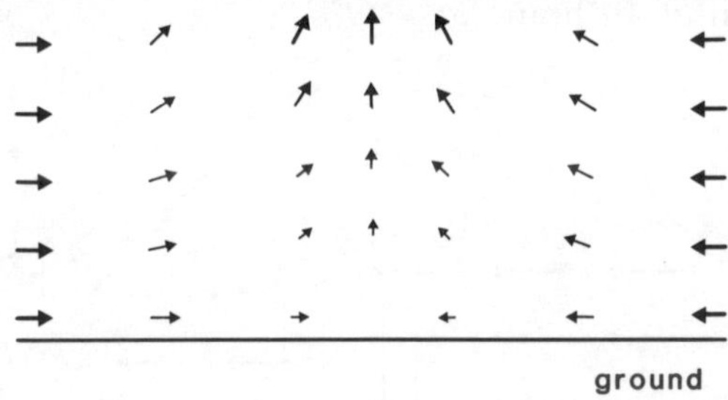

Figure 1. Vertical section through a region of horizontal wind convergence, showing vertical divergence and an increasing upward wind component with height. Arrow length is proportional to wind speed.

Near the ground, horizontal convergence is necessarily accompanied by vertical divergence (figure 1) because there is no significant accumulation of air (increase in air density). Vertical divergence implies an increase in the upward component of the wind with height. (Horizontal divergence is accompanied by vertical convergence and an increasing downward component with height.) Flying insects therefore have a tendency to be carried upwards where there is wind convergence, but their volume density would not increase unless there was also a tendency to resist being taken aloft, particularly where the upcurrent is sloping and there is no circulation to return descended insects into the convergence. Such a tendency would occur if insects actively avoided cold air, where temperatures are below a threshold for flight, varying not only between species but also, to some extent, between individuals within a species. Evidence that this does in fact occur comes directly from laboratory studies on many species and indirectly from trap catches in relation to air temperature. It is also provided by the frequent layering of night-flying insects recorded by radar on clear, quiet nights, once the dusk take-off period has been passed, and there are no persistent updraughts. Layering has been found with locusts (Drake & Farrow 1983), grasshoppers (Riley & Reynolds 1979, 1983; Reynolds & Riley 1988) and moths (Schaefer 1976; Greenbank *et al.* 1980; Drake 1984*b*, 1985*b*; Drake & Farrow 1985; Chen *et al.* 1989). Where there have been temperature soundings, maximum densities have been found in the warmest air, near the top of the nocturnal temperature inversion (Schaefer 1976; Riley & Reynolds 1979; Drake 1984*b*).

Convergence is the instantaneous rate of shrinking of unit horizontal area of a marked volume of air because of net inflow. Where wind direction is uniform over an area, convergence is readily seen to be due to downwind deceleration (figure 2*a*), but where the direction is not uniform (figure 2*b*) it is less readily identified. Convergence is measured in units of change of velocity per unit distance and is typically about 10^{-5} s^{-1}. With such an intensity of convergence, a given area will shrink by a factor of e (about 2.7) in about one day. The concentration rate of insects flying in such convergence is slow: a tenfold increase in volume density, for insects confined to a fixed depth in the atmosphere, would require a few days. However, there are parts of the atmosphere where convergence is much stronger, say 10^{-3} s^{-1}, often in strips known as convergence zones. Flying insects there could be concentrated much faster: a tenfold increase in volume density in about an hour. With such convergence the vertical velocity would increase upwards by 10^{-3} m s^{-1} in 1 m, that is, from zero at the ground to 1 m s^{-1} at a height of 1 km. Upward velocities of 1–5 m s^{-1} are characteristic of convergence zones with such an intensity, and may exceed the fall speeds of insects with their wings folded. Even stronger convergence does occur, as in tornadoes and so-called microbursts, but it is transient, often lasting only a small fraction of an hour.

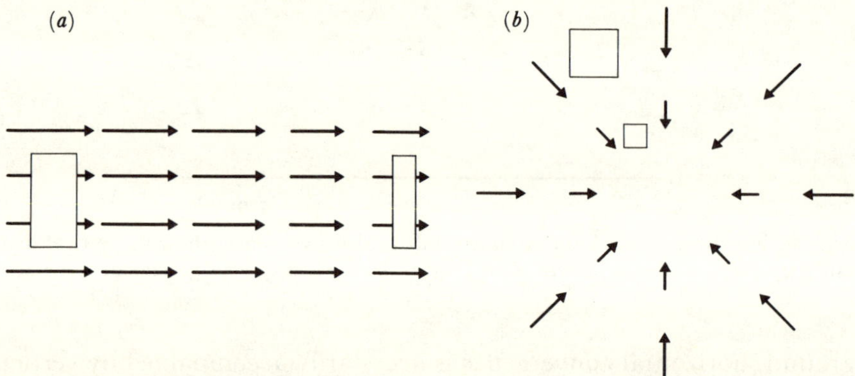

FIGURE 2. Wind arrows in two idealized convergent flow patterns with downwind deceleration, showing contraction of horizontal areas of moving volumes of air: (*a*) uniform direction; (*b*) radial inflow.

There is a tendency for stronger convergence to be associated with smaller and shorter-lived disturbances in the atmosphere (Fujita 1981), with the result that the time needed for a tenfold increase in volume density is comparable with the life-time of the disturbance. Hence an increase in volume density of about tenfold is to be commonly expected due to wind convergence within a single atmospheric system, irrespective of its duration, and only rarely a hundred-fold. Such increases in volume density assume the insects remain in the convergence zone, but winds may carry insects more or less quickly through the zone, with consequently less concentration. However, by comparing vertical profiles of volume density measured by radar in and near echo bands, increases of about fifty-fold have indeed been found (Schaefer 1976, 1979; Pedgley *et al.* 1982; Riley & Reynolds 1983). To compare densities at a given height (as measured by an aircraft, for example) is inadequate as an indicator of concentration when insects are fed from below into the convergence zone. Moreover, neither is an increase in *areal* density necessarily an indicator of concentration because it can be caused by a simple doming up of an insect layer without any increase in *volume* density.

Increased numbers of insects caught in traps that happen to be near convergence zones are sometimes used as evidence that concentration by wind convergence has taken place (Haggis 1971, 1979), but size of catch is also affected by trap efficiency (Douthwaite 1978; McGeachie 1989) and by insect flight behaviour, both of which can be modified by the weather accompanying convergence zones.

4. Atmospheric disturbances and insect concentration

Radar studies have clearly associated echo bands, or insects in line concentrations (the word is used here to describe the observed state, not a process), in the lowest kilometre of the atmosphere at night, with windshift lines and strong horizontal wind convergence. There are zones of strong horizontal wind convergence in three kinds of mesoscale atmospheric motion patterns (i.e. those with horizontal dimensions from a few tens to a few hundreds of kilometres); for a general review, see Atkinson (1981).

1. The leading edges of gravity currents (also known as density currents), where there is undercutting of warm air by cool air produced by evaporation of falling rain or by horizontal variations in the radiation balance of the ground, either on a large scale as at some cold fronts or the Inter-Tropical Front (ITF) or more locally as at coasts and over mountains (and possibly also at the boundaries of wet or cloud-shaded regions, see, for example, Ookouchi *et al.* 1984; Segal *et al.* 1986).

2. Overturning circulations in gravity waves in stably stratified air which has been displaced vertically, either by intrusion of gravity currents (Reid *et al.* 1979; Drake 1984*a*, *b*, 1985*a*), or by flowing over topographic barriers (Pedgley *et al.* 1982).

3. The edges of adjacent convection cells (polygonal or linear) caused by heating of the atmosphere over more or less homogeneous ground.

(a) Gravity currents

The structure and behaviour of gravity currents have been studied extensively in laboratory tank experiments, in the field (first using instrumented towers and dense networks of ground recorders, but later using Doppler radar and aircraft), and by simulation in numerical models. For reviews, see Fujita (1985), Simpson (1987) and Smith & Reeder (1988). Gravity currents in the atmosphere have depths of up to one or two kilometres, and at the leading edge there is convergence and ascent of both airstreams into a domed head containing an overturning circulation, or rotor (figure 3). Flying insects from both sides can be expected to ascend in a sloping curtain with a structure similar to that revealed by duststorms accompanying downdraught squalls in arid country (Lawson 1971). Such a structure is well demonstrated in a cross-section through a cloud of spruce budworm moths at a sea breeze front (Schaefer 1979), and in less detail for grasshoppers (Schaefer 1976) and African armyworm moths (Pedgley *et al.* 1982) at storm outflows. The vertical distribution of volume density found by Schaefer is consistent with a doming up, or even rolling up and mixing, of shallow, dense layers of insects near the ground from both sides of the front, similar to the vertical distribution of water vapour

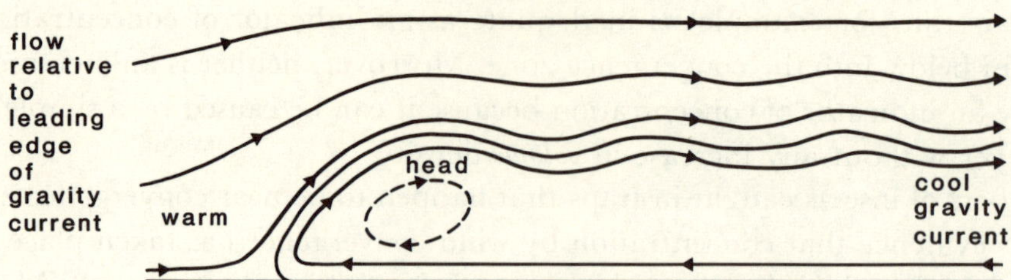

FIGURE 3. Schematic vertical section through a gravity current, showing flow from both sides into a zone of updraughts at the leading edge, and possible overturning in the head.

found by Simpson *et al.* (1977) at a sea breeze front. A result is that the area density increased about tenfold, whereas the volume density changed less. Another example of this rolling-up process is the reconstitution of swarms of the Desert Locust in northern Somalia flying into windshift lines (identified by Sayer (1962) as the ITF, but more likely to be gust fronts from adjacent rainstorms). Similarly, band echoes due to the presence of insect-feeding birds at altitudes of hundreds of metres in a sea breeze front are indicative of the effects of upward motion there on flying insects (Simpson 1967), and the same seems to be true for the leading edges of storm outflows (Harper 1960). Other echo bands due to insects have been reported in association with sea breeze fronts (Neumann & Mukammal 1981; Drake 1982, 1984*b*), storm outflows (Schaefer 1976; Riley *et al.* 1981; Reynolds & Riley 1988), cold fronts (Rainey 1979; Reid *et al.* 1979), and the ITF (Rainey 1976; Schaefer 1976). Gravity currents similar to sea breezes can develop along an escarpment bordering a plateau but their effects on flying insects have not been explored.

It is not known if the lobe-and-cleft structure (Simpson 1987) at the leading edge of a gravity current intensifies the concentration mechanism there. Nor are the effects on flying insects known where increased convergence occurs in vortices that sometimes develop along strongly sheared windshift lines. Such vortices vary in size from as small as dust devils to as large as tornadoes, and their accompanying circulations with dimensions from several kilometres to a few tens of kilometres (James & Browning 1979; Lemon & Doswell 1979; Hobbs & Persson 1982; Carbone 1982, 1983; Parsons *et al.* 1987; Mueller & Carbone 1987). Outflows from neighbouring rainstorms can collide, triggering new rainstorms and hence new outflows.

The process of insect concentration at a gravity current head has been modelled numerically. Mansfield *et al.* (1974) used the observed overturning circulation at a sea breeze front, and simple assumptions about aphid flight behaviour, to show that after two hours almost all the insects were being carried within the circulation. Symmons & Luard (1982) have shown the magnitude of concentration, by using some fairly arbitrary assumptions about both the structure of the wind field and insect behaviour. A two-dimensional (vertical plane), numerical, Eulerian model, based on a pollution diffusion model by using observed insect sources and wind soundings as inputs, is being developed to determine whether convergence simulates the distribution of brown planthoppers near the Bai-u front over China (F. Crummay, personal communication).

(b) Bores and solitary waves

When a gravity current enters a stably stratified fluid, a bore, like that on a tidal river, can be induced, as has been shown in laboratory tank experiments by Maxworthy (1980) and by Rottman & Simpson (1989). There is a sudden and sustained lifting of the streamlines, often in a sequence of waves (figure 4a). Such undular bores are smooth if the lifting is small, but they are progressively more turbulent as lifting increases. Tank experiments and numerical models show that an undular bore propagates away from the parent gravity current and evolves into a set of independently-moving solitary waves (figure 4b); for reviews, see Smith

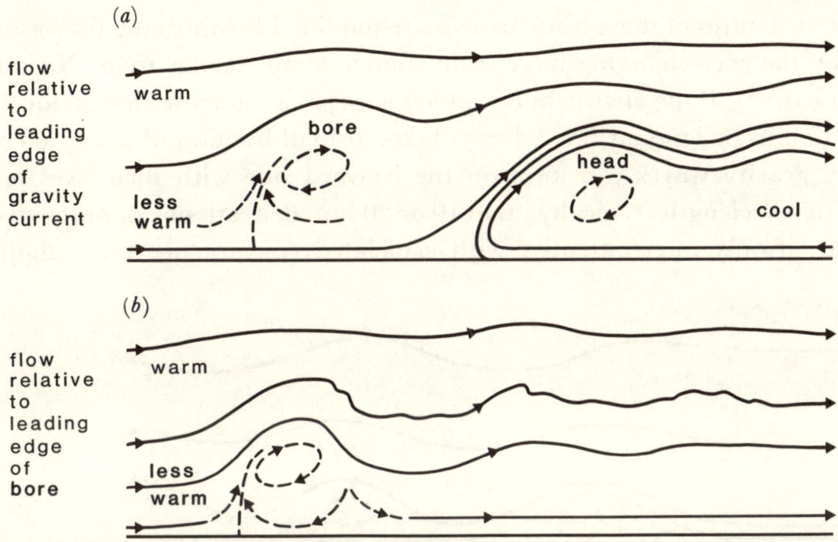

FIGURE 4. Schematic vertical section through an undular bore generated by a gravity current entering a stably stratified fluid: (a) smooth undular bore beginning to move away from the head; (b) turbulent undular bore that has propagated far from its source.

(1988) and Christie (1989). An atmospheric example that has been well documented in recent years is the Morning Glory of northeastern Australia, which seems to be generated by the collision of sea breeze fronts from opposite coasts of the Cape York Peninsula, in Queensland, and travels westwards for hundreds of kilometres at speeds of 5–15 m s^{-1}. Similar disturbances seem to be generated in nocturnal temperature inversions by sea breeze fronts (Simpson *et al.* 1977), by rainstorm outflows, and by horizontal variations in wind accelerations associated with formation of the nocturnal low-level jet. Little is yet known about the conditions for, or even the occurrence of, overturning circulations in such solitary waves, and their significance in concentrating flying insects is also unknown, but they may account for some of the echo bands reported from Australia (Drake 1984a, b, 1985a).

(c) Mesoscale gravity waves

Much larger gravity waves, with wavefront lengths of a few hundred kilometres and speeds of a few tens of metres per second, persisting for ten hours or more, occur occasionally in middle latitudes (Uccellini & Koch 1987). At the ground, these waves are accompanied by wind oscillations (superimposed on the large-scale flow) of amplitude about 10 m s^{-1}, with maxima under the crests blowing in the same direction as that of wave propagation, and under the troughs in the opposite direction. Maximum convergence is of the order of 10^{-4} s^{-1}. Implied vertical circulations extend throughout the depth of the troposphere, for many of the waves seem to have been set up by downdraughts although they are transmitted by marked low-level inversions. Their effects on insect concentration are unknown.

(d) Topographic waves and vortices

The motion occurring in a stably stratified airstream in the presence of a topographical barrier may take many forms, like those in a river flowing over and between boulders, and determined by the upstream profiles of wind and temperature as well as by the shape and size of the barrier. The structures of these disturbances are modified by diurnal patterns of heating and cooling and by the ever-changing large-scale wind systems (Alaka 1960; Nicholls 1973; Smith 1979; Baines 1987). If the airstream flows over a *long ridge*, narrow enough for the effects of the earth's rotation to be ignored (i.e. when it takes a small fraction of a day to cross it), a train of stationary gravity waves can form on the leeward side with their axes parallel to the barrier and with wavelengths typically up to 10 or 20 km. Beneath one or more wave crests, rotors may appear, if only intermittently, with associated convergence lines (figure 5(a)).

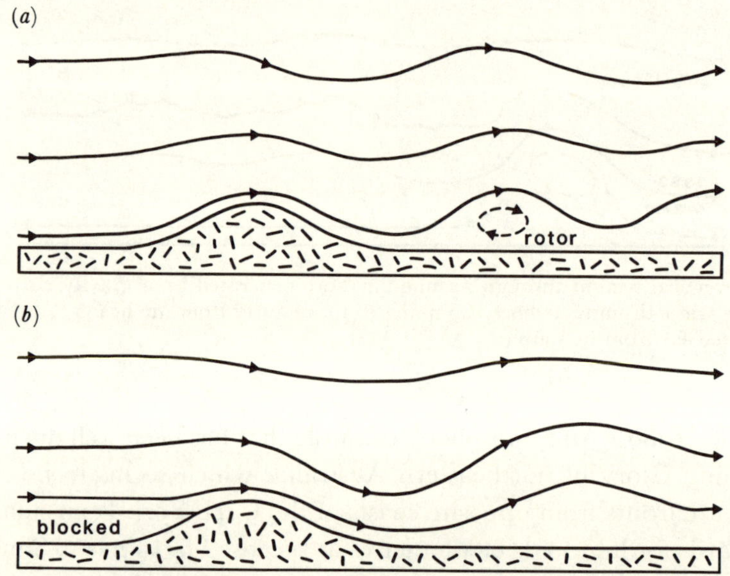

FIGURE 5. Schematic vertical section through a stably stratified fluid crossing a long ridge, showing: (a) lee waves and an overturning circulation (rotor); (b) blocking of the lower part of the upstream flow.

There is radar evidence for concentration of African armyworm moths in rotors (Pedgley *et al.* 1982). On the windward side, the lower part of the airstream may contain a similar rotor, again with an associated convergence line (figure 5a), or it may be brought to rest, or blocked (figure 5b). With parallel ridges, the flow can be more complex, depending on ridge spacing in relation to wavelength (Tampieri & Hunt 1985; Lee *et al.* 1987).

If the barrier is an *isolated hill* then much of the air that would otherwise be blocked is deflected around it, leading to a complex wake (figure 6) (Brighton 1978, Hunt & Snyder 1980). Waves may be induced in an inversion if it is at levels comparable with the top of the hill, perhaps with a horseshoe-shaped spiralling rotor in the first lee wave. Beneath this pattern,

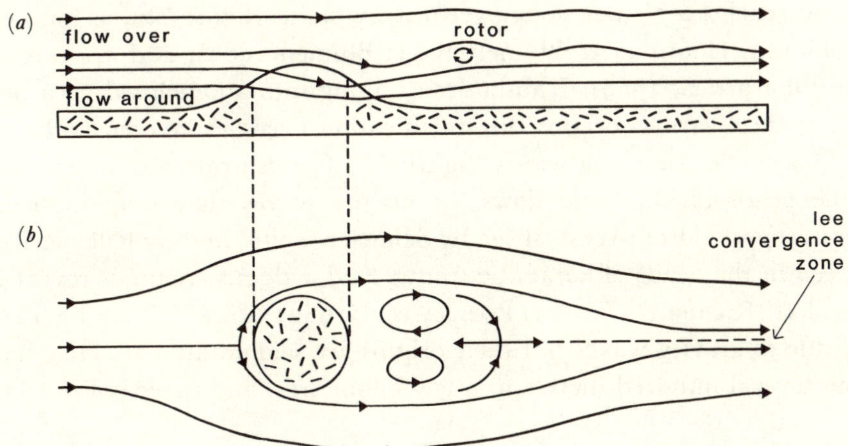

FIGURE 6. Schematic representation of flow of a stably stratified fluid over and around an isolated hill (based on Brighton (1978) and Hunt & Snyder (1980)): (a) vertical section showing the upper part of the stream flowing over the hill, and the lower part around; (b) horizontal plan showing separation of the flow on the upstream side, and a complex wake on the downstream side with vortices and a lee convergence zone.

the deflected flow, accelerating on either side of the hill, can be separated by shear lines from a zone of light winds (wind shadow), or by a lee convergence zone (Edinger & Helvey 1961, Walter & Overland 1982, Mass & Dempsey 1985a), or by contra-rotating vortices with more or less vertical axes and a zone of reversed wind blowing towards the hill. Vortices formed in a stably stratified airstream to the lee of isolated mountainous islands can be periodically shed as a Karman vortex street (Chopra & Hubert 1965). An example of similar shedding over land may be the Denver Cyclone (Wilczak & Glendening 1988), but the effects of such vortices on flying insects are unknown.

Complex interactions can be expected between gravity currents and topographic disturbances. A particularly complex and rapidly evolving wind field would be caused, for example, by downdraughts from adjacent evening storms triggered by a sea breeze front moving into a mountainous region.

(e) Topographic convergence zones

Another topographic effect is the channelling of stably stratified airstreams along valleys, leading to the formation of convergence zones. This has been studied over the Puget Sound (Mass 1981, 1982) and over the Red Sea (Pedgley 1966; Rainey 1976). Swarms of the Desert Locust, and its breeding, are concentrated on the lower land around the central latitudes of the Red Sea from November to January, before seasonal temperature rises are enough to allow daytime flight onto the adjacent plateaux and away from the zone's influence.

(f) Slope winds

When a nocturnal temperature inversion develops on sloping ground, the cooled air tends to drain downhill as a gentle katabatic wind. In a valley, katabatics from both sides can combine into a jet-like mountain wind, blowing down-valley with a speed up to 5 m s^{-1}, sometimes pulsating with a period of oscillation of the order of one hour. Evening onset of the mountain wind in a valley bottom can be sudden (Wilkins 1955; Dickson 1958; Thompson 1967; Tyson 1968) and it resembles a gravity current. The same is probably true where a

640   D. E. PEDGLEY

katabatic wind reaches a gentler slope over nearby plains (figure 7a), as has been shown near
Boulder, Colorado (Hootman & Blumen 1983; Blumen 1984), and appears also to happen
over Tokyo (Ohara *et al.* 1989). Katabatics draining into an enclosed area become ponded,
and because of their various origins the stratification becomes layered and conducive to the
formation of bores and travelling waves (figure 7b). Concentration of insects has not yet been
shown at the heads of katabatic flows, or in any waves they may induce, but the line
concentrations reported from West Africa by Schaefer (1976) and by Riley & Reynolds (1983)
may have been in the heads of katabatic fronts. Radar does sometimes reveal undulations in
insect layers aloft (Richter *et al.* 1973; Pedgley *et al.* 1982; Gossard & Strauch 1983); these are
presumably due to gravity waves but their origins are seldom known. They have separations
varying from several hundred metres to a few kilometres, and speeds of 5–10 m s^{-1}.

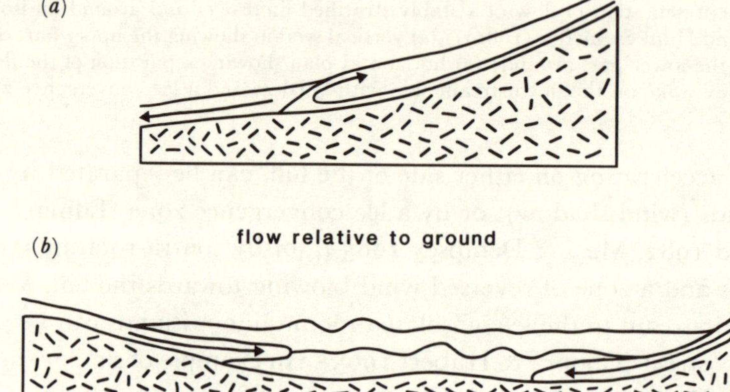

FIGURE 7. Night-time downslope katabatic winds: (a) forming a head at a change in slope;
(b) entering ponded and strongly stratified cold air with generation of gravity waves.

Whereas gravity currents, bores and many waves are mobile, and may travel tens or
hundreds of kilometres, the topographically linked waves and convergence zones are more or
less stationary. There seems to have been little work so far on any breaking or reflection of
travelling waves by topographic barriers, but Black (1979) gives examples of waves in shallow
sea fog reflected from the shores at the head of the Bay of Fundy, Canada.

(g) *Cellular convection*

Daytime convection in the lowest few kilometres of the atmosphere is sometimes organized
into polygonal cells a few tens of kilometres across, or in linear cells, or rolls, a few kilometres
apart but tens of kilometres long. Most studies of polygonal cells have been over the oceans
(Agee 1984, 1987). Similar convection over land seems to be much less common, but it has
been reported from radar studies, in each of which the updraughts in cell walls have been
located by means of the concentrations of insects: in the U.S.A. (Hardy & Ottersten 1969), in
Australia (Schaefer 1976; Reid *et al.* 1979) in Niger (Schaefer 1976) and in India (Mazumdar
et al. 1965). Convective roll vortices more or less parallel to the wind develop when slight
instability is accompanied by wind shear in the vertical; for a review, see Brown (1980).
Convergence of the order of 10^{-3} s^{-1} has been measured beneath upcurrents in such
rolls. Although the effect on flying insects is unknown, some of the daytime radar echoes

reported from Australia were almost linear, suggesting an effect of rolls. Harper (1960) identified echo bands, that we can now reasonably associate with roll vortices, as being caused by birds feeding on insects that were presumably being taken aloft at the convergence lines.

(h) Coastal convergence lines

When a wind over the sea meets a coast on the right-hand side (in the northern hemisphere) at a small angle a coastal convergence line may form due to greater friction over the land (Roeloffzen *et al.* 1986). Such lines have vertical circulations similar to those of sea breeze fronts, and indeed the two mechanisms may interact. Any effects on flying insects have not yet been shown.

5. DISCUSSION

It is often difficult to be sure of the pattern of atmospheric motion that is associated with a given radar echo band. This is because some patterns are still poorly described, and because meteorological observations made during radar field studies are limited, being at best confined to autographic records of surface weather, together with a few soundings of wind and temperature, although aircraft observations are sometimes available. Reliance on surface records to distinguish kinds of night-time patterns is uncertain because temperature inversions more or less uncouple the surface flow from that aloft, even at heights of only a few tens of metres. Additional useful records would be in the vertical plane, obtained from Range Height Indicator radar displays of insect clouds and from frequent wind soundings, perhaps using remote sensing techniques such as Doppler radar.

How can an imperfect understanding of the mechanisms that may concentrate flying insects be used to improve the monitoring of insect populations? If the densities and numbers typical of line concentrations are not worth spraying, for logistic, economic or other reasons, then searching for individual concentrating disturbances is unlikely to be worthwhile. But that may not be true where the same insects are subjected to a sequence of such concentrations. That is most likely to occur where the motion is linked to topography, either coasts (sea breeze fronts and any associated rainstorms) or mountains (complex barrier effects and slope winds, and any associated rainstorms). Away from coasts and mountains, the distribution of convergent wind systems (such as downdraught fronts and bores) is likely to be so erratic in space and time that they would provide little guidance in locating insect concentrations for air-to-air control tactics. An exception is the ITF in Sudan, where it can be as well-defined as the leading edge of a gravity current. In some ways the ITF there resembles a sea breeze front, but on a larger scale. However, the ITF has a north–south diurnal oscillation in position of up to 200 km as well as more irregular movements over a few days of several hundred kilometres, reducing the chances of the same insects being successively concentrated, particularly for those species that fly for only a few days, and for only a part of each day. Another exception is the coastal front of northwest Africa (the 'trade front'), separating cool ocean air from hot continental air, which can persist for days or weeks and trap migrating Desert Locust swarms.

From this it follows that mountains, and particularly coastal mountains, are likely to be the places most favourable to the concentration of flying insects, because of the variety and frequency of suitable atmospheric disturbances. Such disturbances are likely to be most frequent at night, not only because of the static stability necessary for the development of bores, waves and katabatic flows, but also because gravity currents can persist from afternoon sea breezes and rainstorms. But the very complexity of both windfields and topography makes the operation of control aircraft a hazardous one, especially at night. It is difficult to predict the location of convergent wind systems likely to be present in given mountains on a given day or night: partly because the surface observational network is almost always inadequate, except for the larger systems; and partly because fine-mesh numerical models of wind fields have not yet been widely applied to mountainous areas, especially in low latitudes. However, simple diagnostic models, such as that of Mass & Dempsey (1985 *b*) can usefully simulate wind fields in mountainous regions.

Even if a convergent wind system is located we do not know the chances of finding insects in numbers and densities worthwhile to control. Locusts already concentrated into swarms are known to become trapped, for hours or even days, in persistent and slow-moving convergence zones (often topographically linked) which therefore provide suitable sites for monitoring by aircraft. By contrast, short-lived zones (such as gust fronts) and mobile zones (such as cold fronts and the ITF) are much less suitable.

If insects do, in fact, settle on the ground from line concentrations before being dispersed again, and subsequent flight leads to an encounter with another concentrating motion pattern, then a high volume density may result. There is no direct evidence for such mass settling, but storm outflows seem to be a cause of concentration of African armyworm moths that lead to mass egg laying and hence outbreaks of caterpillars, which are known to be associated with night-time rainstorms (Tucker & Pedgley 1983; Pedgley *et al.* 1989). A similar occurrence with the fall armyworm in Canada has been noted (Rose *et al.* 1975), and the distribution of bollworm eggs in the Sudan Gezira may be similarly influenced by the concentration of parent moths (Haggis 1981). Meterological satellites have been used to help locate spreading downdraughts from storms and therefore have potential to help locate any associated armyworm outbreaks.

I am grateful to the following colleagues for their helpful comments on an earlier draft of this paper: P. Burt; C. F. Dewhurst; M. J. Haggis; J. I. Magor; T. Megenasa; P. O. Odiyo; W. W. Page; D. R. Reynolds; D. J. W. Rose and M. R. Tucker.

References

Agee, E. M. 1984 Observations from space and thermal convection. *Bull. Am. met. Soc.* **65**, 938–949.

Agee, E. M. 1987 Mesoscale cellular convection over the oceans. *Dyn. Atmos. Oceans* **10**, 317–341.

Alaka, M. A. (ed.) 1960 The airflow over mountains. *Wld. met. Org. Tech.* Note No. 34.

Atkinson, B. W. 1981 *Mesoscale atmospheric circulations.* New York: Academic Press.

Baines, P. G. 1987 Upstream blocking and airflow over mountains. *A. Rev. Fluid Mech.* **19**, 75–97.

Bergh, J.-E. 1988 Take-off activity in caged desert locusts, *Schistocerca gregaria* (Forsk.) (Orthoptera: Acrididae) in relation to meteorological disturbances. *Int. J. Biometeorol.* **32**, 95–102.

Black, P. G. 1979 Mesoscale cloud patterns revealed by ASTP photographs. In *Apollo-Soyuz Test Project, Summary Science Report. Vol. II* (ed. F. R. El-Baz & D. M. Warner), pp. 617–634. Washington: NASA.

Blumen, W. 1984 An observational study of instability and turbulence in nighttime drainage winds. *Bound.-Layer Met.* **28**, 245–269.

Brighton, P. W. M. 1978 Strongly stratified flow past three-dimensional obstacles. *Q. Jl. R. met. Soc.* **104**, 289–307.

Brown, E. S., Betts, E. & Rainey, R. C. 1969 Seasonal changes in distribution of the African armyworm, *Spodoptera exempta* (Wlk.) (Lep., Noctuidae), with special reference to eastern Africa. *Bull. ent. Res.* **58**, 661–728.

Brown, R. A. 1980 Longitudinal instabilities and secondary flows in the planetary boundary layer. *Rev. Geophys. Space Phys.* **18**, 683–697.

Browning, K. A. & Wexler, R. 1968 The determination of kinematic properties of a wind field using Doppler radar. *J. appl. Met.* **7**, 105–113.

Carbone, R. E. 1982 A severe frontal rainband. Part I: Stormwide hydrodynamic structure. *J. atmos. Sci.* **39**, 258–279.

Carbone, R. E. 1983 A severe frontal rainband. Part II: Tornado parent vortex circulation. *J. atmos. Sci.* **40**, 2639–2654.

Casimir, M. 1987 Plague locusts in New South Wales: studies of migration and displacement of populations. *New South Wales Dept. Agric. Bull.* **91**.

Chen R.-L. & Bao Z.-S. 1987 Research on the migration of oriental armyworm in China and a discussion of management strategy. *Insect Sci. Applic.* **8**, 571–572.

Chen R.-L., Bao X.-Z., Drake, V. A., Farrow, R. A., Wang S.-Y., Sun Y.-J. & Zhai B.-P. 1989 Radar observations of the spring migration into north-eastern China of the oriental armyworm, *Mythymna separata*, and other insects. *Ecol. Entomol.* **14**, 149–162.

Chopra, K. P. & Hubert, L. F. 1965 Mesoscale eddies in wake of islands. *J. atmos. Sci.* **22**, 652–657.

Christie, D. R. 1989 Long nonlinear waves in the lower atmosphere. *J. atmos. Sci.* **46**, 1462–1491.

Cowley, J. M. 1987 Oviposition site selection and effect of meteorological conditions on flight of *Eudonia sabulosella* (Lep.: Scopariinae) with implications for pasture damage. *N. Z. Jl Zool.* **14**, 527–533.

Dickison, R. B. B., Haggis, M. J. & Rainey, R. C. 1983 Spruce budworm moth flight and storms: case study of a cold front system. *J. Clim. appl. Met.* **22**, 278–286.

Dickison, R. B. B., Haggis, M. J., Rainey, R. C. & Burns, L. M. D. 1986 Spruce budworm moth flight and storms: further studies using aircraft and radar. *J. Clim. appl. Met.* **25**, 1600–1608.

Dickson, C. R. 1958 Ground layer temperature inversions in an interior valley and canyon. *Univ. Utah, Dept. Met., Final Report Contract* DA19-129-QM-399.

Douthwaite, R. J. 1978 Some effects of weather and moonlight on light-trap catches of the armyworm, *Spodoptera exempta* (Walker) (Lep.: Noctuidae), at Muguga, Kenya. *Bull. ent. Res.* **68**, 533–572.

Drake, V. A. 1982 Insects in the sea-breeze front at Canberra: a radar study. *Weather, Lond.* **37**, 134–143.

Drake, V. A. 1983 Collective orientation by nocturnally migrating Australian plague locusts, *Chortoicetes terminifera* (Walker) (Orthoptera: Acrididae): a radar study. *Bull. ent. Res.* **73**, 679–692.

Drake, V. A. 1984*a* A solitary wave disturbance of the marine boundary layer over Spencer Gulf revealed by radar observations of flying insects. *Aust. met. Mag.* **32**, 131–135.

Drake, V. A. 1984*b* The vertical distribution of macro-insects migrating in the nocturnal boundary layer: a radar study. *Bound.-Layer Met.* **28**, 353–374.

Drake, V. A. 1985*a* Solitary wave disturbances of the nocturnal boundary layer revealed by radar observations of migrating insects. *Bound.-Layer Met.* **31**, 269–286.

Drake, V. A. 1985*b* Radar observations of moths migrating in a nocturnal low-level jet. *Ecol. entomol.* **10**, 259–265.

Drake, V. A. & Farrow, R. A. 1983 The nocturnal migration of the Australian plague locust, *Chortoicetes terminifera* (Walker) (Orthoptera: Acrididae): quantitative radar observations of a series of northward flights. *Bull. ent. Res.* **73**, 567–585.

Drake, V. A. & Farrow, R. A. 1985 A radar and aerial-trapping study of an early spring migration of moths (Lepidoptera) in inland New South Wales. *Aust. J. Ecol.* **10**, 223–235.

Drake, V. A. & Farrow, R. A. 1988 The influence of atmospheric structure and motions on insect migration. *A. Rev. Ent.* **33**, 183–210.

Drake, V. A. & Farrow, R. A. 1989 The 'aerial plankton' and atmospheric convergence. *Trends Ecol. Evol.* **4**, 381–385.

Edinger, J. G. & Helvey, R. A. 1961 The San Fernando convergence zone. *Bull. Am. met. Soc.* **42**, 626–635.

Fosberg, M. A. & Peterson, M. 1986 Modeling airborne transport of gypsy moth (Lep.: Lymantriidae) larvae. *Agric. For. Met.* **38**, 1–8.

Fujita, T. T. 1981 Tornadoes and downbursts in the context of generalised planetary scales. *J. atmos. Sci.* **38**, 1511–1534.

Fujita, T. T. 1985 The downburst: microburst and macroburst. University of Chicago, Department of Geophysical, Science.

Gossard, E. E. & Strauch, R. G. 1983 *Radar observation of clear air and clouds.* Amsterdam: Elsevier.

Greenbank, D. O., Schaefer, G. W. & Rainey, R. C. 1980 Spruce budworm moth flight and dispersal. *Mem. ent. Soc. Can.* **110**.

Haggis, M. J. 1971 Light-trap catches of *Spodoptera exempta* (Walk.) in relation to wind direction. *E. Afr. J. Agric. For.* **37**, 100–108.

Haggis, M. J. 1979 African armyworm *Spodoptera exempta* (Walker) (Lep.: Noctuidae) and wind convergence in the Kenya rift valley, May 1970. *E. Afr. J. Agric. For.* **44**, 332–346.

Haggis, M. J. 1981 Spatial and temporal changes in the distribution of eggs of *Heliothis armigera* (Hubner) (Lep.: Noctuidae) on cotton in the Sudan Gezira. *Bull. ent. Res.* **71**, 183–193.

Hardy, K. R. & Ottersten, H. 1969 Radar investigations of convective patterns in the clear atmosphere. *J. atmos. Sci.* **26**, 666–672.

Harper, W. G. 1960 An unusual indicator of convection. *Mar. Obs.* **30**, 36–40.

Hendricks, D. E., Perez, C. T. & Guerra, R. J. 1980 Effects of nocturnal wind on performance of two sex pheromone traps for noctuid moths. *Env. Entomol.* **9**, 483–485.

Hirai, K. 1988 Sudden outbreaks of the armyworm, *Pseudaletia separata* Walker, and its monitoring systems in Japan. *Jap. Agric. Res. Q.* **22**, 166–174.

Hobbs, P. V. & Persson, P. O. G. 1982 The mesoscale and microscale structure and organization of clouds and precipitation in midlatitude cyclones. Part V: The substructure of narrow cold-frontal rainbands. *J. atmos. Sci.* **39**, 280–295.

Hootman, B. W. & Blumen, W. 1983 Analysis of nighttime drainage winds in Boulder, Colorado during 1980. *Mon. Weath. Rev.* **111**, 1052–1061.

Hunt, J. C. R. & Snyder, W. H. 1980 Experiments on stably and neutrally stratified flow over a model three-dimensional hill. *J. Fluid Mech.* **96**, 671–704.

Irwin, M. E. & Thresh, J. M. 1988 Long-range aerial dispersal of cereal aphids as virus vectors in North America. *Phil. Trans. R. Soc. Lond.* B **321**, 421–446.

James, P. K. & Browning, K. A. 1979 Mesoscale structure of line convection at surface cold fronts. *Q. Jl R. met. Soc.* **105**, 371–382.

Johnson, S. J. 1987 Migration and life history strategy of the fall armyworm, *Spodoptera frugiperda*, in the western hemisphere. *Insect Sci. Appl.* **8**, 543–549.

Joyce, R. J. V. 1973 Insect mobility and the philosophy of crop protection with reference to the Sudan Gezira. *Pest Abs. News Summ.* **19**, 62–70.

Joyce, R. J. V. 1983 Aerial transport of pests and pest outbreaks. *EPPO Bull.* **13**, 111–119.

Kisimoto, R. 1984 Meteorological conditions including long-distance migration of the brown planthopper, *Nilaparvata lugens* Stal. *Chin. J. Entomol.* **4**, 39–48.

Lawson, T. J. 1971 Haboob structure at Khartoum. *Weather, Lond.* **26**, 105–112.

Lee, J. Y., Lawson Jr., R. E. & Marsh, G. L. 1987 Flow visualisation experiments on stably stratified flow over ridges and valleys. *Meteorol. atmos. Phys.* **37**, 183–194.

Lemon, L. R. & Doswell III, C. A. 1979 Severe thunderstorm evolution and mesocyclone structure as related to tornado genesis. *Mon. Weath Rev.* **107**, 1184–1197.

Li K.-P., Wong H.-H. & Woo W.-S. 1964 Route of the seasonal migration of the oriental armyworm moth in the eastern part of China as indicated by a three-year result of releasing and recapturing marked moths. (In Chinese.) *Acta Phytophylac. Sin.* **3**, 101–110.

Liu C.-H. 1984 Study on the long-distance migration of the brown planthopper in Taiwan. *Chin. J. Entomol.* **4**, 49–54.

Mansfield, D. A., Milford, J. R. & Simpson, J. E. 1974 Redistribution of an insect population by the circulation within a mesoscale front. (Unpublished).

Mass, C. F. 1981 Topographically forced convergence in western Washington State. *Mon. Weath. Rev.* **109**, 1335–1347.

Mass, C. F. 1982 The topographically forced diurnal circulations of western Washington State and their influence on precipitation. *Mon. Weath. Rev.* **110**, 170–183.

Mass, C. F. & Dempsey, D. P. 1985a A topographically forced convergence line in the lee of the Olympic Mountains. *Mon. Weath. Rev.* **113**, 659–663.

Mass, C. F. & Dempsey, D. P. 1985b A one-level, mesoscale model for diagnosing surface winds in mountainous and coastal regions. *Mon. Weath. Rev.* **113**, 1211–1227.

Maxworthy, T. 1980 On the formation of nonlinear internal waves from the gravitational collapse of mixed regions in two and three dimensions. *J. Fluid Mech.* **96**, 47–64.

Mazumdar, S., Bhaskara Rao, N. S. & Gupta, G. R. 1965 Radar and synoptic study of locust swarms over Delhi. *Wld met. Org. Tech. Note* **69**, 162–187.

McGeachie, W. T. 1989 The effects of moonlight illuminance, temperature and wind speed on light trap catches of moths. *Bull. ent. Res.* **79**, 185–192.

McManus, M. L. 1988 Weather, behaviour and insect dispersal. *Mem. ent. Soc. Can.* **146**, 71–94.

McNeil, J. N. 1987 The true armyworm, *Pseudaletia unipuncta*: a victim of the pied piper or a seasonal migrant? *Insect Sci. Appl.* **3**, 591–597.

Miyahara, Y. 1987 Simultaneous trap catches of the oriental armyworm and the diamondback moth during the early flight season at Morioka. (In Japanese) *Jap. J. appl. Entomol. Zool.* **31**, 138–143.

Mueller, C. K. & Carbone, R. E. 1987 Dynamics of thunderstorm outflows. *J. atmos. Sci.* **44**, 1879–1898.

Mueller, E. A. & Larkin, R. P. 1985 Insects observed using dual-polarisation radar. *J. atmos. Ocean Tech.* **2**, 49–54.

Neumann, H. H. & Mukammal, E. I. 1981 Incidence of mesoscale convergence lines as input to spruce budworm control strategies. *Int. J. Biomet.* **25**, 175–187.

Nicholls, J. M. 1973 The airflow over mountains: research 1958–1972. *Wld met. Org. Tech. Note* **127**.

Ohara, T., Uno, I. & Wakamatsu, S. 1989 Observed structure of the land breeze head in the Tokyo metropolitan area. *J. appl. Met.* **28**, 693–704.

Ookouchi, Y., Segal, M., Kessler, R. C. & Pielke, R. A. 1984 Evaluation of soil moisture effects on the generation and modification of mesoscale circulations. *Mon. Weath. Rev.* **112**, 2281–2292.

Pair, S. D., Raulston, J. R., Rumme, D. R., Westbrook, I. K., Wolf, W. W., Sparks, A. N. & Schuster, M. F. 1987 Development and production of corn earworm and fall armyworm in the Texas High Plains: evidence for reverse fall migration. *Southwestern Entomol.* **12**, 89–99.

Parsons, D. B., Mohr, C. G. & Tzvi, G.-C. 1987 A severe frontal rainband Part III: Derived thermodynamic structure. *J. atmos. Sci.* **44**, 1615–1631.

Pasek, J. E. 1988 Influence of wind and windbreaks on local dispersal of insects. *Agric. Ecosys. Environ.* **22/23**, 539–554.

Pedgley, D. E. 1966 The Red Sea convergence zone. *Weather, Lond.* **21**, 350–368, 394–406.

Pedgley, D. E. 1982 *Windborne pests and diseases* Chichester: Ellis Horwood.

Pedgley, D. E. 1983 Windborne spread of insect-transmitted diseases of animals and man. *Phil. Trans. R. Soc. Lond.* B **302**, 463–470.

Pedgley, D. E., Reynolds, D. R., Riley, J. R. & Tucker, M. R. 1982 Flying insects reveal small-scale wind systems. *Weather, Lond.* **37**, 295–306.

Pedgley, D. E., Page, W. W., Mushi, A., Odiyo, P., Amisi, J., Dewhurst, C. F., Dunstan, W. R., Fishpool, L. D. C., Harvey, A. W., Megenasa, T. & Rose, D. J. W. 1989 Onset and spread of an African armyworm upsurge. *Ecol. Entomol.* **14**, 311–333.

Peterson, R. K. D., Higley, L. G. & Bailey, W. C. 1988 Phenology of the adult celery looper, *Syngrapha falcifera* (Lep.: Noctuidae), in Iowa: evidence for migration. *Env. Entomol.* **17**, 679–684.

Rabin, R. & Zrnic, D. 1980 Subsynoptic-scale vertical wind revealed by Doppler radar and VAD analysis. *J. atmos. Sci.* **37**, 644–654.

Rainey, R. C. 1951 Weather and the movements of locust swarms: a new hypothesis. *Nature Lond.* **168**, 1057–1060.

Rainey, R. C. 1963 Meteorology and the migration of desert locusts. *Wld met. Org. Tech. Note* **54**.

Rainey, R. C. 1972 Flying insects as potential targets: initial feasibility studies with airborne Doppler equipment in East Africa. *Aeronaut. J.* **76**, 501–506.

Rainey, R. C. 1974 Flying insects as targets for ultra-low volume spraying. *Br. Crop Prot. Counc. Monogr. No.* **11**.

Rainey, R. C. 1976 Flight behaviour and features of the atmospheric environment. *Symp. R. ent. Soc. Lond.* **7**, 75–112.

Rainey, R. C. 1979 Dispersal and redistribution of some Orthoptera and Lepidoptera by flight. *Bull. Soc. ent. Suisse* **52**, 125–132.

Rainey, R. C. 1980 Possible impact of radar on pest management operations. (ed. C. R. Vaughn, W. W. Wolf & W. Klassen), pp. 81–86. *Radar, insect population ecology and pest management.* NASA Conference Publication. 2070.

Ramana Marty, B. V., Roy, A. K., Biswas, K. R. & Khemani, L. T. 1964 Observations on flying locusts by radar. *J. Sci. Ind. Res.* **23**, 289–296.

Reid, D. G., Wardhaugh, K. G. & Roffey, J. 1979 Radar studies of insect flight at Benalla, Victoria, in February 1974. *CSIRO, Div. Entomol., Tech. Pap.* **16**.

Reynolds, D. R. & Riley, J. R. 1988 A migration of grasshoppers, particularly *Diabolocatantops axillaris* (Thunberg) (Orthoptera: Acrididae), in the West African sahel. *Bull. ent. Res.* **78**, 251–271.

Richter, J. H., Jensen, D. R., Noonkester, V. R., Kreasky, J. B., Stimman, M. W. & Wolf, W. W. 1973 Remote radar sensing: atmospheric structure and insects. *Science, Wash.* **180**, 1176–1178.

Riley, J. R. & Reynolds, D. R. 1979 Radar-based studies of the migratory flight of grasshoppers in the middle Niger area of Mali. *Proc. R. Soc. Lond.* B **204**, 67–82.

Riley, J. R. & Reynolds, D. R. 1983 A long-range migration of grasshoppers observed in the Sahelian zone of Mali by two radars. *J. anim. Ecol.* **52**, 167–183.

Riley, J. R. & Reynolds, D. R. 1986 Orientation at night by high-flying insects. In *Insect flight, dispersal and migration* (ed. W. Danthanarayana), pp. 71–87. Berlin: Springer-Verlag.

Riley, J. R., Reynolds, D. R. & Farmery, M. J. 1981 Radar observations of *Spodoptera exempta*, Kenya, March–April 1979. *Centre for Overseas Pest Research, Misc. Rep. No.* 54.

Riley, J. R., Reynolds, D. R. & Farmery, M. J. 1983 Observations of the flight behaviour of the armyworm moth, *Spodoptera exempta*, at an emergence site using radar and infra-red optical techniques. *Ecol. Entomol.* **8**, 395–418.

Riley, J. R., Reynolds, D. R. & Farrow, R. A. 1987 The migration of *Nilaparvata lugens* (Stal) (Delphacidae) and other Hemiptera associated with rice during the dry season in the Philippines: a study using radar, visual observations, aerial netting and ground trapping. *Bull. ent. Res.* **77**, 145–169.

Roeloffzen, J. C., van den Berg, W. D. & Oerlemans, J. 1986 Frictional convergence at coastlines. *Tellus* **38A**, 397–411.

Rose, A. H., Silversides, R. H. & Lindquist, O. H. 1975 Migration flight by an aphid, *Rhopalosiphum maidis* (Hem.: Aphididae) and a noctuid, *Spodoptera frugiperda* (Lep.: Noctuidae). *Can. Ent.* **107**, 567–576.

Rose, D. J. W., Dewhurst, C. F., Page, W. W. & Fishpool, L. D. C. 1987 The role of migration in the life system of the African armyworm, *Spodoptera exempta. Insect Sci. Appl.* **8**, 561–569.

Rose, D. J. W., Page, W. W., Dewhurst, C. F., Riley, J. R., Reynolds, D. R., Pedgley, D. E. & Tucker, M. R. 1985 Downwind migration of the African armyworm moth, *Spodoptera exempta*, studied by mark-and-capture and by radar. *Ecol. Entomol.* **10**, 299–313.

Rosenberg, L. J. & Magor, J. I. 1987 Predicting windborne displacements of the brown planthopper, *Nilaparvata lugens*, from synoptic weather data. 1. Long-distance displacements in the north-east monsoon. *J. Anim. Ecol.* **56**, 39–51.

Rottman, J. W. & Simpson, J. E. 1989 The formation of internal bores in the atmosphere: a laboratory model. *Q. Jl R. met. Soc.* **115**, 941–963.

Sayer, H. J. 1962 The desert locust and tropical convergence. *Nature, Lond.* **194**, 330–336.

Schaefer, G. W. 1976 Radar observations of insect flight. *Symp. R. ent. Soc. Lond.* **7**, 157–197.

Schaefer, G. W. 1979 An airborne radar technique for the investigation and control of migrating pest species. *Phil. Trans. R. Soc. Lond.* B **287**, 459–465.

Segal, M., Purdom, J. F. W., Song, J. L., Pielke, R. A. & Mahrer, Y. 1986 Evaluation of cloud shading effects on the generation and modification of mesoscale circulations. *Mon. Weath. Rev.* **114**, 1201–1212.

Showers, W. B., Whitford, F., Smelser, R. B., Keaster, A. J., Robinson, J. F., Lopez, J. D. & Taylor, S. E. 1989 Direct evidence for meteorologically driven long-range dispersal of an economically important moth. *Ecology* **70**, 987–992.

Simpson, J. E. 1967 Aerial and radar observations of some sea-breeze fronts. *Weather, Lond.* **22**, 306–316, 325–327.

Simpson, J. E. 1987 *Gravity currents in the environment and the laboratory*. Chichester: Ellis Horwood.

Simpson, J. E., Mansfield, D. A. & Milford, J. R. 1977 Inland penetration of sea-breeze fronts. *Q. Jl R. met. Soc.* **103**, 47–76.

Smith, R. B. 1979 The influence of mountains on the atmosphere. *Adv. Geophys.* **21**, 87–230.

Smith, R. K. 1988 Travelling waves and bores in the lower atmosphere: the 'Morning Glory' and related phenomena. *Earth-Sci. Rev.* **25**, 267–290.

Smith, R. K. & Reeder, M. J. 1988 On the movement and low-level structure of cold fronts. *Mon. Weath. Rev.* **116**, 1927–1944.

Symmons, P. M. 1986 Locust displacing winds in eastern Australia. *Int. J. Biomet.* **30**, 53–64.

Symmons, P. M. & Luard, E. J. 1982 The simulated distribution of night-flying insects in a wind convergence. *Aust. J. Zool.* **30**, 187–198.

Tampieri, F. & Hunt, J. C. R. 1985 Two-dimensional stratified flow over valleys: linear theory and a laboratory investigation. *Bound.-Layer Met.* **32**, 257–279.

Thompson, A. H. 1967 Surface temperature inversions in a canyon. *J. appl. Met.* **6**, 287–296.

Tucker, M. R. 1983 Light-trap catches of African armyworm moths, *Spodoptera exempta* (Walker) (Lep.: Noctuidae), in relation to rain and wind. *Bull. ent. Res.* **73**, 315–319.

Tucker, M. R. & Pedgley, D. E. 1983 Rainfall and outbreaks of the African armyworm, *Spodoptera exempta* (Walker) (Lep.: Noctuidae). *Bull. ent. Res.* **73**, 195–199.

Tyson, P. D. 1968 Velocity fluctuations in the mountain wind. *J. atmos. Sci.* **25**, 381–384.

Uccellini, L. W. & Koch, S. E. 1987 The synoptic setting and possible energy sources for mesoscale wave disturbances. *Mon. Weath. Rev.* **115**, 721–729.

Wada, T., Seino, H., Ogawa, Y. & Nakasuga, T. 1987 Evidence of autumn migration in the rice planthoppers, *Nilaparvata lugens* and *Sogatella furcifera*: analysis of light trap catches and associated weather patterns. *Ecol. Entomol.* **12**, 321–330.

Waloff, Z. 1972 Orientation of flying locusts (*Schistocerca gregaria* Forsk.) in migratimg swarms. *Bull. ent. Res.* **62**, 1–72.

Walter, B. A. & Overland, J. E. 1982 Response of stratified flow in the lee of the Olympic Mountains. *Mon. Weath. Rev.* **110**, 1458–1473.

Wilczak, J. M. & Glendening, J. W. 1988 Observations and mixed-layer modeling of a terrain-induced mesoscale gyre: the Denver Cyclone. *Mon. Weath. Rev.* **116**, 2688–2711.

Wilkins, E. M. 1955 A discontinuity surface produced by topographic winds over the Upper Snake River. *Bull. Am. met. Soc.* **36**, 397–408.

Wilson, A. G. L. 1983 Abundance and mortality of overwintering *Heliothis* species. *J. Aust. ent. Soc.* **22**, 191–199.

Wolf, R. A., Pedigo, L. P., Shaw, R. H. & Newsom, L. D. 1987 Migration/transport of the green cloverworm, *Plathypena scabra* (F.) (Lep.: Noctuidae), into Iowa as determined by synoptic-scale weather patterns. *Environ. Entomol.* **16**, 1169–1174.

Discussion

R. S. Scorer (*Imperial College, London, U.K.*). Meteorologists have been aware of the importance of wind patterns in the transport of any airborne migrating species, mainly through the work of R. C. Rainey and K. Williamson, as I have already described (Scorer 1958, 1978).

The essential mechanism for the concentration of numbers of insects is that horizontal convergence at the surface should support upward motion above, in which the insects resist being carried upwards. It is like the creation of rain in a shower cloud or front, where the amount of rain exceeds the total amount of water in the cloud by a large multiple because it is extracted from a much larger volume of air which flows in at the bottom and out at the top, having dropped rain in the process.

The horizontal convergence maintains the coherence of very large locust swarms, which are far bigger than could be held together by a flight behaviour in which random direction becomes inwards at the edges. Another most effective mechanism is that of shovelling up by a cold front, or any similar phenomenon such as a sea breeze front or cold outflow front from a storm. In this, again, the resistance to being carried upwards causes the collection on the 'shovel' as seen by Harper (1960) when (simultaneously by telescope and radar) he observed swifts soaring on the updraft of fronts and supposed that they were feeding on the insects lifted by the updraft.

The question of how resistance to being carried upwards is effected has several interesting aspects. If the air which is being cooled adiabatically in the updraft falls below what is presumed to be the threshold temperature for flight, the insects would automatically cease to fly actively. This raises problems in the context of layering of insect swarms (or 'clouds') at night. This phenomenon was discovered by Johnson (1957) and Taylor (1958) in the case of aphids already carried aloft by dusk. Their density was measured by traps carried aloft by the Meteorological Research Flight, and was found that after being spread through the full depths of the convection layer on a summer afternoon they often tended to accumulate in the warmer air as a layer just above the inversion produced by the cooling of the ground as the sun sets.

The effects connected with gravity waves comprise a very complex field for investigation. Thus, when potato beetles *Leptinotarsa decemlineata* (Say) were found in Czechoslovakia in the late 1940s, it was alleged that they had been put into the fields by enemy agents because they were found in three or four equally spaced rows. Förchtgott (1950) explained that this could have been because they were in lines where the wind was very light as a result of the occurrence of lee waves, the lines being parallel to the hill ridge which produced the waves. This is scarcely a case of clouds or swarms of insects being formed, and appears to depend on landing conditions preferred by a species that was being studiously hunted all over Europe at the time.

In my view, convergence may be quite alarming even over the sea. Convergence at a front of some sort seems quite capable of forming a locust swarm and the general convergence can be quite enough to prevent the dispersion of a swarm for many days even after the frontal character of the motion has ceased.

Smoke, unlike rain or locusts, is not concentrated in the same way because it has a fall speed much smaller than the updraft speed of the air; and the dispersion of smoke therefore is not a good guide to what would happen to a cloud of locusts, even in random flight. Pollution dispersion models are unreal in that they do not represent the motion of the air, and the models are quite artificial in containing diffusion coefficients designed to produce the observed results from a computation.

Because I see no significant aggregation being produced in the variety of gravity wave motion patterns, I would equally not expect to find them in waves reflected from mountains. Such reflections can occasionally be seen by satellite (Scorer 1986).

References

Förchtgott, J. 1950 Transport of small particles and airborne insects over the Ore Mountains *Bull. Met. Czech.* **4**, 14.
Harper, W. G. 1960 An unusual indicator of convection. *Mar. Obs.* **30**, 36–40.
Johnson, C. G. 1957 The vertical distribution of insects in the air and the temperature lapse rate. *Q. Jl Met. Soc.* **83**, 194–201.
Scorer, R. S. 1958 *Natural aerodynamics.* Pergamon Press.
Scorer, R. S. 1978 *Environmental aerodynamics.* Chichester: Ellis Horwood.
Scorer, R. S. 1986 *Cloud investigation by satellite.* Chichester: Ellis Horwood.
Taylor, L. R. 1958 Aphid dispersal and diurnal periodicity. *Proc. Linn. Soc. Lond.* **169**, 67–73.

D. E. PEDGLEY. Professor Scorer draws attention to two aspects of our ignorance: the behaviour of flying insects encountering a change in air temperature, and the potential for gravity waves to concentrate flying insects. His comments on the former point to the possibility that insects respond to falling air temperature by taking avoiding action rather than simply ceasing flapping flight below a threshold. Such avoiding action might be examined in a wind tunnel. As to the latter, it would be helpful first of all if field observations could establish, beyond reasonable doubt, that radar band echoes are in fact associated with gravity waves as well as gravity currents.

Professor Scorer also asserts that horizontal convergence maintains the cohesion of large locust swarms. I wonder what kinds of wind patterns he has in mind and the order of magnitude of the convergence needed. Micro-scale patterns (smaller than a few tens of kilometres) are too short-lived compared with flying time and, as a form of turbulence, would surely be dispersive. Mesoscale patterns seem to be too rare on the vast plains over which much of the long-distance migration take place for them to play any significant role. Macro-scale patterns (larger than a few hundreds of kilometres), although ubiquitous, have weak convergence. And is there any evidence that coherence is lessened when a swarm encounters divergence?

Concerning the formation of a swarm over the sea through wind convergence, alternative mechanisms need to be considered. For example, it is generally accepted that new-generation locusts form into swarms, some days after they have left widespread fledging areas and up to several hundred kilometres away, because of their behaviour.

J. F. W. PURDOM (*Regional and Mesoscale Meteorology Branch, NOAA/NESDIS*). My comment assumes, as correct, Mr Pedgley's thesis that airborne migrant pests are concentrated within atmospheric convergence zones.

In his opening statement, Mr Pedgley points out that 'to make air-to-air spraying effective, then the possibility arises of improving control by seeking out such patterns.' Among other items, such seeking out would depend strongly on an accurate day-to-day weather forecast followed by an accurate 'nowcast' for the phenomena in question (the location of low-level convergence lines). Satellite and radar imagery are two of the most important tools available (Purdom 1982; Bader & Waters 1987).

Many of the convergence phenomena noted by Mr Pedgley have been routinely observed by satellite for many years, for example, thunderstorm outflow boundaries that appear as arc cloud lines (Purdom 1973, 1976, 1982). These boundaries have been observed to be a major controlling factor in afternoon thunderstorm development over southeast U.S.A. Convection associated with sea breezes and the ITCZ have been located by using satellite imagery since near the beginning of meteorological satellite era in 1961. Indeed, a project during the 1960s routinely used satellite APT imagery in the Desert Locust Control Organization for Eastern Africa in Asmara, Ethiopia. We are now well beyond those early days with once a day APT views of the cloud field. The ability to use frequent interval visible and infrared geostationary satellite imagery to monitor the location of such boundaries allows for their nowcasting, and by using the concept of 'convective scale interaction' makes it possible to nowcast favoured areas for deep convective development along such convergence lines (Purdom 1987). Satellite rainfall estimation techniques permit the location of regions along such convergence lines that should be experiencing heavier precipitation. That knowledge would help to identify regions where insects might land because of rainfall.

Some of the mechanisms that give rise to convergence zones are quite shallow, with a depth of 1–2 km. This is in agreement with aircraft flights through thunderstorm outflow boundary (arc cloud line) as observed by using satellite imagery (Sinclair & Purdom 1983). Those convergence zones were narrow, being some 5–10 km in width. It is well known that convergence in depth contributes to the vertical motion found along such convergence boundaries. However, when the air into which the convergence zone is advancing is stable, there is resistance to vertical forcing and that may contribute to such convergence lines.

As Mr Pedgley said, it is difficult to predict the location of convergent wind systems likely to be present in given mountains on a given day or night, because of inadequate surface observations and the lack of fine-mesh numerical models being applied to the problem. However, by averaging geostationary satellite imagery for the same time of day, each day over a particular season, and stratifying further by flow regime, it is possible to identify regions of preferred low-level convergence (in moist areas) by the higher frequency of convective cloudiness in a particular region. Such studies, undertaken over the mountainous region of western U.S.A., clearly show the influence of the mountains on afternoon convective development (Klitch *et al.* 1985). If such studies were undertaken by using Meteosat imagery, valuable information might be found that would be applicable for both planning and forecasting over the region observed by that satellite. When such satellite image climatologies are prepared for each hour of the day (by synoptic pattern), they may then be sequentially viewed by using an image animation device and compared with actual weather developing over the region at the same time. Although mesoscale models have not been widely applied to mountainous areas, mesoscale modelling of topographically induced thermally and mechanically driven circulations is well within the realm of today's technology. Tripoli & Cotton (1989) used a numerical model to study the development of convective systems over the mountains and their subsequent evolution and movement into the plains at night.

As Mr Pedgley points out, away from coasts and mountains, the distribution of convergent wind systems (such as downdraught fronts and bores) is likely to be so erratic in space and time that they would provide little guidance in locating insect concentrations for air-to-air control. While in some cases this might be true, it must be realized that small-scale features such as lakes, rivers, early cloud cover, and wet versus dry soil are known to give rise to local convergence zones that may trigger thunderstorm formation that in turn may lead to downdraughts that produce new convergence zones. In the mesoscale modelling work of Segal *et al.* (1986), the strength of the convergence zones that developed because of early morning cloudy and adjacent clear regions was found to be nearly as strong as those found along sea breeze fronts. By using knowledge of local effects on thunderstorm development, and the concept of convective scale interaction previously mentioned, nowcasting was done that allowed the aircraft investigations of arc cloud lines of Sinclair & Purdom (1983) to be successfully undertaken. I believe the same might be true for vectoring of flights for air-to-air control; however, having the proper real-time display and analysis equipment for this would be vital.

References

Bader, M. & Waters, T. (eds). 1987 Satellite and Radar Imagery Interpretation, Preprints for a Workshop held at Reading, England, 20–24 July, 1987 (551 pages). *EUMETSAT*.
Browning, K. A. 1982 *Nowcasting* (256 pages). London & New York: Academic Press.
Klitch, M. A., Weaver, J. F., Kelly, F. P. & Vonder Haar T. H. 1985 Convective cloud climatologies constructed from satellite imagery. *Mon. Weath. Rev.*, **113**, 326–337.
Purdom, J. F. W. 1973 Meso-highs and satellite imagery. *Mon. Weath. Rev.* **101**, 180–181.
Purdom, J. F. W. 1976 Some uses of high resolution GOES imagery in the mesoscale forecasting of convection and its behavior. *Mon. Weath. Rev.* **104**, 1474–1483.
Purdom, J. F. W. 1982 Subjective interpretation of satellite imagery for nowcasting. *Nowcasting* (ed. K. A. Browning), pp. 149–166. London & New York: Academic Press.
Purdom, J. F. W. 1987 How satellite data can help a nowcaster predict preferred areas of mesoscale convective development and its evolution. Satellite and Radar Imagery Interpretation, Preprints for a Workshop held at Reading. England, 20–24 July, 1987 (ed. M. Bader & T. Waters) *EUMETSAT*. 263–285.
Segal, M., Purdom, J. F. W., Song, J. L., Pielke, R. A. & Mahrer, Y. 1986 Evaluation of cloud shading effects on the generation and modification of mesoscale circulations. *Mon. Weath. Rev.* **114**, 1201–1212.
Sinclair, P. C. & Purdom, J. F. W. 1983 *The genesis and development of deep convective storms.* CIRA Paper No. 1 Colorado State University, Fort Collins, Colo.
Tripoli, G. J. & Cotton, W. R. 1989 Numerical study of an observed orogenic mesoscale convective system. Part I: simulated genesis and comparison with observations. *Mon. Weath. Rev.* **117**, 273–304.

J. E. Simpson (*Department of Applied Mathematics & Theoretical Physics, University of Cambridge, U.K.*). I would like to underline the probable importance of the effect of atmospheric bores on airborne insect behaviour. Ten years ago such bores seemed to be rare phenomena such as the Morning Glory in north Australia. Since then they have been thoroughly investigated in the laboratory and when looked for they appear to be of common occurrence. All that is needed for their formation is a reaction between a gravity current and a stable layer in the atmosphere. Such layers are almost universal during the night and the gravity current pushes through them, generating a bore disturbance which separates and travels ahead of it. One difference between a bore and a gravity current which is easily detected is the absence of any marked temperature change near the ground; another is the presence of waves, not usually found in gravity currents. Bores are of two types. In a strong bore with an increased height three or four times the undisturbed level, the flow is very similar to that in a gravity current. However in one in which the original depth is less than doubled the bore will take the form of a regular series of waves. This is the form likely to be identifiable in radar observations of insects airborne in bores.

P. G. Wickham (*Meteorological Office, Bracknell, U.K.*). There are many scales of atmospheric phenomena. Many that Mr Pedgley has mentioned, though significant physically, are difficult to observe in practice. Indeed the monitoring of any non-cloudy event presents great difficulties. The most useful scale, both in respect of its physical importance and its operational identification, is the mesoscale (5–50 km). Cloudy systems on this scale can be continuously monitored over the whole of Africa by meteorological satellites.

D. E. PEDGLEY. It is well known that satellites and radar are powerful tools for monitoring clouds and rain, particularly for research on the mesoscale structure and behaviour of weather systems. For day-to-day operations, however, their use is still restricted in many developing countries with arid or semi-arid climates by the limited availability of imagery and by the infrequency of clouds (let alone rain) at many of the convergence zones known or suspected to concentrate flying insects. Nevertheless, a promising start has already been made using half-hourly geostationary satellite imagery at night over eastern Africa to locate rainstorms that may be associated with the concentration of armyworm moths leading to caterpillar outbreaks. Such imagery has recently become available in real time through the introduction of PC-based image capture and processing facilities. Climatologies of such images may well point to preferred sites for outbreaks, but we have yet to learn how to distinguish clouds associated with moth concentration from other clouds. Numerical models, for example, of topographically modified wind fields, would be particularly valuable if simple enough also to be run in the field on PCs, using current synoptic data as input.

R. B. B. DICKISON (*Department of Forest Resources, University of New Brunswick, Canada*). I support Mr Pedgley's judgement that many of the convergence lines leading to concentration of airborne insects would not be accompanied by cloud, and would therefore not be detected on satellite imagery. Marine flows for example are so shallow that the induced vertical motion is usually not sufficient to reach the lifting condensation level of the atmosphere above the layer.

R. M. MORRIS (*Meteorological Office, Bracknell, U.K.*). The only practical way of exploiting operational Numerical Weather Prediction techniques to monitor and forecast the movement and behaviour of flying insects is to use a moveable nested grid mesoscale model, which could simulate the essential topographical and atmospheric details. Such a model is capable of depicting fine shear lines and localized convection. Forecast vertical profiles of wind, temperature and humidity could be exploited given knowledge of the behaviour of the insects under specific weather conditions.

V. A. DRAKE (*Division of Entomology, CSIRO, Canberra, Australia*). It seems to me that ecological and meteorological constraints, together with more general public-health and environmental considerations, may severely limit the use of spraying insecticides into convergence zones as a method of managing populations of insect pests. For most crop/pest systems, targeting of the often relatively few pests that are in a position to cause immediate crop loss seems likely to be more economical, less ecologically damaging, and less likely to induce resistance to the insecticides used than attempting a broad-scale and indiscriminate control of the entire pest population. Increasing concern about spray drift onto populated areas, non-target food crops, and biodiversity conservation reserves will also limit opportunities for emitting insecticide into convergence updrafts, even though total deposition may be reduced when the latter technique is used (Schaefer 1980). Convergence-zone spraying is therefore most likely to be useful over extensive, relatively uniform, low-value agro-ecosystems where human habitation is sparse, biodiversity is relatively low, and heavy conventional applications of insecticide are currently required. Obvious candidate targets include spruce budworm and gypsy moth in North American forests and grasshoppers, locusts, and armyworms in African grasslands.

The practical difficulties of locating and spraying insect concentrations in convergence zones appear considerable, mainly because suitable features occur irregularly, and often also infrequently. In New Brunswick, for example, concentrations of spruce budworm moths in sea-breeze convergence lines, which in many respects appear almost ideal as targets for air-to-air spraying, penetrate far inland only once or twice each moth-flight season (Neumann & Mukammal 1981). If control were attempted only on these occasions, not only would aircraft and personnel be standing idle for long periods, but the few sprays carried out might be insufficient to reduce infestation levels below the economic threshold; moreover, it is possible that in some years no opportunities for convergence-zone spraying would arise. On the other hand, if such sprays are made only as a supplement to some more general insecticidal control scheme, the additional economic benefit obtained from them may be quite small, and the necessary investment in the equipment and patrol flights necessary to locate convergence difficult to justify. Such considerations may have led Schaefer (1980) to emphasize routine air-to-air spraying of insects flying in stable airflows rather than opportunistic spraying into convergence zones.

I see little prospect of convergence-zone spraying finding application in Australia. In coastal eastern and southwestern Australia there are extensive forests, large areas of which are exploited for timber and pulp production, and sea breezes, which frequently penetrate far inland, sometimes accumulate large numbers of insects at their leading edge (Drake 1982). Insects cause few problems in these forests, however, except in the far southwest where the jarrah leafminer (*Perthida glyphopa* Common) perhaps deserves investigation to determine whether adult dispersal flights (Mazanec 1989) ever result in concentrations forming at sea-breeze fronts. In the main agricultural zone further inland, pest problems occur mainly in high-value crops that occupy only a very small proportion of the whole area, and timely conventional spraying of individually monitored field and orchard populations appears more appropriate than any attempt at broad-scale control. Migrating insects frequently become concentrated at convergence zones in this region (Schaefer 1976), but the convergence zones, which arise in mesoscale disturbances of a variety of types, occur irregularly, and many of the resulting concentrations are too weak, too localized, too short-lived, or located too close to storm activity to be practicable spray targets (see my comments on the paper by Riley & Reynolds, this symposium). In addition, ground-survey, phenological, and aerial-sampling observations, indicate that concentrations do not necessarily consist predominantly of economically important pests. In the pasturelands even further inland, where broad-scale population suppression of Australian plague locusts (*Chortoicetes terminifera* (Walker)) is regularly undertaken by spraying of hopper bands and swarms (Symmons 1984), the infrequency and unpredictability of suitable target concentrations would probably make convergence-zone spraying impracticable. Air-to-air spraying in stable airflows might have some application in this region as it could be undertaken as a supplement to daytime swarm control, with the same aircraft flying additional night-time sorties on occasions when conditions were suitable for long-distance migration following a mass take-off flight at dusk (Hunter 1981).

An additional problem with any form of broadscale air-to-air control is that the level of population suppression achieved, and thus the economic or social benefit derived, will be more than usually difficult to establish.

References

Drake, V. A. 1982 Insects in the sea-breeze front at Canberra: a radar study. *Weather, Lond.* **37**, 134–143.

Hunter, D. M. 1981 Mass take-off after sunset in the Australian plague locust. *Australian Plague Locust Commission Annual Report Research Supplement 1979–80*, pp. 68–72.

Mazanec, Z. 1989 Jarrah leafminer, an insect pest of jarrah. In *The Jarrah Forest* (ed. B. Dell *et al.*), pp. 123–131. Dordrecht: Kluwer.

Neumann, H. H. & Mukammal, E. I. 1981 Incidence of mesoscale convergence lines as input to spruce budworm control strategies. *Int. J. Biometeor.* **25**, 175–187.

Schaefer, G. W. 1976 Radar observations of insect flight. *Symp. R. ent. Soc. Lond.* **7**, 157–197.

Schaefer, G. W. 1980 The role of aircraft in an agricultural strategy for the year 2000. *Aeronaut J.* **84**, 131–135.

Symmons, P. M. 1984 Control of the Australian plague locust, *Chortoicetes terminifera* (Walker). *Crop Prot.* **3**, 479–490.

D. E. PEDGLEY. Dr Drake draws attention to a variety of constraints in the strategy for the management of windborne insect pests based on air-to-air application of insecticide in atmospheric convergence zones. These constraints are real; nevertheless, there may be regions and seasons where the strategy is workable with species such as those mentioned by Dr Drake.

Phil. Trans. R. Soc. Lond. B **328**, 655–672 (1990)

Printed in Great Britain

Nocturnal grasshopper migration in West Africa: transport and concentration by the wind, and the implications for air-to-air control

By J. R. Riley and D. R. Reynolds

Overseas Development Natural Resources Institute, Radar Entomology Unit, R.S.R.E.,
Leigh Sinton Road, Malvern, Worcs. WR14 1LL, U.K.

During radar observations of the migratory flight of *Oedaleus senegalensis* and of other grasshoppers in West Africa, we have observed that nocturnally flying insects are sometimes concentrated by mesoscale zones of wind convergence. The concentrations were typically 1.2–2.0 km wide, often more than 20 km long, and were similar to those observed elsewhere. The convergence zones appeared to be usually caused by atmospheric gravity currents. Some of these currents were cold air outflows from rain storms, and others were possibly of katabatic origin. Occasionally zones may also have been caused by bores and gravity waves set off by these currents.

In this paper, we investigate the practicability of controlling populations of sahelian grasshoppers by the air-to-air spraying of insecticide onto such concentrations of insects. Using our data on concentration in convergence zones and a rudimentary model of zone distribution and behaviour, we estimate that less than 30% of the flying population of grasshoppers would be entrained in convergence zones, and that effective search for the concentrations might require the simultaneous use of at least two aircraft per 500 km square.

These results imply that strategic control by air-to-air spraying is unlikely to be practicable. It is necessary to emphasize, however, that the evidence on which this deduction is based is fragmentary. A much more definitive conclusion could be expected from the results of further research with an aircraft equipped with a wind-finding system and a radar able to measure and delineate insect concentrations.

1. Introduction

Recent studies have shown that *Oedaleus sengalensis* (Krauss) and many other species of grasshoppers of the Sahel and Sudan-savanna zones of West Africa, engage in long-distance, windborne, migratory flights at night. Evidence obtained by monitoring changes in ground populations (Popov 1976; Lecoq 1978*a*, *b*; Duranton *et al.* 1979) and, particularly, the direct observations of migration by radar (Riley 1975; Schaefer 1976; Riley & Reynolds 1979, 1983; Reynolds & Riley 1988) have revealed the magnitude and regularity of these flights, which are best viewed as an essential part of an adaptive strategy for colonizing and exploiting the transient sahelian habitat.

One of the consequences of grasshopper migration at heights above the 'flight boundary layer', as defined by Taylor (1974), is that the grasshoppers will inevitably be carried downwind and towards zones of wind convergence (Rainey 1976; Pedgley 1980) where there is a net horizontal inflow of air. Aerial densities will tend to increase if incoming airborne insects are constrained against ascent in the upward divergent air currents which compensate for the horizontal convergence (Rainey 1976). Wind convergence zones occur on various scales, but some of the most active occur at fronts and windshift lines associated with mesoscale (i.e. from tens to hundreds of kilometres) airflows, such as cold outflows from convective storms. Several authors have observed that mesoscale windshift lines are frequently associated with linear accumulations of airborne insects which appear on the radar plan position indicator (P.P.I.) display as a moving 'line-echo' (Schaefer 1976; Reynolds & Riley 1988; Drake 1982 for concentrations of grasshoppers), and it seems clear that these convergent winds can increase the aerial density of flying insects by up to 100 times (Drake & Farrow 1988).

Apart from being of intrinsic meteorological and entomological interest, the concentration of nocturnally migrating insects at windshift lines may have implications for pest control. For example, Joyce (1983) reports that in Sudan, the grasshopper *Aiolopus simulatrix* (Walk.) becomes a pest of economic significance only when concentrated by convergent wind fields. Moreover, he suggests that it is only when the grasshoppers are in these airborne concentrations that control is worthwhile. This has led to proposals for the location of wind convergences by aircraft equipped with doppler navigation systems, followed by the delineation of any associated insect pest concentrations by an insect-detecting radar (Schaefer 1979), and the destruction of the pests by air-to-air spraying (Rainey 1972, 1974*a*, 1976, 1979; Joyce 1976, 1981, 1983). Insecticide droplets are picked up much more effectively by flying insects than by those on the ground (Rainey 1974*a*), so less insecticide would be required with air-to-air ultra low volume (ULV) spraying and reductions in both cost and environmental pollution could be anticipated.

The main purpose of this paper is to interpret available data on grasshopper concentration in a way which may help to assess the practicality of such a control scheme. Most of these data come from radar observations of flying insects made during four seasons' work on grasshopper migration in the sahelian zone of Mali, West Africa. Information is presented on the frequency, extent and aerial density of line-echoes, and we incorporate the data into a simple descriptive model of convergence zone behaviour. We examine the implications of the model for the proposed aerial 'search-and-destroy' control strategy, and discuss other factors which would be of relevance in air-to-air control. The identity of the meteorological mechanisms which may have produced the concentrations is also briefly considered.

2. Methods

Radar observations were made on larger-sized insects (> 100 mg) by using a modified marine, X-band (3.2 cm wavelength) radar at the following places in Mali: Kara (14° 10′ N, 5° 01′ W) in the flood plains of the middle Niger delta; Alfandé (15° 48′ N, 3° 04′ W) in the northeast of the delta; Tin Aouker (16° 48′ N, 0° 08′ E) in the Tilemsi Valley, north of Gao; and Daoga (15° 53′ N, 0° 14′ E) on the River Niger, south of Gao. The observational periods at each site are shown in table 1. The equipment and observational procedures are described in Riley & Reynolds (1979, 1983). Continuous records of wind velocity, and of air temperature and humidity were made at each site.

3. Results

(a) Take-off and emigration

Long-distance migratory flight in sahelian grasshoppers commences just after dusk, when mass take-off of flight-ready, largely post-teneral, individuals occurs in response to decreasing illumination (Riley & Reynolds 1979, 1983; Schaefer 1976). This mass take-off has been

TABLE 1. FREQUENCY OF OCCURRENCE OF LINE-ECHOES OBSERVED
AT FOUR SITES IN MALI, WEST AFRICA

month (dates)	site (year)	number of evenings	number of hours of PPI observations	number of line echoes	number of line echoes per evening	number of hours between line echoes
September (19–30)	Daoga (1978)	12	63.9	3	0.25	21.3
October (1–23)	Daoga (1978)	23	138.0	4	0.17	34.5
October (9–21)	Tin Aouker (1978)	13	120.1	16	1.2	7.5
October (24–31)	Kara (1973)	8	41.5	1	0.13	41.5
October (23–24)	Kara (1975)	2	9.4	0	0	—
November (2–11)	Alfandé (1975)	10	38.4	1?	0.1?	38.4?
November (1–9)	Kara (1973)	6	29.8	0	0	—
November (1–23)	Kara (1974)	2	114.9	0	0	—
November (19–27)	Kara (1975)	9	33.3	0	0	—

routinely observed during radar studies; it occurs to a varying degree every night in the rainy and post-rainy seasons in Mali when suitable grasshopper populations are present and temperatures are above a threshold of about 23 °C. Grasshoppers ascend under their own power (Rainey 1974*b*) at climb rates of *ca.* 0.4 m s^{-1} (Riley & Reynolds 1979) to heights of several hundred metres, when they are usually, but not always (see Riley 1975), above their 'flight boundary layer'.

Radar displays during the dusk take-off of grasshoppers at Tin Aouker typically showed dense plumes and concentrations of multi-target echo appearing from areas of clay known to harbour large numbers (1200–3300 ha^{-1}) of *Aiolopus simulatrix*. The plumes were observed to disperse gradually as the insects in them displaced downwind, presumably because of differences in individual insect orientation and air speed, and because of the effects of wind shear and turbulence. Thus the grasshoppers did not undertake nocturnal migration in a cohesive swarm like those of day-flying, gregarious locusts and it appears necessary to postulate mechanisms that could reconcentrate migrants to the densities seen to invade cultivations.

At the peak of the dusk emigration period (about 50 min after sunset), the volume density at 100 m above ground and averaged over all observational evenings at Tin Aouker and Daoga, gave a mean of about one grasshopper-like insect per 10^4 m^3 (range from about one per 10^3 m^3 to about one per 10^5 m^3). The corresponding densities were lower at the Alfandé site (mean of about 5 per 10^6 m^3). Similarly Schaefer (1976), working on grasshoppers in the Gezira province of Sudan, found that emigration densities ranged from about one per 10^3 m^3 to about one per 10^6 m^3.

Schaefer (1976) reported that large numbers of grasshoppers take off on migratory flights under particular weather conditions in the field, and Bergh (1988) discovered in the laboratory

[139]

that the take-off rate of *Schistocerca* increased with the passage of cold fronts. Other workers have noted that higher numbers of certain acridids often appear at light after storms and other disturbances (Clark 1969, 1971; Farrow (1979) for *Chortoicetes terminifera* (Walk.); L. D. C. Fishpool, personal communication for *O. senegalensis* and other species in Mali), and this may also indicate a relationship between increased flight activity and disturbed weather. In these cases however, it is possible that the observed increase was caused by storm-associated aerial concentration in the vicinity of the light traps, rather than by a stimulation of flight activity *per se*. Any tendency for take-off to be stimulated by disturbed weather would, of course, increase the number of insects available to become concentrated in the mesoscale convergence zones normally present in such weather.

(b) Duration of flight

Our observations have shown that the aerial density of grasshoppers in flight tends to decline as the night develops (figure 1). It could be argued that this decline occurs because habitats progressively further upwind of the radar support progressively lower grasshopper populations, but this explanation seems unlikely because the decline tends to occur at rather similar rates in winds from a wide variety of directions, and at different sites. An alternative, and much more probable cause is that there is a large spread in the flight endurance of the individuals that make up the grasshopper population. Some will begin to descend after only short flights whilst others continue for longer, with the result that aerial density will fall as time progresses. Sometimes the rate of decline is initially rather dramatic, and then becomes less rapid. On other occasions, a more uniform rate is maintained for most of the night (figure 1). The various rates of decline can be interpreted in terms of mean flight period. For example, the

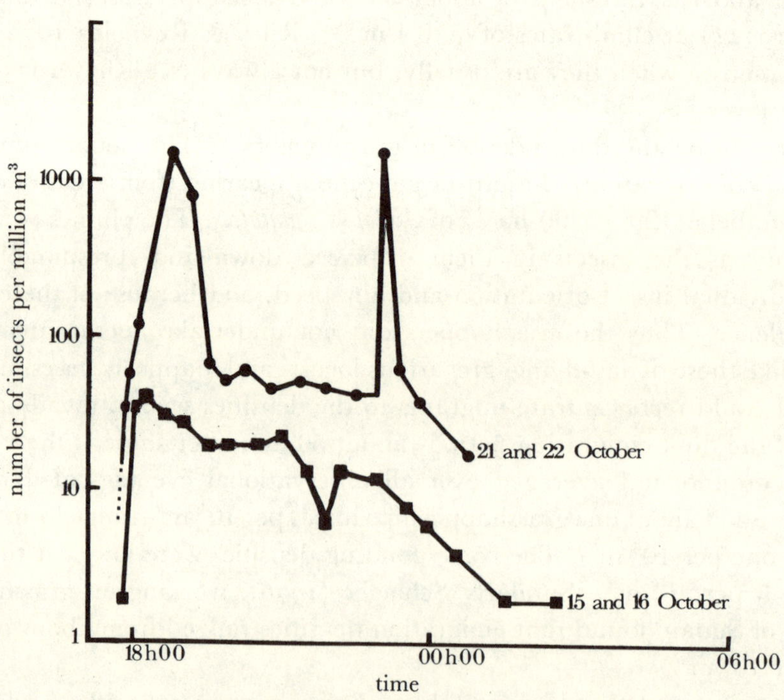

FIGURE 1. The variation in volume density of insects (mainly grasshoppers such as *Oedaleus senegalensis*) flying between 140–170 m above ground at Tin Aouker, during the nights of 15–16 and 21–22 October 1978.

density–time data shown in figure 1 for **15** and **16** October could be explained if it is assumed that the distribution of flight endurance amongst the population was exponential, with a mean flight period of 2.7 h. The more gradual slope found for the **21** and **22** October (after the initial rapid decline) implies a mean period of 6 h for the long-distance fliers. The detection of flying concentrations of grasshoppers 5–7 h after the dusk take-off period (Riley & Reynolds 1983; Reynolds & Riley 1988) provides more direct evidence that a night's migration is usually accomplished in single long flights, rather than in a series of short ones.

(c) Concentration in wind convergence zones

Observations in the rainy season in the sahelian zone of Mali revealed many occasions when dense or relatively dense (namely, 0.003 to 3×10^{-5} insects per m^3) line concentrations passed over the radar. On all the occasions which we investigated, these concentrations were aligned along, and moved with, a windshift line at altitude. The phenomena were usually accompanied by corresponding changes of windspeed and/or direction at ground level, but these changes were sometimes unremarkable and might have passed unnoticed had not the radar results focused our attention on them. Many minor shifts in wind velocity recorded at ground level were not associated with any radar evidence of convergence, so it appears that conventional anemometry did not provide a reliable indication of convergence aloft.

At the altitudes at which most insects were flying, i.e. within the lowest kilometre of the atmosphere, there was always a net horizontal inflow or convergence of air in the features studied. An example of the phenomenon in which the convergence reached a value of $5 \times 10^{-3}\,s^{-1}$ is described in Reynolds & Riley (1988). Another example with a weaker convergence $(1.4 \times 10^{-3}\,s^{-1})$ is shown in figure 2.

Several authors have observed similar moving linear accumulations of insects and have noted their apparently invariable association with windshift lines (see, for example, Schaefer (1976); Riley & Reynolds (1983); Reynolds & Riley (1988) for sahelian grasshoppers). In some cases the cause of the windshift was self-evident, e.g. sea breeze fronts (Drake 1982), storm outflows (Schaefer 1976; Pedgley *et al.* 1982), or the Inter-Tropical Front in Sudan (Schaefer 1976), but in other cases the mechanism was less certain or unknown (Schaefer 1976; Riley & Reynolds 1983).

The concentrations are usually about 1.2–2 km from front to back, but much greater in lateral extent: Schaefer (1976), for example, observed a length of 60 km of a storm outflow in Sudan, and Drake (1982) over 40 km of a sea breeze front in Australia. In these and other cases the concentrations probably extended laterally well beyond the range to which they could be detected by the radar.

Some figures for the aerial density of insects found within line-echoes are given in table 2; the highest values approach those found in (sparse) locust swarms (1 per $10^3\,m^3$) (Waloff 1972).

(d) Frequency of occurrence of line-echoes

The frequency at which we observed the passage of line-echoes varied considerably between month and site (table 1). There appeared to be more line-echoes towards the end of the rainy season (September–October) than after the rains (November), but this could have been partly attributable to the fact that different sites were used in the two periods. At only one of our sites (Tin Aouker) were line-echoes as frequent as one per night. Schaefer (1975) working in the Gezira area of Sudan in October, reported that line-echoes occurred on 12 occasions during

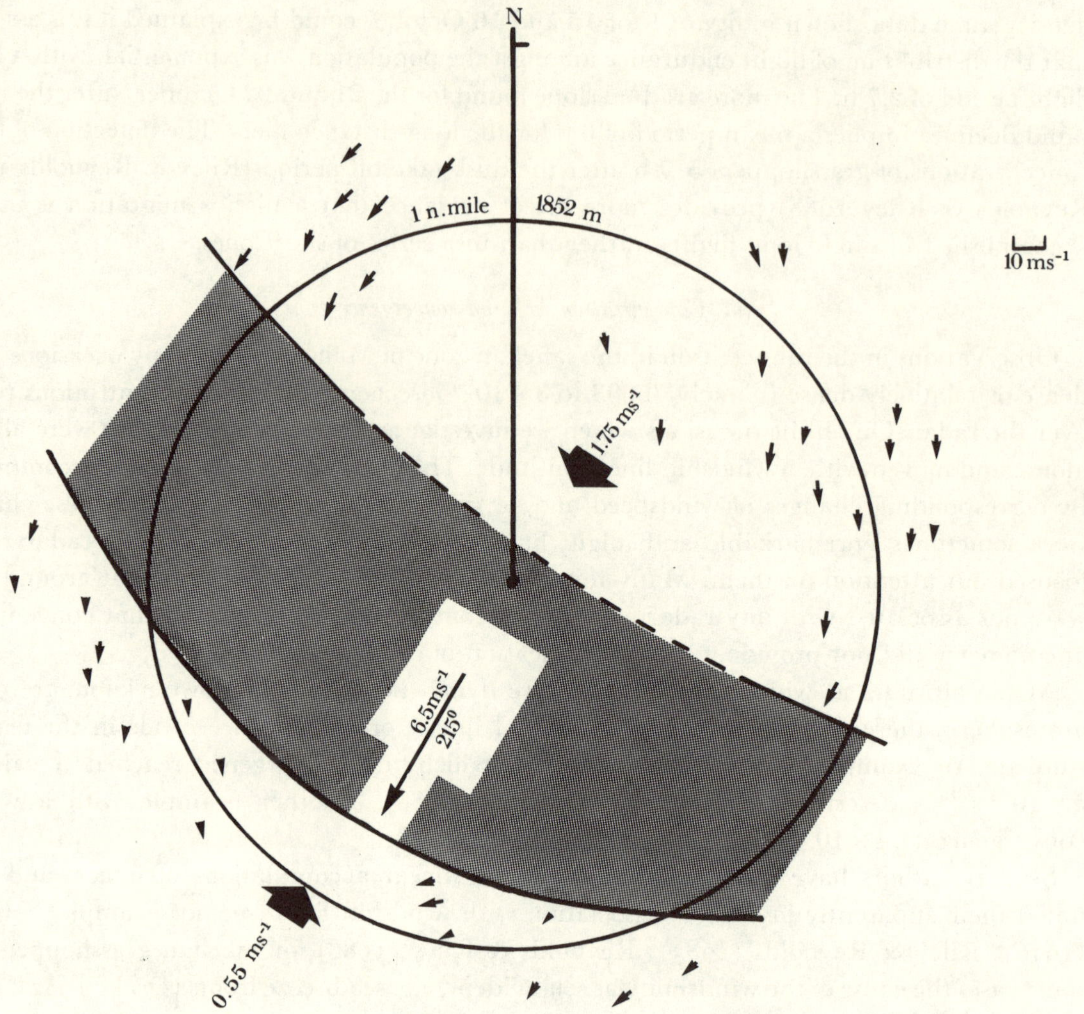

FIGURE 2. Diagram from the radar PPI display showing a line-echo passing over the Tin Aouker site at about 19h22 on 20 October 1978, moving towards 215° at 6.5 m s⁻¹. The small arrows represent the displacement velocity, relative to ground, of a selection of individual insects. The estimated average ground speed and direction of insects in front of the line-echo was 7.0 ± 1.4 (s.d.) m s⁻¹ towards $201° \pm 27°$, and the corresponding values behind the line-echo 9.0 ± 1.5 m s⁻¹ towards $195° \pm 17°$. The large broad arrows show the mean velocity of the insects *relative* to the line-echo and perpendicular to it. The horizontal convergence (Sc) over the 1.6 km width (Lc) of the zone was 1.4×10^{-3} s⁻¹.

96 h of observation over 22 nights. The average interval between events (8 h) was thus very similar to the value (7.5 h) found by us at the Tin Aouker site. There are apparently no data for the frequency of line-echoes earlier in the rainy season (June–August).

(e) *Weather during passage of line-echoes*

Most (18 out of 23) line-echoes studied at Daoga and Tin Aouker were accompanied by an increase in surface wind speed of between 1 and 4 m s⁻¹ (except in one case on 20 October 1978 which was clearly attributable to the gust-front of a storm, where the wind speed increased suddenly by 9 m s⁻¹). Increases in wind speed were sustained for periods ranging from 5–10 min up to several hours. In 10 cases there was a well-marked change in direction of insect movement (reflecting a change in wind direction) at the leading edge of the line-echo (see

TABLE 2. THE CHARACTERISTICS OF SOME LINE-ECHOES OBSERVED ACCUMULATING FLYING GRASSHOPPERS IN MALI AND SUDAN

date and time	volume density in line-echo (number per m³)	increase in volume density at leading edge of line-echo	area density of line-echo (number per hectare)	length (km)	grasshopper species accumulated	place	reference
21-10-78 23h00	Up to 0.0015	*ca.* 50 ×	*ca.* 4000	at least 22 km	mainly *Oedaleus senegalensis*	Tin Aouker, Mali	Riley & Reynolds (1983) and unpublished results
10-11-78 22h15	Up to 0.0001	15 ×	—	at least 8 km	mainly *Oedaleus senegalensis*	Tin Aouker, Mali	Reynolds & Riley (1988) and unpublished results
11-10-78 02h27	Up to 0.003	*ca.* 80 ×	—	at least 16 km	*Diabolocatantops axillaris* & *O. senegalensis*	Tin Aouker, Mali	Reynolds & Riley (1988) and unpublished results
17-11-78 20h28	—	60 ×	*ca.* 1000	60 km	mainly *Aiolopus simulatrix*	Radma, Sudan	Schaefer (1976)

example in Reynolds & Riley 1988); on the other occasions there was an increase in target speed but little change in direction (figure 2). On most occasions there was no change in surface air temperature (14 out of 23 cases) or dew point (13 out of 23 cases) with the passage of the line-echo. In only one case was there a substantial fall in temperature (-8.5 °C) and rise in dew point ($+3°$), and this was clearly attributable to the above-mentioned storm outflow. On the remaining occasions the temperature showed a small rise (*ca.* 1 °C) (or a small fall followed by a small rise), and the dew point usually fell slightly. The line-echoes with an accompanying change in temperature or dew point were also those with a marked shift in wind direction.

Apart from the line-echoes associated with the storm on 20 October, and the line-echo described in Reynolds & Riley (1988), which was probably an outflow associated with patchy rain from middle-level cloud, there was little evidence of precipitation within radar range (at least 40 km) at the time the line-echoes were observed. It would thus appear that if the concentrations are formed by outflow gravity currents they must propagate far from their parent rain-storms.

(f) Speed and direction of movement of line-echoes

The line-echoes moved with ground-speeds in the range 4 m s^{-1} to 14 m s^{-1}. At Tin Aouker, line-echoes came from the southeast quadrant (south to east) or from the north quadrant (N.W.

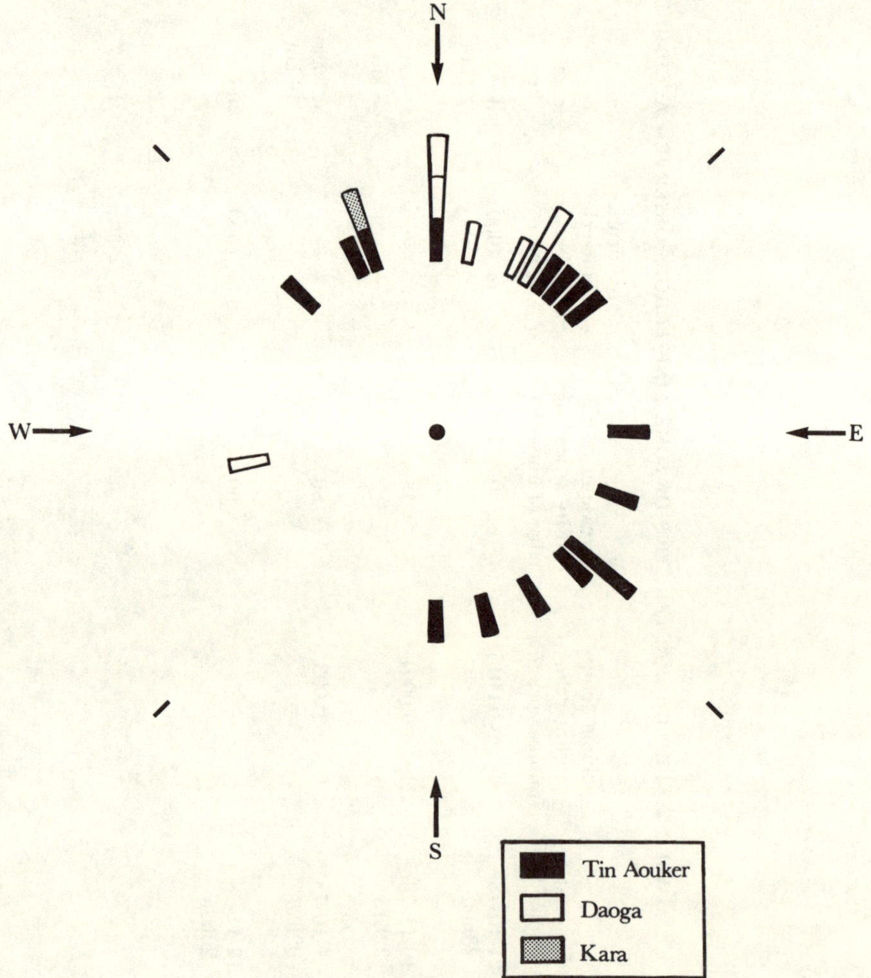

FIGURE 3. Circular histogram showing directions from which the observed line-echoes came.

to N.E.), while at Daoga, they came (with one exception) from the north or northeast (figure 3). We note that line-echoes hardly ever came from the west or S.W., although winds from this direction occurred during the observational period, and, indeed, line-echoes occurred within southwesterly wind fields. Schaefer (1976) working in Niger stated that line-echoes associated with nocturnal windshift lines arrived 'almost invariably from the S.E.'. He attributed them to gravity currents flowing down a shallow drainage slope. Likewise in our study it seems probable that some of the windshift lines approaching from the north or N.E. were katabatic in origin or that their flow was enhanced by the slight gradient from this direction. The line-echoes approaching Tin Aouker from the S.E. could have originated as gust-fronts from storms or disturbances far to the south.

4. Discussion

(a) Factors to be considered in control schemes

There are two main approaches to the problem of controlling migrant insect pests with insecticides. One method is to attempt to reduce the population of the pest species as a whole, over large geographical areas (as in the case of Desert Locust control), and the other is to concentrate on the more limited tactical objective of attacking only those parts of the population which constitute an obvious and direct threat to crops. In the case of grasshoppers migrating over long distances, the direct threat to crops is difficult to predict; proposals for air-to-air spraying of insect concentrations thus appear to be directed more towards large area, strategic control than towards limited tactical defence. If this is indeed the case, their success will depend on whether a significant proportion of the target population is likely to become entrained in convergence zones, and therefore vulnerable, at least in principle, to attack by air-to-air spraying.

(b) Proportion of the population liable to concentration

As we have observed, sahelian grasshoppers such as *O. senegalensis* often spend several hours in migratory flight, usually displacing downwind towards the Inter-Tropical Convergence Zone. During these flights a proportion should be concentrated by mesoscale zones of wind convergence. To compute the size of this proportion we need to have a representative knowledge of the spatial and temporal frequency of the convergence zones and of their strength. Unfortunately, representative data on mesoscale convergence zones in the Sahel are not currently available. This is partly because the density of wind recording stations is so low, but mainly because, in the absence of radar data, conventional anemometry records do not appear to provide a reliable indicator of the presence of zones likely to produce insect concentrations aloft. Data on entomologically significant mesoscale convergence in the Sahel thus appear to be confined to that accrued during the very limited periods of radar operations by Schaefer and by ourselves. Although this information is scarcely adequate to characterize the region as a whole, it does at least provide a basis for a first rudimentary estimate of the proportion of airborne grasshoppers likely to accumulate in convergence zones.

We assume a primitive descriptive model in which linear convergence zones generated from a single source occur randomly but are roughly parallel to each other, and that airborne insects are initially uniformly distributed. It is also assumed that the insects do not change their altitudinal distribution. Then, if the average strength of a zone is Sc, its lifetime Tc, its width

Lc, and the average (perpendicular) distance between zones is L, the percentage, P, of the airborne population likely to be concentrated into a zone (by the end of the zone's life) is:

$$P = (Sc \times Tc \times Lc \times 100)/L \%. \qquad (1)$$

Sc and Lc were directly observable by radar (figure 2), but L must be estimated from the observed line displacement speed and the interval between the passage of successive lines. Thus, suppose that the convergence zones evolve at an average spatial separation of L, at temporal intervals Tg, and that they move with a speed V. If we assume $Tc < L/V$, i.e. that their lifetime is less than the time taken to traverse their separation distance, then the average interval, Ti, at which zones will be seen to pass over a fixed point is:

$$Ti = (L \times Tg)/(V \times Tc). \qquad (2)$$

Substituting for Tc in (1) gives:

$$P = 100 \times ((Sc \times Lc)/V) \times (Tg/Ti) \%.$$

Unfortunately, one of the quantities on the right-hand side of this equation, Tg, is not known, so we cannot evaluate P from our data. However, we can estimate its maximum value because the upper limit of (Tg/Ti) is unity (convergence zones cannot arrive at the radar faster than they are generated). Thus

$$P < 100 \times (Sc \times Lc)/V \%. \qquad (3)$$

Typical values for Sc, Lc and V are, respectively, 10^{-3} s^{-1}, 1.2×10^3 m and 5 m s^{-1}, which suggests that $P < 30\%$, and it may, of course, be much lower. It may be noted that actual values for P, rather than upper limits, could be calculated directly from equation (1) if values for L and Tc could be determined by some other method, for example, by the use of an airborne radar.

It was implicitly assumed in equation (1) that individual grasshoppers remain airborne at least as long as the lifetime of a typical convergence zone, and this assumption merits some scrutiny. Our data suggest that concentrations as high as fifty-fold or more (table 2) occur in the Sahel, and at typical observed convergence rates of the order of 10^{-3}, this implies that zone lifetimes (Tc) are about two to three hours. Our observations of density variation with time, and of discrete overflights imply mean flight lengths are often longer than this in the Sahel, and may be as long as 5–7 h. Schaefer (1976) concluded from time–density data, that in the case of grasshoppers in the Sudan, the average flight duration was probably about 1.5 h during the take-off peak, but about 5 h for individuals which remained in flight after this period. Overall, it seems probable that on many occasions the majority of migrant grasshoppers do indeed remain airborne long enough to be effectively concentrated should they encounter mesoscale convergence phenomena, and so on this count at least, equation (1) should be valid. On some occasions, however, it seems that the bulk of the airborne population lands within an hour or so of take-off (see the data in figure 1 for the night of 21–22 October 1978), and in these cases only a small fraction will remain airborne long enough to be affected by convergence zones.

It must be emphasised that our expression for the maximum value of P was made on the basis of a very rudimentary model of the spatial and temporal distribution of convergence zones. No attempt was made to consider the effect of changing rates of convergence with time, nor that of possible changes in the vertical distribution of insects at the convergence line. Nor was any account taken of the possibility that grasshopper take-off might be preferentially stimulated by disturbed weather conditions. The result should thus at best be taken as a provisional estimate of the maximum fraction of the grasshopper population likely to be concentrated. Much more definitive predictions could be expected if new observations were made with an aircraft equipped with a wind-finding system (to assist in the location, tracking and measurement of convergence zones) and with an appropriate form of radar to monitor insect aerial densities both inside and outside the zones.

(c) *Some practical considerations*

Our analysis of current evidence has indicated that the percentage of grasshoppers likely to be entrained in mesoscale convergence zones in the Sahel is too low to make strategic control by air-to-air spraying worthwhile. However, the evidence on which this conclusion is based is far from perfect, and it is conceivable that future investigations will show that enough entrainment occurs to make effective control at least a possibility. It is thus appropriate to briefly consider here some of the practical implications of the proposed control method.

(i) *Finding the concentrations*

Methods of finding insect concentrations in convergence zones, by using an aircraft equipped with a wind-sensing system and an insect-detecting radar, are described elsewhere in this volume (Rainey & Joyce, this symposium). Here, we examine our ground-based radar data in an attempt to assess the practicality of the search task that such an airborne system would face in the Sahel. In the descriptive model proposed in the preceding section, the average separation between convergence lines is given by the equation:

$$L = Ti \times V \times (Tc/Tg).$$

If it is assumed that the lifetime of a zone is less than the interval between the generation of successive zones, i.e. that $Tc/Tg < 1$, and we take 5 m s^{-1} as a typical value for V, then:

$$L < 18 \times Ti \text{ km}.$$

Our data (table 1) show that Ti varies considerably between October and November. The corresponding average values for L are < 500 km in October, and a larger but undetermined value in November. The lateral extent of the zones we observed was unknown, but we assume (optimistically) that it might approach 100 km. Thus, in a hypothetical search area of 500×500 km, on each search leg perpendicular to the alignment of the zones an aircraft might expect to encounter more than 1 zone per leg in October, but perhaps none at all in November. At a flying speed of 350 km per hour, each search leg would take about 1.5 h, and it would require 4 legs or about 7 h to cover the 500 km square. This is more than twice the estimated average lifetime of a zone, so it appears that half the zones might be missed, and that a search using only one aircraft per 500 km square would thus be only partially effective. Much more reliable conclusions than these could be expected from an experimental trial with an airborne system, because the lateral extent, lifetime, and spatial and temporal distribution of zones could then be measured rather than inferred.

(ii) *Insecticide application*

If the estimate of the time required to conduct a systematic search for concentrations is approximately correct, then it is clear that one aircraft would not have time both to look for concentrations and to spray them. The practical implications of a requirement for simultaneous night-time operation of perhaps several aircraft flying co-ordinated and non-routine flight paths, during disturbed weather are considerable. However, given sufficient investment, the logistic and operational problems could probably be overcome, especially now that accurate navigational data are available from the satellite-based global positioning system (GPS), even in very remote areas. Nevertheless, before considering a full-scale control project, it would obviously be prudent to conduct a research program to establish the optimum droplet size distribution for use in the updraft conditions prevailing in convergence zones, and also to evaluate the efficiency of different spray regimes within the zones.

(iii) *Selection of target concentrations*

It is quite possible that the line concentrations to be found over the Sahel will not always consist mostly of target grasshopper species. If this is the case, then indiscriminate spraying of all detected concentrations would be both wasteful and ecologically irresponsible (see general discussion in Rainey (1976)). Unfortunately, the rather small size of sampling nets that can be carried on aircraft means that determination of the species composition of a concentration by aerial trapping is a slow process and would sometimes be impractical. It is thus very desirable that the radar carried by the aircraft should have the capacity at least to distinguish between, say, moths and grasshoppers, and between grasshoppers of substantially differing sizes. There is some prospect that this will be possible in future generations of entomological radars.

(d) *Origins of the convergence zones*

Mechanisms that produce wind convergence zones are described in detail elsewhere in this volume (Pedgley, this symposium); we concentrate briefly here on clues that might identify the origins of the zones we detected. The majority of line concentrations observed by us had probably been formed by the head of an advancing gravity (or density) current. Here a flow of cool, dense air displaces or undercuts warmer air and so produces a narrow zone of horizontal convergence. Another type of disturbance of the nocturnal boundary layer which could concentrate flying insects is an internal solitary wave of large amplitude with a closed circulation (these have been observed to form line-echoes in Australia (Drake 1985)). In a circulation, the horizontal components of flow at some level are opposed to the overall wind direction and similarly produce linear zones of (horizontal) convergence.

Two sources of mesoscale cold air flow to be expected in the Sahel are the spreading downdrafts from rainstorms, and the downslope (katabatic) drainage of air which has cooled during the evening. At first sight one might expect that the passage of a convergence zone produced by a gravity current would be clearly revealed by a decrease in air temperature. However, because the arrival of a gravity current usually disturbs the nocturnal temperature inversion, a slight temperature decrease may be masked by mixing within the inversion layer (at least at ground level), and so it may not provide a clue to the origin of the flow. In fact, we almost never detected a substantial fall in surface air temperature during the passage of a line-echo. It thus appears that positive identification of gravity currents is often not possible from surface temperature and humidity data alone.

The scarcity of line-echoes outside the rainy season could indicate that the source of the convergences were gravity currents of the spreading downdraft type. However, as mentioned above, rainfall was seldom detected on the radar behind an advancing convergence zone. The tendency of many of the line-echoes to move in the general direction of the large-scale topographic gradient, i.e. from the north or northeast (figure 3), strongly suggests that some were produced by katabatic flows, or possibly by internal bores or solitary waves set off by these flows (Simpson 1987).

References

Bergh, J. E. 1988 Take-off activity in caged desert locusts, *Schistocerca gregaria* (Forsk.) (Orthoptera, Acrididae) in relation to meteorological disturbances. *Int. J. Biometeorol*, **32**, 95–102.

Clark, D. P. 1969 Night flights of the Australian plague locust, *Chortoicetes terminifera* Walk., in relation to storms. *Aust. J. Zool.* **17**, 329–352.

Clark, D. P. 1971 Flights after sunset by the Australian plague locust, *Chortoicetes terminifera* (Walk.), and their significance in dispersal and migration. *Aust. J. Zool.* **19**, 159–176.

Drake, V. A. 1982 Insects in the sea-breeze front at Canberra: a radar study. *Weather, Lond.* **37**, 134–143.

Drake, V. A. 1985 Solitary wave disturbances of the nocturnal boundary layer revealed by radar observations of migrating insects. *Bound.-Layer Met.* **31**, 269–286.

Drake, V. A. & Farrow, R. A. 1988 The influence of atmospheric structure and motions on insect migration. *A. Rev. Ent.* **33**, 183–210.

Duranton, J. F., Launois, M., Launois-Luong, M. H. & Lecoq, M. 1979 Biologie et ecologie de *Catantops haemorroidalis* en Afrique de l'Ouest (Orthopt. Acrididae). *Annls Soc. ent. Fr.* (N.S.), **15**, 319–343.

Farrow, R. A. 1979 Population dynamics of the Australian plague locust, *Chortoicetes terminifera* (Walker), in central western New South Wales. I. Reproduction and migration in relation to weather. *Aust. J. Zool.* **27**, 717–745.

Joyce, R. J. V. 1976 Insect flight in relation to problems of pest control. *Symp. R. ent. Soc. Lond.* **7**, 135–155.

Joyce, R. J. V. 1981 The control of migrant pests. In *Animal migration* (ed. D. J. Aidley), pp. 209–229. Cambridge: Cambridge University Press.

Joyce, R. J. V. 1983 Aerial transport of pests and pest outbreaks. *EPPO Bull.* **13**, 111–119.

Lecoq, M. 1978*a* Biologie et dynamique d'un peurplement acridien de zone soudanienne en Afrique de l'Ouest (Orthoptera, Acrididae). *Annls Soc. ent. Fr.* (N.S.) **14**, 603–681.

Lecoq, M. 1978*b* Les déplacements parvol à grande distance chez les acridiens des zones sahelienne et soudanienne en Afrique de l'Ouest. *C.r. hebd. Séanc. Acad. Sci., Paris* D **286**, 419–422.

Pedgley, D. E. 1980 Weather and airborne organisms. World Meteorological Organisation *Tech. Notes Wld. met. Org.* **173**.

Pedgley, D. E., Reynolds, D. R., Riley, J. R. & Tucker, M. R. 1982 Flying insects reveal small-scale wind systems. *Weather, Lond.* **37**, 295–306.

Popov, G. B. 1976 The 1974 outbreak of grasshoppers in western Africa. In *Report of the Sahel Crop Pest Management Conference, held at the invitation of the Agency for International Development, United States Department of State, Washington DC, USA, December 11–12, 1974* (ed. R. F. Smith & D. E. Schlegel), pp. 35–43. Berkeley: University of California (for United States Agency for International Development).

Rainey, R. C. 1972 Flying insects as potential targets: initial feasibility studies with airborne Doppler equipment in East Africa. *Aeronaut. J.* **76**, 501–506.

Rainey, R. C. 1974*a* Flying insects as targets for ultra-low-volume spraying. *Br. Crop Prot. Counc. Monogr.* **11**, 20–28.

Rainey, R. C. 1974*b* Biometeorology and insect flight: some aspects of energy exchange. *A. Rev. Ent.* **19**, 407–439.

Rainey, R. C. 1976 Flight behaviour and features of the atmospheric environment. *Symp. R. ent. Soc. Lond.* **7**, 75–112.

Rainey, R. C. 1979 Dispersal and redistribution of some Orthoptera and Lepidoptera by flight. *Mitt. Schweiz. Ent. Ges.* **52**, 125–132.

Reynolds, D. R., & Riley, J. R. 1988 A migration of grasshoppers, particularly *Diabolocatantops axillaris* (Thunberg) (Orthoptera: Acrididae), in the West African Sahel. *Bull. ent. Res.* **78**, 251–271.

Riley, J. R. 1975 Collective orientation in night-flying insects. *Nature, Lond.* **253**, 113–114.

Riley, J. R. & Reynolds, D. R. 1979 Radar-based studies of the migratory flight of grasshoppers in the middle Niger area of Mali. *Proc. R. Soc. Lond.* B **204**, 67–82.

Riley, J. R. & Reynolds, D. R. 1983 A long-range migration of grasshoppers observed in the Sahelian zone of Mali by two radars. *J. Anim. Ecol.* **52**, 167–183.

Schaefer, G. W. 1975 Radar studies of the flight activity and the effects of windfields on the dispersal of Sudan Gezira insects. *Ciba-Geigy Seminar on the strategy for cotton pest control in the Sudan Gezira, Wad Medani, Sudan.*

Schaefer, G. W. 1976 Radar observations of insect flight. *Symp. R. ent. Soc. Lond.* **7**, 157–197.

Schaefer, G. W. 1979 An airborne radar technique for the investigation and control of migrating pest insects. *Phil. Trans. R. Soc. Lond.* B **287**, 459–465.

Simpson, J. E. 1987 *Gravity currents: in the environment and the laboratory.* Chichester: Ellis Horwood.

Taylor, L. R. 1974 Insect migration, flight periodicity and the boundary layer. *J. Anim. Ecol.* **43**, 225–238.

Waloff, Z. 1972 The orientation of flying locusts, *Schistocerca gregaria* (Forsk.), in migrating swarms. *Bull. ent. Res.* **62**, 1–72.

Discussion

S. E. HOBBS (*Cranfield Institute of Technology, Bedfordshire, U.K.*). Drs Riley and Reynolds have produced a valuable attempt to quantify insect concentrations in mesoscale convergence zones, based on their field work in West Africa. I can add no more to their model, but I agree that it is desirable to have more information about the convergence zones. In particular, it would be useful to have firmer information about the distribution of the zones (how and when they are generated), and the distribution of insects within them.

Our recent experience at Cranfield is concerned with entomological radar systems for both ground-based and airborne work. The airborne entomological radar system is described by Hobbs & Wolf (1989) with examples of the information available from recent field experiments with the United States Department of Agriculture. We have designed and built signal processors to control the radar and data collection, and have developed associated signal analysis techniques. Recent work should allow field radars to measure individual insect radar cross-sections and so provide some degree of target identification (although not as far as species in general). Based on this experience, these comments relate mainly to the radar component of the project.

The most suitable radar for locating insect concentrations would have a wide area coverage, as provided by conventional weather radars for aircraft. The low altitude of some insect concentrations compared with typical cloud heights may limit the useful radar range, but such a tool would be a very powerful way of mapping out the concentrations. Once such concentrations have been found, the Cranfield airborne radar would be suitable for measuring their height and thickness and the number of insects involved, and could provide some indication of species composition (by measuring an insect's mass).

Entomological radars similar to current systems are able to detect individual insects the size of grasshoppers at a range of the order of 1.5 km. The airborne system uses a pencil beam about 40 m wide pointing vertically below the aircraft. With suitable signal analysis techniques, the insect's range is known to a resolution of 15 m or better and its orientation can be measured to within a few degrees. It should also be possible to measure its mass to within a factor of 2–3 by using more recent analysis techniques. In dense concentrations (greater than about 100 per 10^6 m^3) it is difficult to identify signals from individual insects so that individual size and orientation information is lost. However, individuals should be measurable at the fringes. In general, current radar techniques provide all the information required for locating insect concentrations accurately enough for air-to-air control and should be able to distinguish target species from benign species if the insects differ significantly in mass. Entomological radar systems are generally based on standard technology (e.g. marine X-band radars, with standard microcomputers for data collection and analysis) so that implementing such a system does not require inordinate investments in equipment.

Radar, both ground-based and airborne, should help to quantify the importance of insect concentrations at mesoscale convergence zones, and to exploit them if they prove to be significant.

Reference

Hobbs, S. E. & Wolf, W. W. 1989 An airborne radar technique for studying insect migration. *Bull. ent. Res.* **79**, 693–704.

P. J. MASON (*Meteorological Office, Bracknell, U.K.*). First, I would like to ask what is known about the efficiency of retention of insects within a convergence line. Is there any evidence for varied insect response and turbulence causing significant numbers of insects to leave the region of observed concentration? Might this factor cause your estimates of capture to be optimistic?

Second, I think it is worth noting the interest of air pollution meteorologists in these phenomena. Convergence lines do not of course concentrate passive pollution but they do significantly affect its destination. The convergence lines under stable conditions may be especially relevant as in these conditions pollution hazards are often greatest.

Finally, I must support Drs Riley and Reynolds particular note of the need to formulate a spraying tactic for use in convergence lines. In the case when a free flying swarm is sprayed, I presume that material will tend to move with the swarm and turbulence will mix it through the swarm. In the case of the convergence line the insects are 'trapped' but the spray material will probably not be trapped and will pass through the insects. Spray introduced alongside the insects might only encounter a small fraction of them before being swept away. I think it may require some thought and study to find how to deposit spray in an effective manner so that it is both mixed and encounters the swarm.

J. R. RILEY. Very little is known about the efficiency of retention of insects within convergence lines. Large insects with high air speeds (~ 4.5 m s^{-1}) make progress against the slow winds associated with weak convergence, and hence could presumably fly 'through' a weak convergence zone. Small insects would not be able to do this, but on the other hand, their relatively low falling speeds (~ 1 m s^{-1}) would make them susceptible to being carried involuntarily out of strong zones by the vigorous updrafts associated with them. Nevertheless, the regular occurrence of 50–100-fold density concentrations suggests that, at least in the early lifetime of a convergence zone, the rate of loss is very much lower than the rate of accumulation. The rate of loss could be established by a systematic study using airborne radar. Such studies might also help to characterize the air flow regime in mesoscale convergence zones occurring under stable conditions and, as Dr Mason suggests, the results might well prove useful in understanding the distribution of airborne pollutants in such conditions.

In an ideal spray regime, the droplet size would be such that their rate of fall more or less matched the average strength of updraft, so that they would remain airborne in a convergence zone for a long time. The insects, by virtue of their (horizontal) air speed, would fly through this 'dynamic suspension' of insecticide and the efficiency of collection could be very high. A fairly detailed knowledge of updraft strength distributions would have to be established before a drop size distribution could be selected, and then there would be the problem that updraft strength tends to increase with height. Thus rising droplets would encounter faster updrafts and hence rise even faster, whilst falling ones would fall faster the lower they got, i.e. the 'suspension' mechanism would be unstable.

R. J. V. JOYCE (*Cranfield Institute of Technology, Bedfordshire, U.K.*). Dr Riley has argued that air-to-air spraying of grasshoppers concentrated in zones of wind convergence would be unlikely to make a significant contribution to control. In the sense that 'control', as in the case of the Desert Locust, is the overall regulation of populations, he is surely correct, but it is unlikely that this objective is desirable or economically possible. I think that the most we should aspire to is the regulation of the potentially dangerous populations, i.e., those that are an immediate threat to crops.

My experience with grasshoppers is confined to Sudan, where *Aiolopus simulatrix* (Walker) is of greatest economic importance. In years when this species endangers crops, it has completed two generations of successful breeding, the second generation in the northerly extremity of its distribution where it occurs at densities of the order of 10^3 ha^{-1} over vast areas. To cause crop loss, this density must increase to at least 10^5 ha^{-1} and it is only when airborne in zones of wind-convergence that this increased density has been observed to occur, and it is only when the flight of these insect concentrations terminates in cropped areas that the species is of economic importance. It is also only when the species is present in such zones of wind convergence that they are an economic target for attack. Here they could be destroyed with the minimum amount of insecticide (say, of the order of < 100 ml active ingredient ha^{-1}) with no danger of spray residues on the crop and with no hazard to farmers. Spraying would have to be carried out at night by aircraft instrumented to find the zones of wind convergence and identify the species composition of any insect concentrations. All the elements for such performance are contained in the Airborne Radar System I have described (Rainey & Joyce, this symposium).

J. J. SPILLMAN (*Cranfield Institute of Technology, Bedfordshire, U.K.*). I think that it would be possible to spray within convergence zones, as it is very unlikely that the insects are actually in the continuous updraft. I think you will find they are in the rotary region associated with this, rather than in the updraft itself. Aircraft can cope with the velocities we are talking about. There will be a considerable amount of turbulence, and we would use very small droplets which would diffuse in quite small scale turbulence. I certainly agree that some experiments are needed to demonstrate whether it can be done in practice, but I don't envisage a great deal of difficulty.

K. A. BROWNING, F.R.S. (*Meteorological Office, Bracknell, U.K.*). Do I understand Professor Spillman to say that he is assuming that the insects would not be in the strong updraft? I would have thought that it's almost inevitable that if they are anywhere, that is where they would be.

J. J. SPILLMAN. They will be there but they will be on the edges as well. They would be falling, so one would have to match the spray conditions to get the relative motion between the droplets and the insects.

V. A. DRAKE (*Division of Entomology, CSIRO, Canberra, Australia*). This paper represents a welcome first attempt at assessing, for a specific pest and geographic region, the feasibility of controlling populations by air-to-air spraying into mesoscale convergence zones. The authors' simple estimates of the proportion of the airborne population accumulating in wind convergence lines and the search effort required to locate the resulting concentrations, based on observations for a number of localities and years, are generally discouraging, although high variability between sites and seasons (table 1) leaves room for optimism that the technique might be practicable in particular areas, e.g. the Tilemsi Valley in Mali.

Although a comparable series of radar observations exists for the agricultural zone of inland southeastern Australia, I have not yet subjected them to any comprehensive analysis. However, it is my impression that line concentrations in this region arise mainly from storm outflows or from deeply penetrating and katabatically reinforced sea-breezes. In terms of their suitability as targets, the latter have the advantage of being potentially predictable, of probably having a long (hundreds of kilometres) lateral extent, and, in marked contrast to outflows, of being associated with generally settled weather which presents few hazards to aircraft operation. On the other hand, in the more inland areas this type of disturbance does not arrive until around midnight, by which time the intensity of insect migration has often fallen well below its early evening peak. In any case, intense line-echoes with confirmed lateral extents of tens of kilometres occur perhaps only once per week on average in this region. Even when weaker or smaller-scale concentrations are included, the frequency of 'one to three times per night' recorded by Schaefer (1976) in his pioneering Australian observations has rarely been approached in subsequent studies. It therefore appears likely that conditions are no more favourable for routine convergence-zone spraying in southeastern Australia than they are in Mali.

Some more general observations about the practicability of this technique are made in my comments on Pedgley's paper (this symposium).

Reference

Schaefer, G. W. 1976 Radar observations of insect flight. *Symp. R. ent. Soc. Lond.* 7, 157–197.

M. E. Irwin (*Agricultural and Economic Entomology, University of Illinois and Illinois Natural History Survey, 607 E. Peabody, Champaign, IL 61820, U.S.A.*) and S. A. Isard (*Department of Geography, University of Illinois, 607 S. Mathews, Urbana, IL 61801, U.S.A.*). Drs Riley and Reynolds described the use of radar for locating linear echo zones containing large concentrations of grasshoppers within the planetary boundary layer to initiate air-to-air chemical control. That Riley and Reynolds were unable affirm the potential of the technique because their calculations suggested low spatial concentrations and temporal difficulty in locating the zones of convergence, should not dissuade us from investigating this process further.

The inability to locate these zones of convergence and their associated linear echoes, according to Riley and Reynolds, is due to the length of time the zones remain active and their spatial distribution over the Sahel. This points to a considerable knowledge gap concerning the modelling of the planetary boundary layer and to our lack of sufficient instrumentation to detect these zones early during their short-lived existences. One can only wonder if finding these zones might be easier over areas such as the southwestern deserts of North America, a region with a more extensive and intensive meteorological grid.

Riley and Reynolds broached a crucial issue facing our understanding of the dynamics of insect migration. A conceptual model of insect migration often divides the migratory event into three equally important components: take-off and ascent, horizontal translation, and flight termination and descent (Hendrie *et al.* 1986). Riley and Reynolds focus on the horizontal translation components which dictate flight pathways and, consequently, an insect's destination. Several analytical methods have been developed to estimate source regions and pathways of migrating insects in the planetary boundary layer (Scott and Achtemeier 1987). Further improvement of these techniques to forecast the long-distance movement of insects requires the identification of factors that govern the vertical distribution of insects in the planetary boundary layer (Isard *et al.* 1990).

Observations by Riley & Reynolds of grasshopper concentrations in cold air outflows from rain storms, possibly associated with gravity currents, also implicates meteorological factors in the descent component of migration, perhaps the most difficult phase to investigate, and thus understand and predict. In other papers, zones of convergence have been associated with flight termination and descent of insects (Rainey 1976, Schaefer 1976, Pedgley 1982). Other observations suggest that a similar phenomenon is associated with flight termination and descent of weakly flying insects such as aphids (Irwin & Thresh 1988). Thus, an important working hypothesis governing the descent component of migration has been formulated, one that demands rigorous testing. If the hypothesis withstands these tests, the emerging principle will permit a much greater degree of predictability concerning migratory insect descent, something that is sorely lacking in our current abilities to forecast migratory episodes.

References

Hendrie, L. K., Irwin, M. E., Liquido, N. J., Ruesink, W. G., Mueller, E. A., Voegtlin, D. J., Achtemeier, G. L., Steiner, W. W. M. & Scott, R. W. 1986 Conceptual approach to modelling aphid migration. In *The movement and dispersal of agriculturally important biotic agents* (ed. D. R. MacKenzie, C. S. Barfield, G. C. Kennedy, R. D. Berger & D. J. Taranto), pp. 541–582. Baton Rouge, Louisiana: Claitor's Publishing Division.

Irwin, M. E. & Thresh, J. M. 1988 Long-range aerial dispersal of cereal aphids as virus vectors in North America. *Phil. Trans. R. Soc. Lond.* B **321**, 421–446.

Isard, S. A., Irwin, M. E. & Hollinger, S. E. 1990 Preliminary analysis of the vertical distribution of aphids in the planetary boundary layer. *Env. Entomol* (In the press.)

Pedgley, D. E. 1982 Windborne pests and diseases. Meteorology of airborne organisms (250 pages). New York: Halsted Press.

Rainey, R. C. 1976 Flight behaviour and features of the atmospheric environment. *Symp. R. ent. Soc. Lond.* **7**, 75–112.

Schaefer, G. W. 1976 Radar observations of insect flight. *Symp. R. ent. Soc. Lond.* **7**, 157–197. Blackwell Publishing, Oxford.

Scott, R. W. & Achtemeier, G. L. 1987 Estimating pathways of migrating insects carried in atmospheric winds. *Env. Entomol.* **16**, 1244–1254.

Phil. Trans. R. Soc. Lond. B **328**, 673–677 (1990)

Printed in Great Britain

Training: the missing link in pest control

By Kit R. A. Kitenda

Desert Locust Control Organization for Eastern Africa, Nairobi West Airport,
P.O. Box 30023, Nairobi, Kenya

[Abstract only]

Today when scientific development has so much to offer, the Desert Locust still poses as much threat to many parts of the world as it did many decades ago. When I was first exposed to aerial control of Desert Locust in the late 1960s, everything was being done to tackle the problem scientifically, with a collective and Regional approach. However, the recent past points towards a negation of this, because of a total disregard of the necessary manpower training. Unless a remedy is found soon the war will be won by the pests.

Because of the long intervals between plague recurrences, usually 7–10 year lapses, those involved in locust control either retire by the next time around or just completely forget reading through the amount of ample literature available to update their knowledge. There is no effort to conduct refresher or mock exercises during recessions because it is assumed that anyone can fight the threat as the locust has been with us from time immemorial. There must be an effort to conduct symposia on locust control as there are on other migratory pests, which would involve people in real field work. Great emphasis must be laid on training and planning must begin for future needs in the effective control of all migratory pests by: (*a*) training of aircrew; (*b*) scouts, drivers and ground support.

The assumption that anybody with a flying licence or anyone who sits in an aircraft can identify a locust is a fallacy, as regards aerial control and observation. Training is the most potent instrument in producing a confident airman capable of conducting a serious campaign against a locust plague. Safe and efficient operations in this kind of flying can only be delivered by properly trained crews, as experience has shown. It is only a properly trained pilot who is likely to prevent pollution of the environment by avoiding overkill. Scouts and ground-support training is vital for a successful campaign. There is every danger of causing harm to human and animal life, as well as the environment, by using untrained personnel. A major cause of unnecessary wastage is often untrained manpower. It is therefore impossible to put enough emphasis to manpower training.

In an area covering over 6 million km^2 with a population of about 121.5 million in eastern Africa, one may be hard-pressed to find 30 properly trained locust scouts today. A properly trained scout and driver would be a great asset in such a vast and inhospitable region, as he may be the most likely person to recognize the onset of an infestation. It is also very important for every administrative staff member to appreciate the enormity of all the elements involved in conducting a successful campaign. It calls for dedication and involvement. Looking at and considering the vastness of the locust area, one question comes to mind. What is the most effective way to combat the menace when it occurs? Aerial control is the answer. But, faced with the present types of aircraft, one sees a big gap between available migratory pest control aircraft and our future needs. The typical agricultural aircraft is inadequate. The converted models of normal utility aircraft cover some role but are still unable fully to meet our needs.

There is therefore a great need for all of us; operators, manufacturers and researchers, to make a concerted effort to develop a suitable aircraft. An aircraft to cover the long distances fast enough not to miss the target, to be capable of carrying sufficient loads to control an average sized swarm without having to return to base for reloading, and yet be manoeuvrable comfortably, working from poorly maintained airfields and strips and be economical to run. We will need to fit the future aircraft with very advanced navigation equipment to enable more accurate long range survey and control. Presently, there are no such aircraft on the market.

Discussion

N. KINVIG (*CKS Ltd, Southend Municipal Airport, U.K.*). I have to congratulate Captain Kitenda on his very good talk. I agree with what he said about training being the most important subject. Proper locust pilots are becoming like Dodo birds, extinct. So something has to be done. I think that if the average experienced agricultural pilot is put into the field for locust spraying he won't know what to do, but it is very difficult to train when locusts are not present. However, the average agricultural pilot can make a good locust pilot if he has a good observer. As Captain Kitenda said, if you don't know what you are looking for you won't find it.

I don't entirely agree with Captain Kitenda's comments on the unsuitability of aircraft types. Some of the present aircraft are suitable, but I agree that a lot of the purpose-built agricultural aircraft are not. If they are loaded up, as the manufacturers suggest, then you can't fly more than 50 or 60 feet above the ground with a full load, so I think it will be necessary to use utility aircraft. I know it is a compromise, but to develop an aircraft totally suitable for locust spraying is out of the question. So, in conjunction with the new ideas on radar, we have to look for aircraft with a reasonable range and good manoeuvrability. Of existing aircraft, single-engined aircraft of the turbo-prop type such as Pilatus–Porter or Beaver, converted to turbo-prop, would make ideal aircraft for the purpose.

R. J. V. JOYCE (*Cranfield Institute of Technology, Bedfordshire, U.K.*). Captain Kitenda, with his great experience at the sharp end of locust control, has ably directed attention to the increasingly large gap that exists between those who see insect pest species as ones whose behaviour and sudden changes in numbers and location is of exceptional scientific interest, and those whose job it is to prevent crop loss. It is often forgotten in the scientific world that money is allocated to research into pest biology in the expectation that this will lead to saving crops from damage. In my view, the development of locust control has been successful because most of the knowledge concerning locusts was derived from questions raised by the needs of control: such as where have the swarms come from, where will they be tomorrow, and how can I transfer enough insecticide, with minimum waste, in the time they are in range, to destroy the bulk of them? It was questions such as these that generated Dr Rainey's convergence hypothesis, that have enabled locust populations to be studied quantitatively, and that stimulated search for an understanding of the physical principles which determine how efficiently insecticide released from an airborne tank can be transferred to its site of action within the locust. For several decades in the past, research workers have worked hand-in-hand with those engaged in control, to their mutual benefit. It is sad to learn from Captain Kitenda that this is no longer so.

Captain Kitenda is right to stress that it is a fallacy to assume that a flying licence is all that is necessary for successful survey and control of locusts, and recent experience has shown that this assumption has led to gross inefficiency and horrific environmental contamination. With no means of checking the efficiency of kill, it is impossible to establish whether the emergency use of even experienced crop-spraying pilots, paid as they are by the size of the area they treat, has been cost-effective. Locust survey and control is a highly specialized job that demands experience, skill, dedication and courage. Moreover, it also requires expertise in the use of instrumentation rarely employed in agricultural work. We are fortunate in having with us today a pilot who has been a pioneer in the development of the use of navigation systems for track guidance of agricultural aircraft, Noel Kinvig with recent experience in locust control, and I am sure he will agree how important is such instrumentation.

What, I think, is needed to surmount the difficulty to which Captain Kitenda has drawn our attention, namely the intermittent nature of locust upsurges, is the formation of a 'fire-brigade' unit, consisting, perhaps of no more than two aircraft, flown by an experienced crew under the flag and general direction of FAO; during recessions, conducting surveys for swarms in the most suspect areas, and, during upsurges, providing assistance to local control teams and training and updating aircrew in control techniques.

J. D. GILLETT (*London School of Hygiene and Tropical Medicine, U.K.*). Captain Kitenda makes a plea for new types of aircraft for anti-locust work; has he, I wonder, considered small modern airships? Airships have several advantages over both helicopters and fixed-wing aircraft: airships are cheap both in initial cost and in operation; airships have considerable superiority in endurance; airships can carry a far greater payload; airships can operate from poorly maintained or even non-existent airfields.

A high technology airship with an endurance of between 20 and 30 h and with sophisticated navigation and advanced radar, such as Airship Industries' SK600 class, would, I understand, cost in the order of £2000000; a smaller, low technology airship with an endurance of some 3–5 h, such as those now being offered by Thunder and Colt, would cost about £300000.

The disadvantages of airships for this sort of work are: (*a*) their relatively slow *economic* cruising speed, limited to between, say, 65 and 40 km h^{-1} (although the larger ships could have a sprint capability of somewhere well over 100 km h^{-1}) and (*b*) the necessity for a relatively large handling crew; both disadvantages are offset by the superior flight endurance and payload.

Finally, most of the locust problems that we have been discussing involve desert or semi-desert regions. If we move with the times and consider a solar powered airship (as at present under serious consideration at Imperial College), then the advantages in these sunny regions would be even greater. An airship, with its relatively huge surface area, is ideal for this type of development.

G. H. LEE (32 *Moreton End Lane, Harpenden, U.K.*). As an Aeronautical Engineer by profession, I was interested in the remarks made by Captain Kitenda concerning desirable aircraft characteristics. It is obvious that more than one type of aircraft would be needed: in addition to the 'strike' aircraft of short range suggested, there is also a need for a long-range aircraft for transport and survey duties.

I was surprised at the preference shown for single-engined aircraft. If aircraft cost was the deciding factor, I would have thought that the avoidance of a forced landing due to a single engine failure was a desirable feature of a twin-engined machine and might justify higher purchase and operating costs. Any aeroplane used must be propeller driven to give the operational flexibility required. The fuel type available might be limited and it is for this reason, as well as for lightness of engine, that a gas turbine might be preferable. Possible twin engined types might be later versions of the Short 'Skyvan' (the **360** series) or of the 'Jetstream' (formerly made by Handley Page and now manufactured by British Aerospace at Prestwick).

As it is well known for aircraft to fly through locust swarms, they would need special modifications to air intakes (especially engine air intakes) and to oil coolers to prevent clogging by flying insects.

Professor Gillett raised the question of the possible use of airships. As a consultant to an airship company, I would like to point out that despite obvious operational advantages (such as very low flight speed, e.g. hovering), airships are expensive to buy and to operate; the latter would require the services of a ground crew of, say, 10 people, with considerable associated ground equipment (mobile mast, etc.) that have to accompany the airship. This is for an airship having a useful load (flight crew, operational crew plus operating equipment and load, together with the necessary fuel and oil) of about 2500 kg. An endurance of 30 h was suggested; this is possible (and has in fact been achieved when demonstrating the possibility of Channel traffic control to the French Navy), but certification might require a night landing to be done to cope with engine failure; is that possible 'in the bush'?

Regarding helicopters, I agree that they have desirable flight characteristics, but they are more expensive to buy and operate than aeroplanes (and are probably more costly than airships). Their limited endurance and range restrict their usefulness.

R. LE BERRE (*WHO, Geneva, Switzerland*). I thank Captain Kitenda for so clearly drawing our attention to the need for training. To give two examples from the experience of WHO: since 1964, more than 160 Africans of high or good standard have been trained in medical entomology and vector control in excellent European and African centres. Only 40 are still working in the field, which means a loss of 75 %. The reasons for the losses are the difficulty of working under field conditions, and lack of a career structure at National/Regional levels. By contrast within The Onchocerciasis Control Programme (OCP) in West Africa again, 160 or more people have been trained in all disciplines involved. Most of them are still associated with the OCP because of a good structure, career insurance and excellent, although demanding, conditions of work. So I emphasize the need for providing these as well as training.

D. RIJKS (*WMO, Geneva, Switzerland*). At the request of the interdisciplinary workshop on Meteorology for locust control, Tunis, July 1988, a small working group with representation from Food and Agriculture Organization (FAO), the Overseas Development Natural Resources Institute (ODNRI), Programme de Recherches Interdisciplinaire Français sur les Acridiens du Sahel (PRIFAS) and the World Meteorological Organization (WMO), prepared a syllabus for training of observers and operators in acrido-meteorology. I fully support Captain Kitenda's suggestion for training of spray pilots. In fact, many of those involved in the past campaign have indicated the value they themselves would have gained from timely training, as outlined by Captain Kitenda. If it really proves difficult to develop an aircraft that will be able to both prospect and spray efficiently, thought might be given to using a combination of aircraft. Such a combination might consist of a group of five or six spray airplanes and one reconnaissance aircraft, with 10–12 h autonomy and equipped with Global Positioning System (GPS) and High Frequency (HF) communication.

R. J. Courshee (*The Meadows, Plough Lane, Ewhurst, Surrey U.K.*). It might be clear from our discussions by now that swarm control with chemicals is the main practical way we have of dealing with major locust infestations in 1989. In my view, a bigger missing link in locust control than training, is survey during recession periods. For this, I think, helicopter transport will be needed to enable surveys to be carried out on foot, so I would ask Captain Kitenda to consider the same helicopters for control of these swarms. If such control helicopters were also supplied by a transport helicopter, they could stay with the swarms and get over a main problem of fixed wing aircraft which cannot always stay in contact with them, economically that is, without airstrips and ground support.

Phil. Trans. R. Soc. Lond. B **328**, 679–688 (1990)
Printed in Great Britain

A Meteosat motion picture of a weather situation relevant to locust reports

By C. Zick

Freie Universität Berlin, ZEAM, Malteserstrasse 74-100 *D*-1000, *Berlin* 46 *F.R.G.*

[One plate]

The motion picture, of Meteosat VIS and IR half-hourly image sequences, covers the North African area with close-ups of northwest Africa from 28 to 31 March 1985. A cold front and horizontal vortex rolls, with their convergent upward atmospheric motions, visualized through dust being transported south from the Atlas mountains, may have transported locust swarms from the Algerian breeding area down to southeast Mauritania, where specimens were observed a few days after the event. This followed more than a year of no locust observations at all. Observation of dust in the atmosphere may give some insight into the transport and behaviour of locusts, both on the synoptic and mesoscale.

1. Introduction to Meteosat data

The use of satellites for monitoring sequences of weather favourable for reproduction and gregarization of Desert Locust *Schistocerca gregaria* (Fôrskal) has been discussed by Abdallahi *et al.* (1979), Müller (1976) and Hielkema & Howard (1976). Since 1978, the European satellite Meteosat has been used for the same purpose.

Meteosat transmits every half hour, that is, with a very high time resolution, pictures of the earth's disc, with Africa in the centre of view, in three different channels:

(i) VIS (0.4–1.1 μm) (the visible part of the electromagnetic spectrum). The intensity of the signal is proportional to the albedo of the objects (water, land, clouds, dust);

(ii) IR (10.5–12.5 μm) (the thermal infrared window of the electromagnetic spectrum). The intensity of the signal is proportional to the radiation temperature of the objects.

(iii) WV (5.7–7.1 μm) (the water vapour absorption band in the infrared spectrum). The intensity of the signal is inversely proportional to the upper tropospheric water vapour content.

The latter channel has little direct use for migrant pest research and operation. VIS and IR, however, are quite useful. The VIS data provide information on daytime cloud coverage with a spatial resolution of about 2.5 km. Processing the VIS data of cloud-free areas provides information about soil moisture, water content of lakes (ESA–EOQ 1989) and vegetation. The IR data, with a spatial resolution of about 5 km, provide information on cloud height and coverage 24 h per day, and information on convective precipitation (Turpeinen *et al.* 1987). IR data of cloud-free areas, after atmospheric correction, provide information on ground temperatures.

A further aspect of meteorological analysis with Meteosat imagery is the possibility of displaying time sequences of the half-hourly data in motion, which makes it possible to:

(i) calculate wind speeds and directions, through cloud motions;

(ii) make very short-term forecasts of weather, by using pictures showing cloud motion and change;

(iii) discriminate ground-fixed patterns from moving ones;

[161]

(iv) illustrate the nature of large-scale as well as mesoscale weather systems and their interactions, through their cloud motions and changes.

The possibility of obtaining wind data from cloud displacements observed in consecutive images is not discussed here in any detail. For a survey of such, the reader is referred to MOAG (1979). In the subtropics and tropics, the 'low level' cloud base (stratocumuli, cumuli and larger convective clouds) is generally around 1000 m or higher. Below this level the planetary boundary layer winds are the result of local small-scale atmospheric phenomena, which are usually not represented by the cloud winds. The 'low level' cloud winds represent the lowest level of the free atmosphere with their synoptic and mesoscale phenomena. They are useful for nowcasting (up to a few hours) and short-range (up to about 20 h) forecasting of subsynoptic to large-scale atmospheric motions. In the absence of other wind reports, such cloud winds can be useful for anti-locust operational measures and field research.

2. THE METEOSAT MOTION PICTURE OF 28–31 MARCH 1985 AS RELATED TO LOCUST OBSERVATIONS

On 29 March 1985, the Meteosat imagery shows large hazy patterns over North Africa, which are easily identified as dense dust clouds, both from their origin, their texture and their meteorological context (figure 1, plate 1 and figure 2). This strong dust event, which was certainly rich in details, was chosen as the subject of a Meteosat cloud motion film, produced by C.D.Z., in close cooperation with the European Space Agency (ESA) and the Free University of Berlin. The scene covers an area from the Mediterranean to 15 °N and from the Canary Islands to 25 °E. It shows the entire life cycle of a weak depression developing over north Algeria until its cloud patterns disintegrate over Libya. On its rear side, masses of dust are mobilized from the Atlas mountains (see mountain station report of 29 March 1985, 00h00 G.M.T., in figure 2), where copious fine soil weathering debris is available. This dust is transported south (10 °N) and east, and some is also westwards, over to the Canary Islands. This dust transport happens spatio-temporally in the same way in which mid-latitude cyclones mobilize cold air masses with their unstable lapse rates, stronger winds and convective clouds.

At first, this event was not connected with research on migrant pests. Later on, however, it was realized that in April 1985, i.e. only a few days after the period covered by the film, four isolated, but gregaricolour, Desert Locusts were observed in S.E. Mauritania, after a whole year (1984) with no reports of Desert Locust swarms anywhere (Rainey 1989). In the following September and October, observations from Mauritania and the tropical Atlantic west of Africa seem to support the idea that these April 1985 observations indicated the beginning of an upsurge (Rainey 1989). Thus it was of interest to investigate whether there was any relation between the upsurge and the satellite observation of huge dust clouds; not that the dust is important in itself, but rather the larger scale atmospheric environment and context. There are also meso-scale details observable in the motion picture which may contribute to the understanding of how swarms are mobilized and maintained in the air for considerable lengths of time.

3. LARGE-SCALE FEATURES

On the large scale there are two distinct types of weather systems relevant to successful breeding of locusts. On the one hand the tropical weather systems of the Inter-Tropical

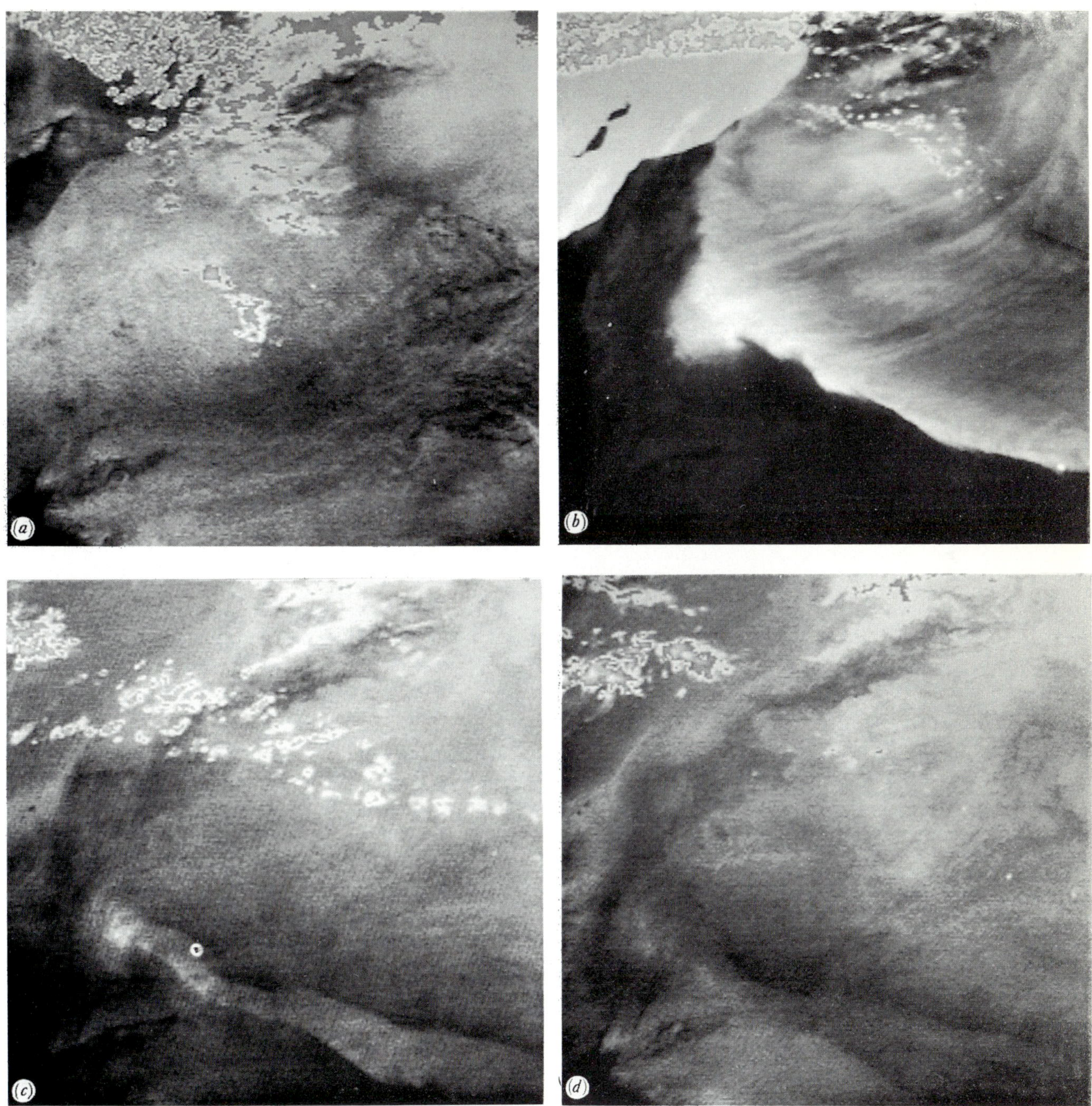

FIGURE 1. Meteosat IR pictures during 29 March 1985 ((*a*) 02h30 G.M.T. (*b*) 13h00 G.M.T. (*c*) 19h00 G.M.T. (*d*) 21h00 G.M.T.). Dust is transported southward from the Atlas mountains, the leading edge of it is the 'dust cold front'. Different contrast enhancements have been applied in these pictures to display the ground (with its highly varying temperatures) against the dust features. During the night (d), the 'dust-free' ground has a lower temperature than the dust concentrated aloft in the 'dust front'.

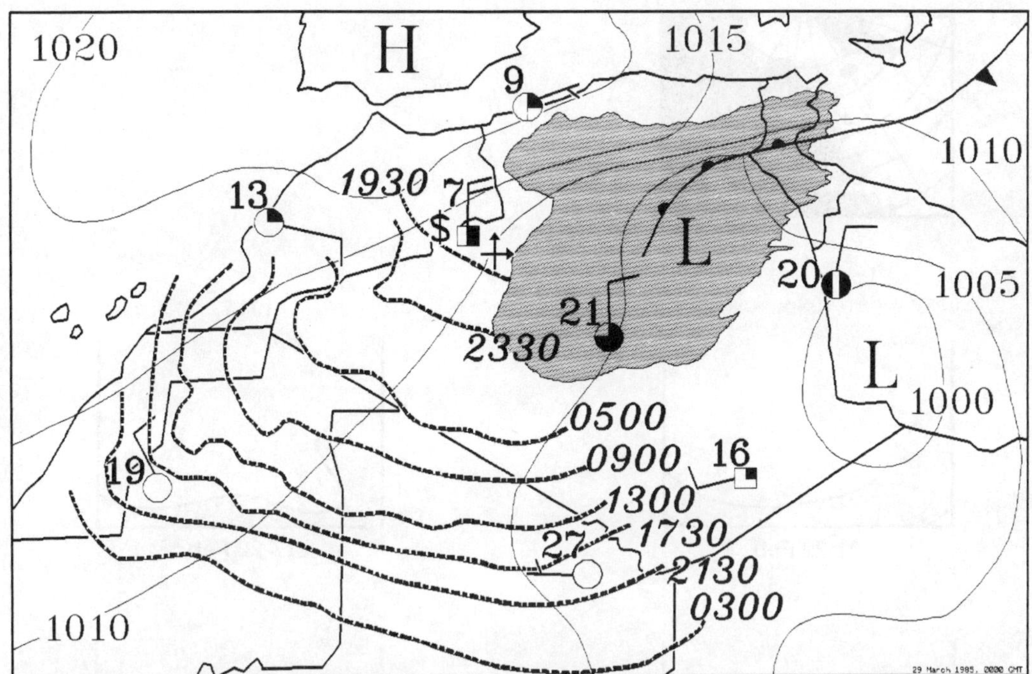

FIGURE 2. Surface weather map 29 March 1985, 00h00 G.M.T. (after German Weather Service, DWD). The central cloud deck of the developing lee depression is shown in a grey tone. The positions of the 'dust cold front' between 28 March, 19h30 G.M.T., and 30 March, 03h00 G.M.T. (times shown in italics) are dashed. Note the strong surface pressure gradient in the lee of the Atlas mountains; the mountain station Colomb-Béchar reports a north wind of 20 knots and dust storms continuing for at least 6 h.

Convergence Zone (ITCZ); on the other, troughs, with disturbances penetrating south but originating from the west. The main geographical distribution of the Desert Locust over Africa reflects this differentiation (Rainey 1963). The case presented here relates to the latter type of large-scale weather system.

To see any large-scale relation between the weather at the end of March 1985 and the 1985 locust upsurge after this date, one may study the analyses published monthly by the German Weather Service (DWD 1984, 1985), each containing a sequence of 500 hPa maps showing several characteristic weather periods ('Witterungsabschnitte') over Europe during that month. Each map is an average of the 500 hPa maps of a varying number of days (2–6 generally), showing the characteristic 500 hPa patterns during that period of weather. A simplified presentation of the significant features of these maps of relevant periods in 1984 and 1985 is shown in figure 3. Clearly, these maps do not extend far into North Africa. The author hopes, however, that the reader will accept that these available analyses for the edge of the area of interest are sufficient for the requirements of this work.

For our case, the relevant map (figure 3) comprises the days of 25–29 March 1985. It shows a large-scale westerly current over North Africa, with a short trough over Algeria. Set into its temporal context, we see that it marks the end of a rather quiet homogeneous 'cold' period beginning on 6 March with a deep trough over the same area. It is comparable to almost the same period (11–28 March) of the previous year. It is followed by a rapid build-up of a ridge and consequently warmer and drier weather in the area.

With regard to the observation soon afterwards of four isolated gregaricolor locusts in S.E.

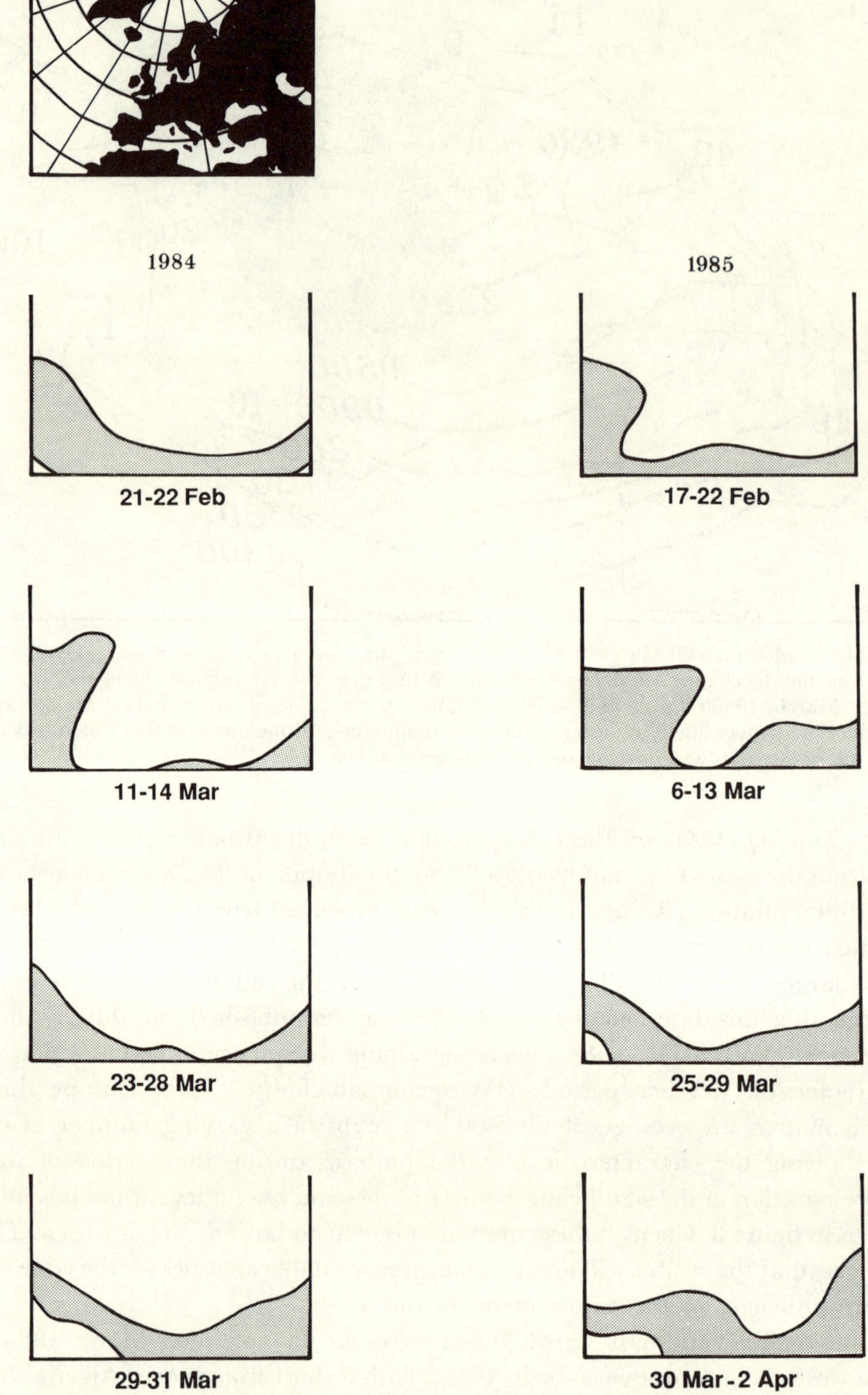

FIGURE 3. The change of characteristic periods of 500 hPa patterns over southern Europe, for periods in 1984 and 1985. The diagrams were generated from the German Weather Service 500 hPa maps showing characteristic weather periods ('Witterungsabschnitte') over Europe, by isolating maps, the 564 gpdm and the 580 gpdm lines to emphasize the southernmost 500 hPa patterns. The geographical background is shown once (top left-hand corner of diagram).

Mauritania, the large-scale weather situation of our case should be considered as the end of a prolonged period of depressions passing over northwest Africa and providing, through their cold fronts and convective clouds, the intermittent rainfall necessary for successful breeding. Indeed, several weak depressions brushed the Atlas mountain range during that time, as can be seen from the daily Meteosat images (European Space Agency 1985). In March 1985 there was over 200 % (90–279 mm) of the average monthly precipitation on the north side of the Atlas, but it was slightly colder than the monthly average. The preceding month, however, was 2–3 degrees warmer than average, in the same area.

A possible explanation for the unsuccessful breeding during the similar 1984 period (11–28 March) in the North African area is suggested by the climatological data for 1984. Both February and March 1984 were colder than the monthly average, and only February 1984 had a local Tunisian maximum of precipitation (200–400 % of the climatological average). The following month of March remained very dry (25–50 %).

The period 22 December 1984 to 18 January 1985, although otherwise similar to the March 1985 period, was colder. Also, during the period 1–20 February 1984, there was a comparable large-scale trough over Egypt and Libya, but both periods were probably not favourable to successful breeding because of the winter season: northwest African locusts need the warmer spring temperatures and the precipitation of spring depressions for successful breeding.

4. Synoptic and mesoscale features

Rainey (1963) has emphasized the importance of depressions for the transport of locusts from southern Morocco to Tunisia, and this may indeed have happened during February–March 1985: the periods 17–22 February and 6–13 March (see figure 3) with their large-scale southwesterly flow in that area would have been favourable. The depression in question here, however, is an Atlas lee depression travelling east. On its rear side, huge masses of dust were raised by strong winds over the Atlas mountains during the night of 28–29 March (figures 1 a and 2). During daytime (figure 1 b) the leading edge of the cool, i.e. bright, dust cloud becomes a clear-cut feature against the diurnally heated 'dust-free' ground. Relative to the cloud-covered vortex centre this 'dust front' is positioned and shaped like a typical cold front of a depression in the westerlies. And in our case it is quite reasonable to speak of a cold front, as it is dynamically defined rather than appearing only as a belt of clouds. The cold front is the leading edge of the cooler, drier air mass being transported equatorwards on the rear side of the depression. Along the front, dynamic mass convergence forces lifting of the air.

The dust, in particular the dust concentrated along the cold front, shows clearly the southwestward trade wind flow on this day (29 March, see figure 2). The dust is transported in an atmospheric layer between about 500 m and 1000 m. In the motion picture one can see that by night (figure 1 c, d) the surface has undergone strong nocturnal radiative cooling, while the dust, in particular in the 'dust front', remains at its level of transport and consequently keeps a constant temperature, resulting in a temperature slightly higher than that of the adjacent ground (figure 1 d). Thus this rapidly moving frontal system remains clearly identifiable as a thermal pattern and, as we know, a dynamic synoptic scale convergence pattern, for about two days. Within this atmospheric system, locusts of Algerian origin may well have been carried across the Sahara to their place of detection in Mauritania the following April.

During 29 March 1985 the Meteosat picture (figure 1 b) shows long streamers of dust aligned with the mountain ranges of the Atlas as well as on each side of the Mcherrah/Aftout massif with its huge Ergs on either side. These streamers of dust result from mesoscale convergent patterns at the level of the dust, about 500 m to 1000 m. This effect, when observed with small cumulus clouds, is called cloud streets. It is due to the formation of longitudinal vortex rolls with a horizontal axis in the atmospheric boundary layer (Brown 1980); they are aligned more or less with the direction of the mean wind in that layer, and they usually develop through thermal (convective) instability of the boundary layer due to surface heating and/or cold air advection. The destabilization of the lowest 1000 m of the atmosphere during daytime because of heating of the ground, well observable in the IR motion picture, and the consequent upward motion of air provide the energy necessary to mobilize swarms of locusts. Also the mesoscale 'dust street' convergences within the air mass may well keep the swarms together and in the air during daytime.

5. Conclusions

The Sahara and the Ergs have been an area of dust deposition for many thousands of years, with the dust being raised and transported in a manner similar to the case described here. This case, with its cloud and dust patterns in motion, is a clear example of a regularly occurring interaction between the atmosphere and ground, which raises and transports dust from the Atlas mountains southwards, southeastwards and southwestwards. The existence of the long-term regular repetition of this process is shown by the huge sand dunes within the Ergs, having the same orientation and shape as our 'dust street' convergences. On this large scale the regular transport mechanisms of dust across northwest Africa, by physical mechanisms, which are also favourable to the mobilization and transport to locusts, may provide a clue to the link between the otherwise widely separated northwestern Africa areas of locusts (Rainey 1963) and the tropical belt of West Africa. Similar to conclusions drawn by Rainey about the systematic swarm displacements from northwestern Africa to Mauritania, the specimens observed in April 1985 may have been carried there from breeding areas in Morocco and Algeria, by the synoptic scale features observed in the cloud motion picture. Subsequently, the ITCZ over Mauritania has been claimed, with good reason, to be the appropriate feature governing their movement.

A similar coincidence of dust and locust reports occurred on 23 September 1985. Over the ship 'Ile Maurice' in the tropical Atlantic numerous groups of 'criquets' in flight were reported, and four days later isolated locusts over a distance of some hundred kilometers, some 600 km further out to sea (Rainey 1989). These reports also coincided with a huge dust cloud over that area in the Meteosat VIS pictures of the same dates (European Space Agency 1985).

It is not claimed that dust may serve as a tracer for locusts in the atmosphere; such a correlation would have been well-known since biblical times. But by observing the mechanisms for transporting dust, and its 'behaviour' in the atmosphere we may gain some insight into the similar mechanisms and behaviour of locusts, both on the mesoscale and on the very large synoptic scale.

I thank Dr K. A. Browning for stimulating this work, and Dr R. C. Rainey for providing literature and data about migrant pests and insect flight.

References

Abdallahi, O. M. S., Skaf, R., Castel, J. M. & Ndiaye, A. 1979 OCLALAV and its environment: a regional international organization for the control of migrant pests. *Phil. Trans. R. Soc. Lond.* B **287**, 269–276.

Brown, R. A. 1980 Longitudinal instabilities and secondary flows in the planetary boundary layer: a review. *Rev. Geophys. Space Phys.* **18**, 683–697.

Deūtscher Wetterdienst 1984 Die Grosswetterlagen Europas. *Offenbach* **37**, 1–12.

Deūtscher Wetterdienst 1985 Die Grosswetterlagen Europas. *Offenbach* **38**, 1–6.

European Space Agency 1985 METEOSAT image bulletin, 3, 9. Darmstadt.

ESA–EOQ 1989 METEOSAT applications in African flood monitoring. *Earth Obs. Quart.* **26**, 4–5.

Hielkema, J. U. & Howard, J. A. 1976 Pilot project on the application of remote sensing techniques for improving Desert Locust survey and control. AGP/LCC/76/4. Rome: FAO.

MOAG 1979 METEOSAT cloud winds quality. Darmstadt: ESA–ESOC.

Müller, W. A. 1976 Seasonal locust development potential in the Niger area analysed through ecoclimatological interpretation of satellite images. *Int. J. Biomet.* **20**, 249–255.

Rainey, R. C. 1963 Meteorology and the migration of desert locusts. *Tech. Notes Wld met. Org.* **54**.

Rainey, R. C. 1989 *Migration and meteorology*. Oxford University Press.

Turpeinen, O. M., Abidi, A. & Belhouane, W. 1987 Determination of rainfall with the ESOC precipitation index. *Mon. Wea. Rev.* **105**, 2699–2706.

Discussion

R. S. Scorer (*Imperial College, London, U.K.*). The phenomenon of 'rope clouds' shows how well the dividing front between two air masses can remain preserved long after cloud and rain systems have ceased to function, and may sometimes never have existed.

Rope clouds can be seen in satellite pictures in many parts of the world: a good example is the N.W. Pacific where cold fronts crossing Japan remain marked by rope clouds for hundreds of miles into the ocean (Scorer 1990). The same can be seen in the north Atlantic, but it is a less common feature there, possibly because the large dry inland area of Asia is not replicated west of the Atlantic (Scorer 1986; 1990; Scorer & Verkaik 1989). The boundary of an area of dust collected over the Gobi Desert is a classic example in the Japanese literature (Scorer 1986).

Rope clouds frequently occur in the Mediterranean, where it is quite usual for a stable layer of air to be enclosed by the surrounding mountains, with occasional invasions from the N.W. the boundary being marked by rope. But these are usually too shallow to cross the Atlas mountains, as in Zick's case.

Sea breezes do not show rope clouds normally, because they would be evaporated by surface heating. But the fronts where convergence occurs can sometimes be seen in IR satellite pictures by the temperature contrast on, or in moisture in air just above, the ground (Scorer 1990). A very spectacular case of rope at the front of the reverse flow at the inversion out to sea was recorded by the Gemini crew orbiting in September 1966, although the phenomenon is more common in the quieter weather of March and April (Scorer & Verkaik 1989).

The very sharp boundary of the upper level outflow is not usually included in discussions of sea breezes, but it might be important over the Arabian Sea, which is sometimes crossed by locust swarms that experience convergence and arrive in Pakistan as dense swarms (for example, as in January 1962).

There should be no surprise when convergence lines are found around the Red Sea and the Gulf of Aden, and inland in Africa near mountain ridges or plateaux.

In the remarkable case shown by Dr Zick it is worth noting that at the front fresh dust is being picked up all the time so that individual particles do not travel the whole distance traversed by the dust front.

References

Scorer, R. S. 1986 *Cloud Investigation by satellite*. Horwood.
Scorer, R. S. 1990 *Satellite as microscope*. Horwood.
Scorer, R. S. & Verkaik, A. 1989 *Spacious skies*. David & Charles.

J. R. MILFORD (*Department of Meteorology, University of Reading, U.K.*). The film of well-defined fronts marked by dust, and also the illustrations of rope cloud over Europe provided by Professor Scorer are of considerable general interest, but are essentially anecdotal. The question is how such ideas can be put to use in dealing with migrant pests in practice.

The first method is through training operational personnel so that they are aware of the importance of mesoscale convergence regions and are therefore looking out for them. Such training may need to be extensive, as few cases are as clear as those chosen as illustrative material, and many are hard to detect on satellite pictures available in real time.

Before providing training in recognition of these particular features, we should also be able to quantify their importance: are they the prime mechanism for concentrating swarms? Such a question asks for an answer based on statistics: it does not appear that anyone has devised a viable means of gathering such data since no such results have been published, or referred to in this meeting.

I would welcome comments, particular from those involved in the field, which would tell us whether we need to quantify the effects of mesoscale convergences on insect populations. Some 15 years ago Dr J. E. Simpson, Dr D. A. Mansfield and I used a numerical model to calculate the concentration of aphids over southern England resulting from the passage of a sea-breeze front. This was not published because there seemed no way to verify the results, and also because no-one placed much reliance on the few figures we found for flight characteristics. I observe that models of this type do not seem to have developed and ask whether this is again because of the absence of reliable flight data on migrant pests. Such models, with stochastic features built in, appear to be necessary if the effects of the features shown here in films and slides are to be fully understood.

I would also wish to voice my disappointment at the absence of quantitative modelling from the topics of this meeting in general, and ask whether this is related to the absence of any section in the programme, which could be headed 'progress', with regard to the Desert Locust, and similar pests.

P. WICKHAM (*Meteorological Office, Bracknell, U.K.*). As a meteorologist, I was excited to see on Mr Zick's pictures of March 1985, that the structure of the event showed up so clearly on both the visible and infrared (IR), particularly when highly enlarged. Of course the IR shows temperature contrast, and I wonder what that was on these pictures. Was it entirely dust that we were looking at, coming down behind the main frontal zone and in those streaks in the airflow? What was the temperature contrast that showed so clearly there? Could water vapour have been involved?

C. ZICK. I did not see clear structures such as showed on the film, in the water vapour images; the Meteosat photographs don't show any streaks behind the leading edge. You can conclude from the visible images that the streaks were not clouds. If they had been, they would have shown as white spots, at least around midday. Dust behaves, in terms of radiation theory, very much as semi-transparent cirrus does: the visible short wave radiation simply goes through it, without being absorbed. It's only the longer wave thermal radiation that is absorbed, by dust in the atmosphere for example, and emitted again at the relevant wave length. That is why you can see it only in the IR and not in the visible.

M. S. BOULAHYA (*National Meteorological Office, Algiers*). I would like to confirm that it was dust in March 1985. I was travelling in western Algeria at the time and all the southwestern airports were closed for three days, so I had to go by road. Even then, going around the mountains from Algiers to Bechar, the car's engine became completely filled with sand.

R. M. MORRIS (*Meteorological Office, Bracknell, U.K.*). Satellite imagery needs to be exploited much more thoroughly to improve the conventional weather analysis in the region. For example, the movement of the dust cloud needs to be related to the boundary layer wind speed and direction for consistency; conceptual models of frontal structures can be invoked to improve the analysis of temperature, in the horizontal and in the vertical.

R. J. V. JOYCE (*Cranfield Institute of Technology, Bedfordshire, U.K.*). Mr Zick has shown a most illuminating and impressive film of two Meteosat case studies of two periods important in relation to the recent upsurge of the Desert Locust.

The dates enabled Dr Rainey to secure further information on locust reports, which he regarded as critical for the understanding of the initiating events. Dr Rainey has drawn attention to the fact that, after a whole year (1984) with no reports of Desert Locust swarms anywhere, the only slender evidence that the species might be somewhere in West Africa was a report of four locusts showing the pinkish colouration that is manifest only by locusts from swarming populations, which were seen in April 1985 near Aioun-el-Atrouss in southeastern Mauritania. Further reports were lacking and, by June, it was suggested that Desert Locust numbers might be at their lowest ebb since sytematic collection of data started some 60 years earlier. In late September, however, reports from three ships recorded scattered flying locusts seen at sea off the coast of Mauritania and Senegal, extending N.N.E.–S.S.W. over a distance of some 500 km. In early October many hundreds of square kilometres of advanced-instar hoppers and pink adults were discovered in southwestern Mauritania, necessitating control on a substantial scale. By the end of the month, there was the first report of a young swarm of the next generation. It was clear that there had been undiscovered breeding on a large scale on the heavy rains which had been recorded in the area in July and early August, and that the locusts seen at sea would have represented some of the last parent population.

Mr Zick's Meteosat image sequences show that the weather system, which developed over Mauritania during the last days of March 1985, may well have brought together locusts from a wide area for breeding on the rains associated with this disturbance. His subsequent sequences also fit the inferred later movements of the locusts of the new generation.

D. E. PEDGLEY (*ODNRI, Chatham, U.K.*). Mr Zick has shown us a fascinating use of satellite imagery to reveal remarkable detail in a synoptic-scale disturbance over northwest Africa. He must compare his results with the hourly observations available from all the synoptic observing stations over which the disturbance passed. Perhaps Mr Boulahya would be able to provide observations from Algeria, and also help to get others from neighbouring countries. Such a comparison would go towards answering the query on temperature contrasts raised by Mr Wickham. Should it be thought that the use of weather maps and satellite imagery is something new in the forecasting of locust movements, let me say that both have been employed routinely for nearly 30 years. Strong winds behind cold fronts, sometimes accompanied by dust storms but usually less dramatic than the one shown by Mr Zick, are well known to be involved in the long-distance displacements of locust swarms. The cold front associated with Mr Zick's storm would appear very clearly in a synoptic analysis of all the available records. The analysis presented from the daily European Meteorological Bulletin was deficient, no doubt through the sparsity of observations, as often happens when there are widespread dust storms, owing to poor radio communications.

Phil. Trans. R. Soc. Lond. B **328**, 689–704 (1990)

Printed in Great Britain

Monitoring of rainfall in relation to the control of migrant pests

By J. R. Milford and G. Dugdale

Department of Meteorology, University of Reading, Whiteknights, Reading RG6 2AU, U.K.

Of all climatic parameters, rainfall has the greatest variability in space as well as time. It also has the greatest influence on the breeding and behaviour of migrant pests, supplying the moisture needed both for their development and for the growth of vegetation to sustain a population. Soil moisture controls both processes directly but cannot be adequately surveyed even by remote sensing techniques, so we rely upon water balance models to interpret rainfall measurements and to forecast pest populations.

Both terrain and rainfall are very inhomogeneous on the kilometric scale, a scale which is not matched by observations either from conventional raingauges or current meteorological satellites. Modelling the effects of rainfall must take account of these inhomogeneities, and the processes involved will be discussed. However, it is clear than more detailed studies of rainfall–habitat interaction are needed to derive soil moisture from rainfall estimates by using knowledge of the microtopography. Such relationships must be capable of being generalized so that future monitoring by satellites, essential to give complete and uniform coverage, can be realistically interpreted.

1. Introduction

Knowledge of rainfall, or any other meteorological variable, is only one among many information inputs to the complex decision-making system which is involved in the monitoring and control of migrant pests. The aim of the system is to find, at the earliest possible time, those populations of pests which will increase to significant proportions. Locating such populations is the major problem in operational crop protection against migrant pests (World Meteorological Organization 1988), and for this purpose the meteorological information, together with a variety of biological and geographic information, must be fed into a model which incorporates what is known of the processes governing the population dynamics and also migration. The aim of such a model is to predict the pests' behaviour, and so to maximize the chance of discovering swarms with the minimum of effort.

To make the complex information/decision-making system as efficient as possible, we must first identify the information requirements, and the characteristics of the information which is available. As these never fit the initial statement of the ideal requirements an iterative process ensues, in which supply and demand are progressively modified until they match as well as possible. This paper discusses this matching process for information on rainfall in relation to the search for, and the prediction of migrant pests, with particular reference to the Desert Locust in Africa and Arabia. In this case rainfall is the meteorological factor whose variations are predominant: temperature and humidity are also important, but they vary less dramatically and are in any case often closely correlated to rainfall variations. In the migratory periods wind observations are clearly vital, but they play a relatively minor part in the dynamics of the build-up of local plague populations.

[171]

2. Requirements for rainfall information

Rainfall itself is not, of course, a very important parameter in pest prediction; what we generally need to know is the moisture distribution in the uppermost 300 mm or so of the soil, and, if it were possible to map this, we would have the information we need directly. However, this is not a realistic prospect, for even the most ardent proponent of microwave remote sensing will not promise to provide information with the required discrimination and spatial resolution from operational satellites within the next decade or two. For the foreseeable future we must continue to rely on observations of rainfall amounts as the input to soil moisture models, although there may be special circumstances where information on surface wetness may be both attainable and of interest.

As has been known for a long time (Popov 1958), water equivalent to about 20 mm of rainfall must be present in the typical, sandy soils of the relevant regions, to induce female Desert Locusts to lay their eggs, and then for these to absorb sufficient moisture for hatching to take place. The rate of development of the eggs decreases quite rapidly with temperature, but so does the rate of evaporation, and the initial quantity of water required in the soil to lead to successful hatching may not vary greatly. The other water requirement before a population can expand is for the growth of vegetation to sustain the emergent hoppers. Where no substantial vegetation is present at the time of laying, a minimum available soil moisture of 20 mm or so of rainfall is again a suitable threshold level to support vegetation development sufficient to give a risk of population expansion.

3. Rainfall observations

Until recently, the only method available for the measurement of rainfall was the raingauge. If it is well sited and maintained a raingauge can tell us the amount of rain which has fallen at that point with an accuracy of around 5% (HMSO 1981; Robinson & Rodda 1969). What is less well known is what the gauge can tell us about the rainfall distribution in the surrounding area. In detail, this depends on the climatological characteristics of a region, in particular the nature of the rain-producing systems. From observations made with a dense, regular network in Niger we derived figure 1, which shows the best estimate of the rainfall from a single storm, averaged over a square of increasing area centred on a gauge. Also shown are the confidence limits associated with this estimate which are substantially independent of the rainfall amount (Flitcroft et al. 1989). If the gauge departs from the centre of the area, or the area is enlarged, the 'corrections' to be applied increase. However, this result cannot be extrapolated far because of the assumption included in the underlying model that all points in the area considered have experienced some rainfall.

Up till now, it has been customary to estimate the rainfall at a particular point within a raingauge network by interpolation between the observed values, more or less regardless of the time and space scales involved. Given the variability within individual storms, and the rarity of rain events in many of the areas of our current concern, the assumption of continuity that lies behind this isohyetal analysis is dangerous even for seasonal total rainfall. For individual rain events, the rate at which the correlation between station rainfalls decreases with separation (figure 2) shows that for sites more than 20 km or so from a gauge the best estimate of a daily fall may be the mean rain per rainday, established from climatology, and not the fall recorded

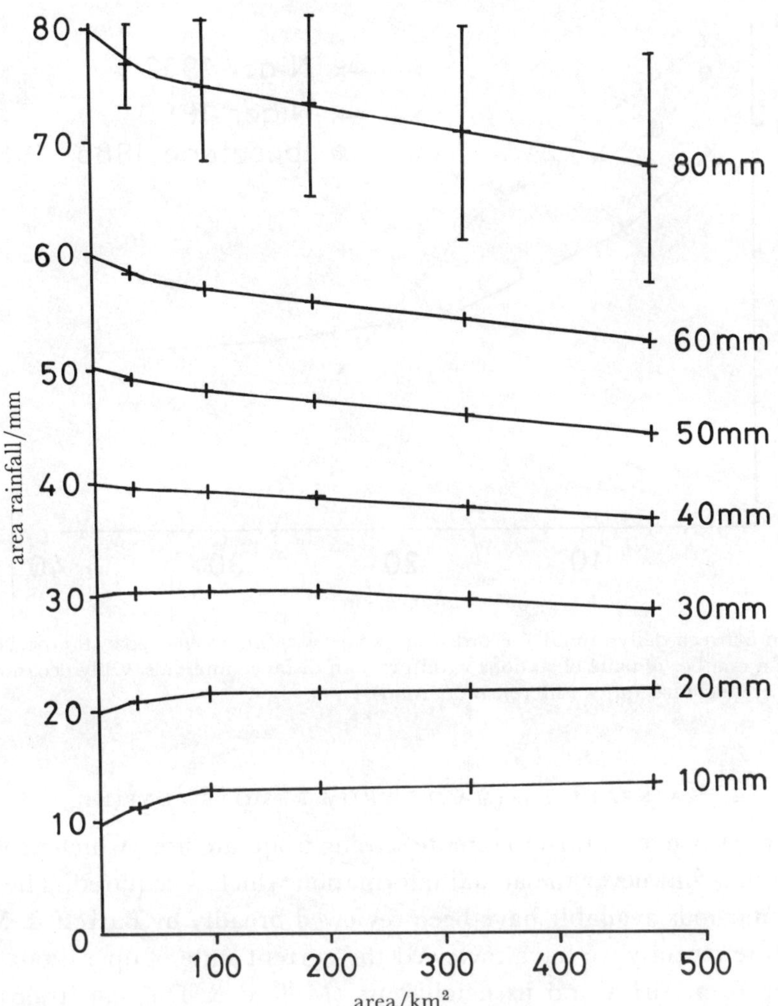

FIGURE 1. The best estimate of area average rainfall for a single storm for an area centred on a single raingauge, shown as a function of the size of area and the rainfall. The error bars include 90% of cases for recorded falls of 80 mm (after Flitcroft *et al.* 1989).

at the nearest gauge. Some of the results of comparisons between satellite estimates and gauged amounts of rain confirm this warning (Snijders 1990). The rate of decay of correlation with distance, shown in figure 2, affects the quantitative results, including those of figure 1: however, calculations (not reported here) based on samples of data from other dense networks suggest that figure 2 is not typical of seasonally or semi-arid regions of Africa, where the majority of the rain comes in heavy falls from deep convective cloud cells (Sumner, 1983). An accurate account of the sporadic rainfall, whether for monitoring water resources or migrant pest potential, thus requires an impracticable number of gauges, higher even than the 1 per 1000 km² recommended by WMO (1983). An additional requirement is for daily reports, implying also an unrealistic network of communications.

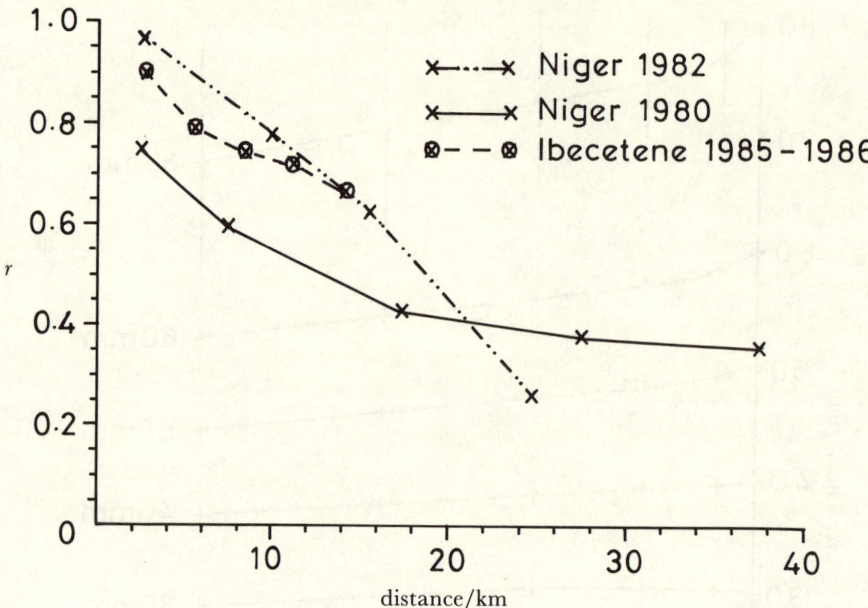

FIGURE 2. Correlation between daily rainfalls recorded at gauges with increasing separation in Niger. Each point is the average of a number of pairs of stations within certain distance intervals. Only occasions when rain fell at both stations are included (after Flitcroft *et al.* 1989).

4. RAINFALL ESTIMATES FROM REMOTE SENSING

To gain an overview we must turn to remote sensing from satellites, which provide impartial and simultaneous data but never the actual information which is required. The potential and limitations of the methods available have been reviewed broadly by Barrett & Martin (1981) among others. More recently we have reviewed the current state of operationally established rainfall estimation over Africa and its neighbours (Milford & Dugdale 1990), which relies almost entirely upon the statistical relationships between cold, deep cloud and rainfall. The optimum threshold to use for classifying a cloud as 'cold', and the regression parameters to use, depend on the region of interest, the time of year, and possibly other variables such as cloud type and characteristics of the season so far, which have yet to be parameterized effectively. In spite of this, such rainfall estimates are being used to provide hydrological information as a routine, and also inputs to food early warning systems, such as that in the Sudan. We now describe the information which may be obtained from such systems if they are optimized for migrant pest monitoring and control purposes.

The requirement that a system be fully operational makes it very desirable to acquire the data from the satellite and process it automatically, and to pre-calibrate all numerical steps. In the case of short period rainfall, frequent views are needed, and this limits us at present to geostationary satellites, such as METEOSAT which is located some 36000 km above 0° N, 0° W. Currently the only input to our rainfall estimates is the thermal infrared (TIR) channel on Meteosat: the addition of data from the water vapour band (5.7–7.1 μm) has made only small improvements to the accuracy of the final outputs (Turpeinen *et al.* 1987), and too much of the storm activity occurs at night for the visible channel to make a useful contribution. While Meteosat provides the area coverage and hourly images which we need, well registered by European Space Operation Centre (ESOC) before they are retransmitted, the spatial resolution is limited to 5 km at nadir, and about 7 km near the horizon in Saudi Arabia, for instance. The latter area is also covered by the Indian satellite INSAT, a similar satellite around 70° E: if the INSAT data were available for general use, combining the two data sets would give improved information in principle.

Although the rainfall estimates are being made within the Africa Real Time Environmental Monitoring using Imaging Satellites ARTEMIS system in FAO, Rome, a project primarily designed to help in migrant pest control (Hielkema, this symposium), the calibrations have not yet been optimized for this purpose for any particular area. Up till now the calibration process has consisted of collecting the durations of cold cloud (CCD), by using three different threshold temperatures, over a number of raingauge stations. Ten-day totals of CCD are compared with measured falls, first in contingency tables to select the best threshold and then by linear regression to provide a conversion from CCD to a rainfall estimate. We have found that both the threshold and the regression coefficients vary geographically. Operational programs include smoothing procedures to eliminate abrupt changes in the outputs appearing at the edges of the calibration zones, and interpolation to compensate for missing data (within limits).

The credibility of the system depends on its ability to provide useful estimates of rainfall in real time, and we therefore spend much effort on validating these in years subsequent to those which have provided the calibration data. Table 1 shows an example of the results. Even though about half the residual variance is due to the non-representative nature of the raingauge (see Flitcroft *et al.* 1989), it is still so large than the estimates are not likely to be useful as inputs to localized crop production models. Here, however, we are concerned with discriminating falls within 1–3 days above a certain threshold and figure 3 shows that we may indeed be able to provide such information with significant success. A substantial amount of time is needed to optimize the information and then to test that any preliminary result is robust enough to be used operationally.

TABLE 1. VALIDATION OF 10-DAY RAINFALL ESTIMATES, NIGER, JULY 1985, 1986 AND 1987

(Calibration used: $R = 0$ when $D = 0$, otherwise $R = 4.52D + 5.1$ mm where D is CCD in hours below -60 °C. Rainfall categories used: 0, 1–10, 11–20, 21–30, 31–40, 41–60, 61–80, 81–100, etc. in mm.)

year	number of obs	exact category	1 category out	2 & 3 categories out	> 3 categories out	% ≤ 1 categories out
1985	327	69	106	132	20	54
1986	309	68	106	109	26	56
1987	318	60	121	109	28	57

We must recognize that the statistics based on a single radiometer channel are fundamentally limited by their inability to distinguish convective from layer cloud, specifically cirrus. Averaged over large enough areas or lengths of time this may not be too important if the ratio of the types remains constant. Texture analysis could help, but we may well have to wait for the advent of additional infrared radiometers on geostationary satellites to deal with this problem in a fully automatic system. Meanwhile, we have shown that hourly images are needed to obtain the maximum useful information from the method (Milford & Dugdale, this symposium). Direct validation of area average of daily rainfalls is difficult because of the need for many gauges in the area. Some indirect confirmation of the estimates is gained from our use of the satellite data as input to hydrological catchment models, but this is also at a relatively early stage (Hardy *et al.* 1989).

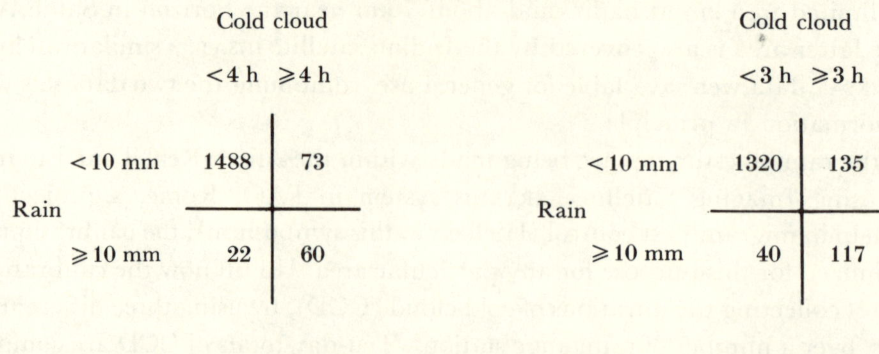

FIGURE 3. Sample contingency tables to show discrimination between falls < or > 10 mm. In July, rainfall ≥ 10 mm was measured within 1.5 % of pixels showing less than 4 h cold cloud (below −50 °C) and 45 % of those showing ≥ 4 h. Both tables are for stations north of 14° N and threshold −50 °C. Rainfall over a period of three days including the day of cold cloud is included.

Considerable research has been carried out on other methods of observing rainfall by remote sensing, and some of these may have useful applications in particular situations. For example, radar monitoring of rainfall can be used in the study of individual systems such as squall lines, but the limited coverage of a radar set prevents the method being viable as a general monitoring tool. Microwave instruments, whether active or passive, mounted on aircraft or satellites, are also under development but will, for the foreseeable future, give too low resolution and too infrequent views for operational use where large areas have to be monitored for brief, intense storms.

Reverting to current meteorological satellites, the Advanced Very High Resolution Radiometer (AVHRR) data from the National Atmospheric and Oceanic Administration (NOAA) satellites has spatial resolution of 1.1 km (in place of 5 km or more) but this does not make up for the poorer frequency (four observations per day at most, in place of 48). These data were the basis for an interactive scheme for anti-locust operations in northwest Africa (Barrett 1979), but the system was not used for long enough to provide independent evaluation of its accuracy. More recently, the Bristol group (Barrett *et al.* 1989) advocated the use of varying amounts of data from geostationary and/or orbiting satellites, according to the area and time span to be covered. For large areas and 10-day periods days are classified as having rain/no rain and a climatological rain/rain-day figure applied. The interactive interpretation of images, together with detailed local knowledge, may well improve localized estimates, particularly where topographic effects are substantial, for instance in the coastal strips on either side of the Red Sea.

Finally, we should refer to the possibility of observing where rain has fallen through the reduction of the amplitude of the diurnal temperature wave at the earth's surface. Much effort has been expended on the general methodology, but it has not yet been shown to be operationally viable anywhere. However, locust recession areas may offer the most favourable sites for testing the technique because of their small amounts of vegetation. The main drawbacks are the difficulty of establishing when an area is free of cloud, a small amount of which will reduce the temperature observed as an average over a whole pixel, and also the fact that the method can discriminate water in the surface layers of a sandy soil up to a few mm (Milford 1987) but cannot differentiate between 10 and 20 mm. Quantification of the reduction in amplitude is also hampered by uncertainty over the moisture content of the atmosphere, and hence in the correction to be applied to the apparent surface temperature reported by the satellite.

For an automatic system the images must be processed to reveal the changes from the minimum reflectivity in the visible, and the maximum midday temperatures, both mapped to include variations across the scene due to the terrain (rock, soil or sand for example). If information on the areas where it was probable that some rain had fallen in the previous 24 h were particularly helpful, it might now be possible to use a dedicated image processing system to alert an operator to areas which merited closer study. A combination of such a system with the cold cloud statistics may improve the chance of locating the areas most at risk from a locust population explosion. At the least, some areas may be shown with some certainty to have received no rain.

5. Moisture in the soil

We have said that it is the moisture in the soil, which is important for both locust eggs and vegetation growth. If it is known how much rain has infiltrated into the soil, and what the physical properties of the soil are at different depths, there is now little difficulty in using a numerical model to compute the distribution of water with depth at some later time (Campbell 1985).

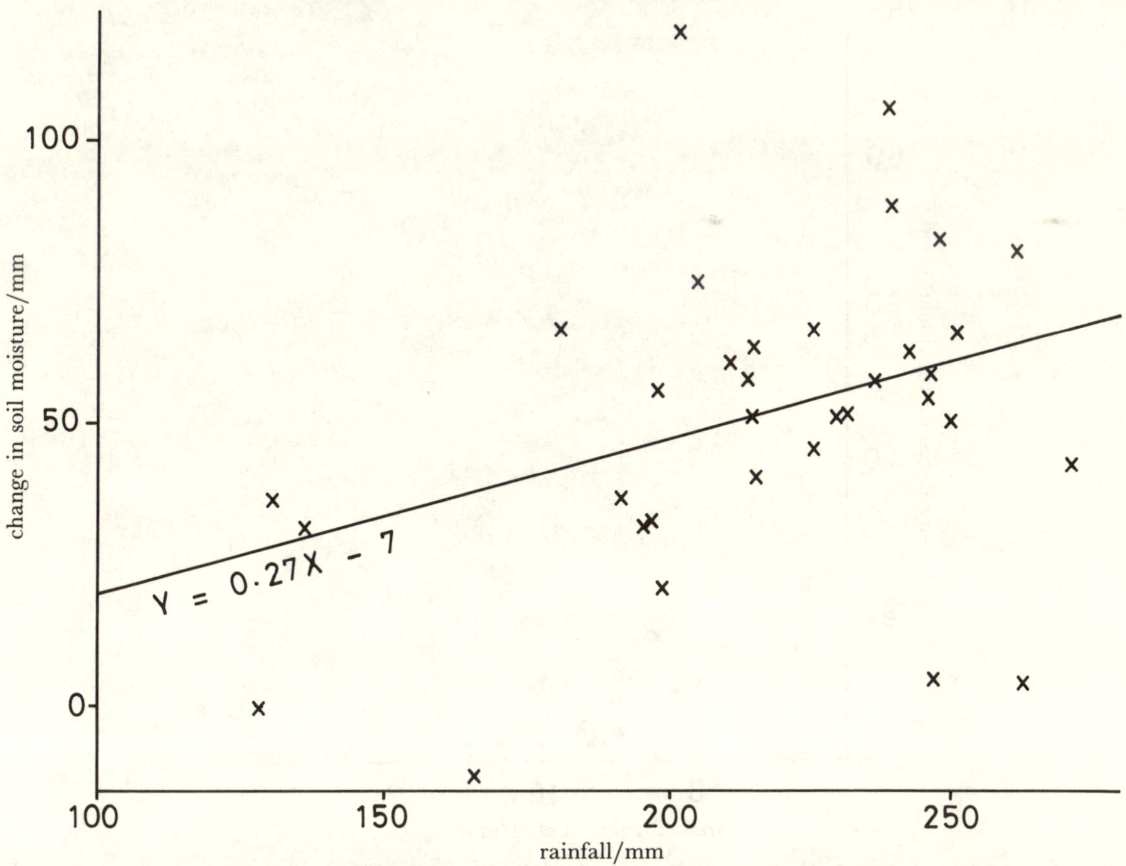

Figure 4. Change in soil moisture storage against total rainfall over an eight-week period. Observations are from a regular grid, with 2 km spacing. Ibecetene, Niger, July–August 1985.

At this stage we must consider the problems of representivity again: all quantities derived from satellite observations are averages over the pixel of the particular radiometer in use, with some blurring because no registration of images can be better than to the nearest pixel. Even if the rainfall were uniform across a pixel we would find that the amount infiltrating into the soil was not uniform. This is shown in figure 4, where the change in soil moisture storage (as measured with a neutron moisture probe) bears little relation to the rainfall measured at exactly the same spot. A subsequent experiment confirmed the importance of the local slope and surface characteristics: local run-off within the pixel scale was indeed substantial, but was reduced by the presence of vegetation, or by breaking up the 'pan' which tended to form on the surface of the sandy soil at this particular station.

6. Interpretation of rainfall estimates

We have discussed the use of contingency tables to select criteria which delimit areas where there was little chance that significant rain has fallen, or there was a substantial chance that rain has fallen. More detailed interpretation of the rainfall estimates from satellites is based on analyses of the spatial distribution of falls of different amounts. On a large scale, comparable to the grid squares of general circulation models of the atmosphere, we may use the variability of the cloud itself. For increasing CCD (and hence estimated area average rainfall) figure 5

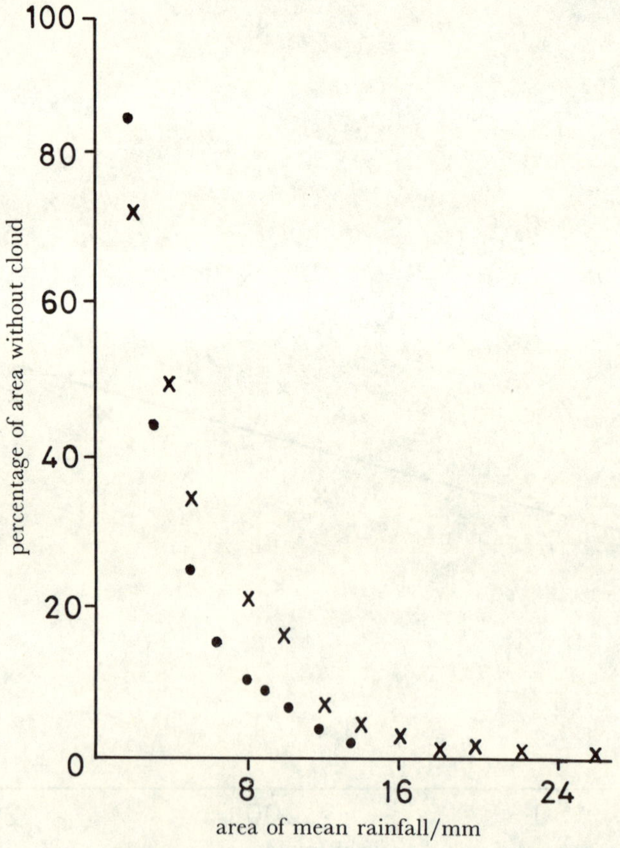

FIGURE 5. Proportion of a grid square without cloud as a function of area mean rainfall estimated from daily cold cloud duration. 200 km grid, Niger, July 1985, threshold −60 °C. 100 km grid, Niger, July 1987, threshold −50 °C (after Dugdale 1989). (×, 200 × 200 km grid (1985); ●, 100 × 100 km grid (1987).)

shows the fraction of an array of 20×20 Meteosat pixels which experience no cold cloud over the 24-h period, and, we assume, no rain. Figure 5 also includes figures for a 40×40 array (Dugdale 1989).

With each pixel cold cloud duration we associate a distribution of point rainfalls, as deduced from our dense gauge networks. The resultant distribution of point falls corresponding to mean falls over a grid square of 3, 7.5 and 15 mm are shown in figure 6 (Dugdale 1989). We see, for example, that for an area mean of 15 mm some 20% of the area would experience a fall of 20 mm or more, sufficient to meet the criterion for locust proliferation. This result applies regardless of the source of information on the area mean.

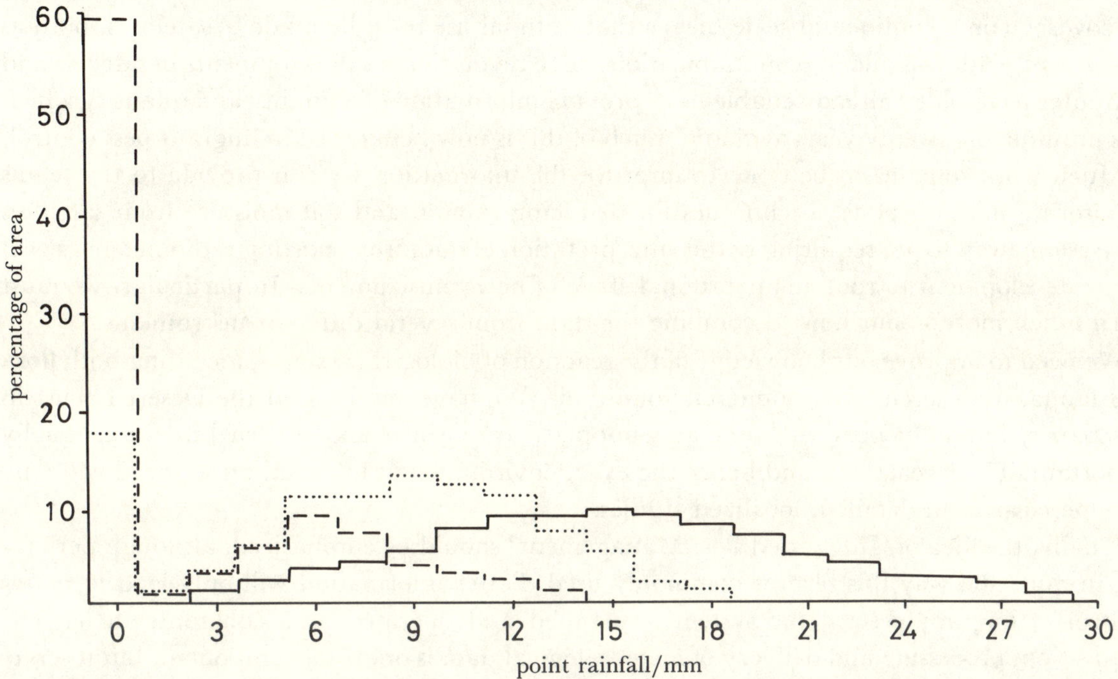

FIGURE 6. Proportion of a 100 km grid square receiving different daily rainfall amounts when area average is 3 mm (---), 7.5 mm (....) and 15 mm (—). Distributions are calculated from the distribution of cold cloud amounts for given average, and the distribution of rainfall observed for each cloud amount (after Dugdale 1989).

A further important factor which has yet to be quantified is the extent to which rainfall is concentrated into a limited part of an area by the topography, for instance where we find soil in wadis between impervious rocks. Even a light average fall over a satellite pixel may thus provide 20 mm of soil water over a fraction of the area, and therefore conditions suitable for Desert Locusts to lay, and also for vegetation growth. Relations between the rainfall, whether observed with a dense gauge network or estimated from satellite, and the range of soil moistures to be found in particular sites must be established. Vast areas have to be covered, so no doubt we must turn to remote sensing techniques to survey them. However, the interpretation of earth resource satellites with this particular purpose in mind has not yet been carried out: it is therefore apparent that studies of such rainfall–habitat interactions are needed if we are to predict the effects of rainfall in specific areas, particularly the fringes of the desert.

7. Summary and prospects

We have been discussing the means available for obtaining information on the rainfall and soil moisture content, a vital component in the prediction of the development and movement of populations of migrant pests. All the discussion on rainfall can be applied to other meteorological data to some extent, although the spatial variability of temperature abnormalities, for example, is much less extreme than that of rainfall. The difficulties of providing the actual information required for the predictive models are manifest, and there is a continuing and major task to match the information that can be supplied on a regular and reliable basis to the decision-making processes of the control operations. In every case the need for coverage on a continental scale means that optimal use must be made of satellites, both as instrument platforms and for communications. The revolutionary developments in satellite and computer technology already enable us to provide information of a quality and quantity which was unthinkable twenty years ago, and much of this is now being used in migrant pest control.

Much work remains to be done to improve the information we can provide to the teams monitoring migrant pests. Techniques for deducing rainfall and soil moisture from satellites have a long way to go, requiring better interpretation of data from existing radiometers as well as the development to routine operational stage of newer instruments. In particular, we must learn much more about how to combine the data from several different instruments.

We need to improve our knowledge of the reaction of biological systems, including both flora and fauna, to different environmental conditions. We have emphasized the Desert Locust in this paper, but influences on other grasshoppers, armyworm and quelea birds are equally important. These reactions, and hence the exact environmental information we need will only become clear from detailed, localized studies.

Finally, the idea of 'Integrated Pest Management' should be emphasized, although perhaps not in quite the way this term is commonly used. Better information will only lead to better control of migrant pests if the system is planned and operated as a coherent whole. The acquisition, processing and delivery of meteorological data is one vital component, but it serves no useful purpose unless it is matched to field data, modelling and guidance to pest control operations which are maintained at an adequate level of readinesss. Systematic study of the overall process is at least as important as research into any individual component.

The research reported here from the Department of Meteorology, University of Reading, has been supported by the U.K. Overseas Development Administration, the UN Food and Agricultural Organization, and the EEC. We thank the substantial team which has contributed over the years.

References

Barrett, E. C. 1979 Satellite rainfall estimation by cloud indexing methods for desert locust survey and control. *Am. Water Res. Ass. Mem. Symp.* In *Satellite hydrology*, pp. 92–100.

Barrett, E. C., Beaumont, M. J. & D'Souza, G. 1989 Hierarchical system for operational rainfall monitoring by Meteosat. *Proceedings of the 7th Meteosat Scientific Users' Meeting, Eumetsat EUM P*04, pp. 393–407.

Barrett, E. C. & Martin, D. W. 1981 The use of satellite data in rainfall monitoring (340 pages). London: Academic Press.

Campbell, G. S. 1985 Soil physics with BASIC – transport models for soil-plant systems. (150 pages). Amsterdam: Elsevier.

Dugdale, G. 1989 The influence of variability of rainfall and soil moisture on evaporation from semi-arid areas. *ECMWF Workshop Proceedings, October 1988*, pp. 63–78.

Flitcroft, I. D., Milford, J. R. & Dugdale, G. 1989 Relating point to area average rainfall in semi-arid West Africa and the implications for rainfall estimates derived from satellite data. *J. appl. Met.* **28**, 252–266.

Hardy, S., Dugdale, G., Milford, J. R. & Sutcliffe, J. V. 1989 The use of satellite derived rainfall estimates as inputs to flow prediction in the River Senegal. *Proc. Balt. Symp. May 1989. IAHS* **181**, 23–30.

Her Majesty's Stationary Office 1981 *Handbook of meteorological instruments*, vol. 5: *Measurement of precipitation and evaporation*. London: HMSO.

Milford, J. R. 1987 Problems of deducing the soil water balance in dryland regions from Meteosat data. *Soil Use Mgmt.* **3**, 51–57.

Milford, J. R. & Dugdale, G. 1990 Estimation of rainfall using geostationary satellite data. In *Applications of remote sensing in agriculture* (ed. J. A. Clark & M. D. Steven) London: Butterworth.

Popov, G. B. 1958 Ecological studies on oviposition by swarms of the Desert Locust (*Schistocera gregaria* Förskal) in Eastern Africa. *Anti-Locust Bull., Lond.* **31**.

Robinson, A. C. & Rodda, J. C. 1969 Rain, wind and the aerodynamic characteristics of raingauges. *Met. Mag.* **98**, 113–120.

Snijders, F. 1990 Rainfall monitoring based on Meteosat data. *Int. J. Remote Sensing.* (In the press.)

Sumner, G. N. 1983 Daily rainfall variability in coastal Tanzania. *Geogr. Annlr* **65A**, 53–66.

Turpeinen, O. M., Abidi, A. & Belhouane, W. 1987 Determination of rainfall with the ESOC precipitation index. *Mon. Weath. Rev.*, **115**, 2699–2706.

World Meteorological Organization 1983 *Guide to climatological practices*, 2nd edn. Geneva: WMO

World Meteorological Organization 1988 Agrometeorological aspects of operational crop protection. *Tech. Notes wld. met. Org.* **192**. Geneva: WMO.

Discussion

K. A. Browning, F.R.S. (*Meteorological Office, Bracknell, U.K.*). Dr Milford described the use of cloud top temperature as a proxy for surface rainfall measurements. I'm sure he would be the first to acknowledge that this approach must be applied with caution: not only does the relationship need to be tuned for different regions and/or seasons, but in many areas it breaks down altogether. Thus in very arid regions a lot of the rain may evaporate before reaching the ground, whilst in some other regions much of the high cloud may be frontal cirrus and unrelated to heavy rain. A study by Turpeinen *et al.* (1987) has shown, for example, that a useful relationship exists for Kenya but that the technique does not work for Morocco and Tunisia.

Reference

Turpeinen, O. M, Abidi, A. & Belhouane, W. 1987 Determination of rainfall with the ESOC Precipitation Index, *Mon. Weath. Rev.* **115**, 2699–2706.

E. C. Barrett (*University of Bristol, U.K.*). The use of regional regressions in the ('Tamsat') method described in some detail seems to work reasonably well in the western Sahel, a region of relatively simple topography and climatic zonation, but seems likely to be less than effective where topography, climate and weather are more complex.

The problem of cirrus recognition is important, for thick cirrus can masquerade as deep rain cloud. The authors say that '...we may well have to wait for the advent of additional infrared radiometers on geostationary satellites to deal with this problem in a fully automatic system', but the Adler/Negri technique is well suited to this task, although it would require local calibration for efficient application.

The potential of passive microwave as a complement to visible and/or infrared data at least for cirrus discrimination and visible/infrared technique calibration is not afforded the recognition it richly deserves.

The use of collateral data (e.g. from raingauges and radar) should be promoted wherever possible. The best results are likely to stem from the systems which use the most, not the least, types of input data.

Twenty years after the first satellite rainfall monitoring schemes were proposed, much has been learnt about the physical and practical problems with which this field is fraught. However, progress towards operational implementation of such schemes has been much slower and more piecemeal than the science and technology could sustain. It has been estimated that each year one third of the total global food crop production is damaged or destroyed by insects. We owe it to mankind to increase our efforts to reduce such undesirable effects by using satellite-based techniques wherever this seems likely to be beneficial.

J. R. MILFORD AND G. DUGDALE. We entirely agree with Dr Barrett that information derived from satellites has been underexploited. This is partly because the time and effort required to move from the research and case study stage to the incorporation of any technique in an operational system has been consistently underestimated, and hence underfunded. In addition, collaboration between local scientists in developing countries has often been poor when they find themselves competing for aid projects and prestige: in this context the donor governments and international agencies have not always set a good example.

We welcome the opportunity to clarify the rationale of our current work in collaboration with the U.K. Overseas Development Administration. We are convinced that if information on rainfall is to be used by non-meteorologists we must provide a virtually automatic system which will deliver easily intelligible products with the utmost reliability. This is particularly true where the information is needed on a limited number of occasions each year, and over a huge area, as in the monitoring of migrant pests in Africa and Asia. Interpretation of the products must be straightforward, and the limits to their accuracy well defined.

We make absolutely no claim to have invented a technique for rainfall monitoring: rather, we are using existing ideas in the engineering of operational systems to match the requirements outlined above. We chose to use the cold cloud statistics method and we are trying to see how far it can usefully be refined. Our main research is on the comparison with other methods, including surface gauges, and on the interpretation of the products. In the process we must educate the technicians who will operate the system, the meteorologists who interpret it and the final users. Only when a truly operational system is established can we decide which of many possible additional techniques would be viable for delivering further information to specific users. Until then we are likely to remain limited to the simplest, and admittedly simplistic techniques.

In the context of this particular meeting our paper emphasized the interpretation of the primitive technique because it is already in operational use, or about to be so, in a dozen centres in Africa. We are already including collateral data where it is appropriate, particularly in the hydrological applications.

With regard to other techniques that were mentioned in the comments, none of these seem to be ready for operational implementation for migrant pest monitoring, for a variety of reasons. Radar is invaluable for research but cannot cover the extensive areas involved in monitoring. Satellite-borne microwave has a great long-term future, but the 20% accuracy predicted for TRMM hardly meets the need as it refers to monthly totals averaged over an area of 10^5 km^2 or more. Turpeinen *et al.* (1987) used a very small sample of data, and we hope to follow up their work with a larger sample in the near future. Finally, our interpretation of Negri and Adler's papers differs from that of Dr Barrett: for example they state that 'One-parameter models were as effective in explaining the variance of cloud volume rainrate as multiparameter methods...' (Negri & Adler 1987). None the less, we hope that texture, pattern and information from other satellites will improve the overall accuracy of operational systems in due course.

Reference

Negri, A. J. & Adler, R. F. 1987 Infrared and visible satellite rain estimation. Part II: a cloud definition approach. *J. Clim. appl. Met.* **27**, 1565–1576.

C. G. COLLIER (*Meteorological Office, Bracknell, U.K.*). This paper provides a useful summary of the importance of rainfall in monitoring the behaviour of migrant pests. The authors note the particular usefulness of satellite data in providing estimates of rainfall over wide areas. Indeed, they stress that conventional raingauge observations cannot hope to satisfy the requirements for widespread continuous coverage.

Although the paper does stress the potential and limitations of satellite methods of measurement, the main method employed, that of estimating the duration of cloud tops below a certain temperature threshold, is presented rather uncritically. It is correctly pointed out that the optimum threshold used depends upon location and season; in addition it will depend upon the type of cloud observed.

Recent work by Brown & Cheng (1989) supports the earlier conclusions of Lovejoy & Austin (1979) that a correlation of visible and infrared data is overall a better measure of rainfall than using visible alone, which in turn is better than by using infrared alone. Of course, visible data are not available at night and substantial rainfall may occur at night. Recently, Vejen (1989) has found that use of water vapour imagery may enhance rainfall estimation techniques. Likewise, the differences between channel 3 (3.7 µm, near infrared) and channel 4 (11 µm, infrared) of the AVHRR instrument flying on the NOAA polar orbiting satellites may be used to differentiate between water and ice cloud. Although data from channel 3 are not available on the current Meteosat, it is possible that Meteosat Second Generation will provide measurements at this wavelength.

Perhaps the most exciting development recently has been the use of microwave frequencies from the Special Sensor Microwave Image (SSM/I) instrument flying on the U.S. Department of Defence Meteorological Satellite Programme (DMSP) satellite. Early results (Barrett *et al.* 1988) by using the different polarizations available at a frequency of 85.5 GHZ are very encouraging. Unfortunately it is unlikely that such an instrument will fly on a geostationary satellite for many years. However, plans are well advanced to fly an active radar system in low earth orbit to provide data for use with such passive microwave radiometers. This project, known as Tropical Rainfall Measuring Mission (TRMM) (Simpson *et al.* 1986), aims to provide monthly estimates of rainfall in the tropics with an accuracy over land of 20%.

Although these new developments may, in the long run, provide more accurate estimates of rainfall than existing satellite techniques, the use of geostationary satellite data will remain the basis of practical operational procedures for measuring rainfall. However, the use of infrared thresholding alone should be regarded as only a first step. The use of procedures based upon multi-spectral analysis are likely to bring worthwhile improvements.

The next ten years or so will see a significant increase in the number of ground based digital weather radars in Africa. Systems are already being planned in Botswana, Malawi, Lesotho and Zimbabwe, and some north African countries already have radar networks. These systems should provide additional data with which to tune satellite measurement techniques. I concur with the authors that the inhomogeneity of rainfall has a major impact upon the way in which its quantity and distribution are inferred from satellite data. Further effort must be put into improving our understanding of exactly what the satellite instrumentation is measuring if inferences are to be made about soil moisture and hence vegetation growth and pest development.

References

Barrett, E. C., Kidd, C. & Bailey, J. O. 1988 The Special Sensor Microwave Imager. A new instrument with rainfall monitoring potential. *Int. J. Remote Sensing* **12**, 1943–1950.

Brown, R. & Cheng, M. 1989 Real-time combination of radar and satellite data for very-short-period precipitation forecasting. *Preprint Vol. COST-73 Int. Seminar on Weather Radar Networking, 5–8 September, Brussels, Commission of the European Communities, XII/475/89-EN*, pp. 353–360.

Lovejoy, S. & Austin, G. L. 1979 The delineation of rain areas from visible and IR satellite data for GATE and mid-latitudes. *Atmos. Ocean.* **17**, 77–92.

Simpson, J., Adler, R. F. & North, G. R. 1986 A proposed Tropical Rainfall Measuring Mission (TRMM) Satellite, *Bull. Am. met. Soc.* **69**, 278–295.

Vejen, F. 1989 Meteosat-2 som nedbormaler verificeret med radar data. Hvordan Kan Water Vapour Kanaler aniverides? M.Sc. thesis, University of Copenhagen, Denmark.

D. P. ROWELL (*Meteorological Office, Bracknell, U.K.*). This contribution leads on from Dr Milford's excellent paper on the techniques and problems of monitoring rainfall, by detailing the current skill of short-range numerical forecasts of tropical rainfall. If skilful forecasts were possible, this would increase the lead time at which significant increases in pest population could be predicted. This contribution also links in with the discussion after Mr Boulahya's paper, concerning the skill of model wind (and precipitation) forecasts in the area surrounding Algeria, and also with the need to predict areas of convergence which are of course closely linked to rainfall.

The results presented here concern the prediction of rainfall in the Sahel at the height of the wet season. Verification of model forecasts was performed for three grid boxes in southern Niger, chosen because of the availability of data from a 'reasonable' number of raingauges within each box. The prediction errors shown in table 1 are for the rainfall which accumulates during the first 24 h of each of the 31 forecasts made from 12h00 G.M.T. during August 1985. It is clear that both models are unable to provide forecasts of rainfall amounts, which improve upon simply by using the climatological values. The ability of the model to forecast simply whether or not rainfall will be above or below a given threshold (2 mm) was also found to lack any useful skill (although a higher threshold is appropriate for the development of pest populations, it is thought that this is unlikely to alter the result of low skill).

Forecasts for at least a few days ahead are the main requirements, and these are clearly out of question, given the skill shown here for just the first 24 h. However, it must be emphasized that the level of skill varies with season and between climatic regions.

The quality of the European Centre for Medium-range Weather Forecasting (ECMWF) and U.K. Meteorological Office 850 mb wind forecasts up to 4 days ahead were also investigated. This study covered the north African region between latitudes 5° and 25° N, again for August 1985. Prediction of wind speed was very poor, but limited skill was found in

TABLE 1. PREDICTION ERRORS FOR THE RAINFALL ACCUMULATING DURING
THE FIRST 24 H OF FORECASTS MADE IN AUGUST 1985

| | | | mean absolute error (mm/day) |
grid box	number of gauges	model forecast	forecast of the climatological amount
		ECMWF	
15° N, 5.6° E	15	4.9	2.6
13.1° N, 7.5° E	10	4.7	6.5
13.1° N, 9.4° E	15	6.3	6.9
mean		5.3	5.3
		U.K. Meteorological Office	
15° N, 5.6° E	13	3.9	2.8
13.5 °N, 7.5° E	11	4.6	6.4
13.5° N, 9.4° E	16	5.5	6.8
mean		4.7	5.3

predicting the position of African waves (affecting wind direction) up to two or three days ahead. This skill may or may not be high enough for the purposes of forecasting pest migration. Again, it should be emphasized that skill varies with season and between climatic regions, and indeed Dr Boulahya has found the Meteorological Office forecasts to be useful up to 5 days ahead in the northern Sahara during November.

I conclude with a brief discussion of future prospects for short-range tropical weather prediction by using numerical models. There is reason to believe that significant improvements will be seen. The main problem at present is to produce a good analysis from the small quantity of data available in regions such as north Africa. Three factors will help with this: the use of automatic weather stations in data-sparse areas; better communications to enable more of the observations to reach the GTS (Global Telecommunications System) in time to be incorporated into the model analysis; the availability of satellite data more appropriate to the needs of meteorologists, and the undertaking and use of research into its objective interpretation. This is likely to be the most important factor leading to improved tropical analyses.

Finally, the parameterization of physical processes is particularly important for the evolution of tropical forecasts, and further research in this area may also lead to greater predictive power.

G. SZEJWACH (*EUMETSAT, Darmstadt, F.R.G.*). The paper by Milford & Dugdale stresses the importance of satellite rainfall monitoring as an indirect means of obtaining information on soil moisture at the considered scale. At the present time, it is indeed true that the method is indirect and this will remain the case until much more sophisticated remote-sensing data (including microwave imagery) become available from satellites. In this context, it is noted that satellite data obtained from a radiometer (visible, infrared, and water vapour channels) provide information on the upper part of the cloud cover from which rainfall at the surface is indirectly estimated using statistics or a physical model. Limitations are well known and most of them are indicated. They include presence of cirrus clouds in the vicinity or convective tops, global representativeness of the relationship between rainfall and cloud top temperature, as well as local, regional, topographical, temporal, and seasonal variability. Research is ongoing to try better to understand, and calibrate this relationship. One should note that most publications (including this one) address the problem of convective rain but very little has been done concerning the much more difficult problem of stratiform rain. It is possible that in the proposed application stratiform rain could play an important role in the regions considered here or in other parts of the globe. One other important parameter not considered here (or by others) related to rain and soil moisture is atmospheric water vapour.

Despite these limitations, the authors attempt to arrive at an operational and useful product related to rainfall. The research conducted by Milford & Dugdale to address the problem of rainfall estimates from satellite, and to correlate this to soil moisture, is relevant to a wide variety of meteorological and climatological studies and applications such as the one indicated in the present paper.

Phil. Trans. R. Soc. Lond. B **328**, 705–717 (1990)

Printed in Great Britain

Satellite environmental monitoring for migrant pest forecasting by FAO: the ARTEMIS system

By JELLE U. HIELKEMA

Food and Agriculture Organization of the United Nations Remote Sensing Centre,
Via delle Terme di Caracalla, Rome 00100, *Italy*

(*Paper introduced by G. B. Popov, FAO, Rome*)

Since 1975, the Food and Agriculture Organization of the United Nations (FAO) has been pioneering the development of the use of satellite remote sensing techniques for improving the surveillance and forecasting capabilities of the centralized Desert Locust Reporting and Forecasting Service at FAO Headquarters and, indirectly, those of Regional Organizations and National Plant Protection Services.

On the basis of findings from experimental activities on the use of Landsat and NOAA satellite AVHRR data for Desert Locust habitat detection and monitoring through vegetation assessment, and the use of Meteosat data for rainfall monitoring, FAO defined an operational system for satellite environmental monitoring in support of the FAO Desert Locust Plague Prevention Programme and the FAO Global Information and Early Warning System on Food and Agriculture.

The system, African Real Time Environmental Monitoring using Imaging Satellites (ARTEMIS) is an advanced computer hardware and software configuration, equipped for direct acquisition of hourly Meteosat digital data and for automated thematic processing of Meteosat and NOAA AVHRR data for large area precipitation and vegetation condition assessment, being the key environmental factors for supporting Desert Locust population development.

Since August 1988, the ARTEMIS system has generated a number of operational products documenting the occurrence of rainfall and vegetation development in the Desert Locust recession area on a 10-day and monthly basis at spatial resolutions varying from 7.6–1.1 km.

These products are being used by the FAO Emergency Centre for Locust Operations (ECLO), along with synoptic weather and locust data, for the preparation of the bulletins containing the Desert Locust situation summaries and forecasts.

For making ARTEMIS output products and other relevant data available in a timely manner at regional and national levels, a dedicated satellite communications system, Data and Information Available Now in Africa (DIANA), is currently being developed by the European Space Agency in cooperation with the FAO Remote Sensing Centre.

The DIANA system will, by mid-1991, provide a capability for high speed (64 kb s^{-1}) two-way transfer of facsimile images of documents and maps, character-coded text documents and digital images in raw or processed form from computers at FAO Headquarters to personal computer based terminals of recipients, initially in Africa, by using the commercial Intelsat satellites.

1. INTRODUCTION

During the 1970s several earth resources and environmental satellites were launched from the United States and Europe, which offered new prospects for mapping the preferred habitats of the Desert Locust (*Schistocerca gregaria* (Fŏrskal)) and for monitoring key environmental parameters for the successful development of Desert Locust populations, that is, precipitation and vegetation development, on the scale of the Desert Locust recession area, which covers some 16 million km² between the West African coast and northwest India.

After a successful demonstration of the use of Landsat imagery, by using visual image analysis techniques, for detecting potentially active Desert Locust breeding habitats on the Tihama plains of Saudi Arabia by the former Centre for Overseas Pest Research (COPR) (Pedgley 1973), the Food and Agriculture Organization (FAO) of the United Nations has since 1975 been actively involved in the development and testing of various satellite remote sensing techniques for improving Desert Locust monitoring and forecasting on an operational basis (Roffey 1975; Hielkema & Howard 1976).

The experience gained in various developmental programmes in this respect, undertaken by the FAO Plant Protection Service and FAO Remote Sensing Centre, with the cooperation of the FAO Commissions for Controlling the Desert Locust in northwest Africa, the Near East and Southwest Asia as well as the Regional Locust Organizations Desert Locust Control Organization for Eastern Africa (DLCO–EA) and Organisation Commune de Lutte Antiacridienne et de Lutte Anti Aviaire (OCLALAV), resulted in the formulation, development and implementation of the Africa Real Time Environmental Monitoring by using Imaging Satellites (ARTEMIS) system. The formulation and development of this system was a cooperative effort between the FAO Remote Sensing Centre, relevant FAO user divisions, NASA Goddard Space Flight Center (GSFC), the National Aerospace Laboratory of the Netherlands (NLR) and the Universities of Reading and Bristol, U.K., financially supported by the Government of the Netherlands through an FAO Trust Fund. The environmental information requirements of the FAO Global Information and Early Warning System (GIEWS) on Food and Agriculture were also taken into account.

This paper summarizes the development of the use of satellite remote sensing techniques for operational Desert Locust monitoring and forecasting from the mid-1970s to the mid-1980s. Furthermore, it describes the technical characteristics of the ARTEMIS system, its present data-processing capabilities and output products and telecommunications aspects as developed in the Data and Information Available Now in Africa (DIANA) project of the European Space Agency (ESA) in cooperation with FAO.

2. Satellite remote sensing systems for Desert Locust monitoring

The satellite systems carrying sensors with spectral, spatial, temporal and radiometric characteristics suitable for precipitation and vegetation detection and monitoring, including habitat mapping, can be broadly divided in two major groups.

(a) Environmental satellites

This group consists of both geostationary and polar orbiting satellites which are characterized by relatively low spatial (1–5 km) and high temporal (30 min–12 hrs) resolutions. The European Meteosat, U.S. Geostationary Operational Environmental Satellite (GOES) and National Atmospheric and Oceanic Administration (NOAA) Satellites and the Japanese Geostationary Meterological Satellite (GMS) belong to this group. The sensors of these satellites were designed for observing atmospheric processes to support weather forecasting, including snowcover and precipitation assessment. Moreover, the Advanced Very High Resolution Radiometer (AVHRR) sensor on the NOAA satellites has spectral and radiometric characteristics suitable for vegetation detection and monitoring. Environmental satellite data allows monitoring of dynamic atmospheric and terrestrial phenomena. Facilities for receiving and processing this type of data are relatively low cost and the spatial characteristics of the data, combined with the orbital characteristics of the satellites, permit simultaneous observations on regional and continental scales.

(b) Earth resources satellites

This group of satellites is characterized by relatively high spatial (10–80 m) and relatively low temporal (16–18 days) resolutions. The U.S. Landsat, French Satellite Probatoire d'Observation de la Terre (SPOT) and Japanese Marine Observational Satellite (MOS) polar orbiting satellites belong to this group. Their sensor characteristics make them suitable for Desert Locust habitat mapping and for detailed vegetation monitoring at a local scale.

At present, some of the above satellite systems, for example, Meteosat, NOAA, Landsat, SPOT, provide data on a fully operational basis generally within the time constraints of applications requiring realtime or near-realtime data to satisfy user information needs.

3. REMOTE SENSING TECHNIQUES AND METHODOLOGIES FOR DESERT LOCUST MONITORING

The strategy of Desert Locust plague prevention, which is coordinated and supported by FAO, is based on detection of areas in the vast recession area where enough rain has fallen to create suitable soil moisture conditions and vegetation development for egg laying, hatching and subsequent population development, using the vegetation as a source of food and shelter as well as an agent in the insect's phase transformation into gregarious behaviour. Early detection of these areas, which are highly variable in time and space, allows for effective aerial and/or ground surveys to be undertaken and preventive control of potentially dangerous populations, thus minimizing the use of environmentally dangerous insecticides and preventing the spread of locust populations to agricultural production areas.

Remote sensing of soil moisture in desert regions is possible with microwave sensors (Schmugge 1981) and by thermal techniques (Wetzel & Atlas 1981; England et al. 1983). However, no satellite-based soil moisture remote sensing system, using microwave techniques, is currently available on an operational or semi-operational basis for an area the size of the Desert Locust recession area; thermal remote sensing techniques that use Meteosat data are still in the research phase.

Remote sensing of meteorological events that cause soil moisture changes relevant for locust population development, can be undertaken from geostationary environmental satellites, e.g. Meteosat for Africa and the Near East. The high temporal frequency of this type of data in the visible and thermal infrared wavelengths permits detailed monitoring of weather systems likely to produce the rainfall necessary to initiate locust breeding.

Various schools have developed and tested remote sensing techniques for estimating rainfall quantitatively from Meteosat data. FAO has been closely involved in these developments as undertaken by the Universities of Bristol (Barrett 1977, 1980; Hielkema 1980; Barrett & Harrison 1986) and Reading (Milford & Dugdale 1987). The technique developed by the University of Reading, based exclusively on the use of Meteosat thermal infrared data, which was selected by FAO for implementation on the ARTEMIS system, is discussed in detail by Milford & Dugdale (this symposium).

TABLE 1. CHARACTERISTICS OF FOUR SATELLITE SENSOR SYSTEMS WITH CAPABILITIES FOR VEGETATION MONITORING

characteristic	Landsat MSS	Landsat TM	NOAA/AVHRR	SPOT
spatial resolution	80 m	30 m	1100 m	20/10 m
spectral resolution	4 bands	7 bands	5 bands	3 bands
useful bands for vegetation monitoring	2	4	2	2
radiometric resolution (quantitative levels)	64	256	1024	256
temporal resolution	16 days	16 days	2–3 days every 9 days	9 days
swath width	185 km	185 km	2700 km	92 km
single frame cover	34 000 km²	34 000 km²	2.10⁶ km²	8500 km²
scenes/orbits required for Desert Locust recession area coverage	700 scenes	700 scenes	7 orbits	2800 scenes

Remote sensing of green vegetation biomass is a well developed technique for which several satellites frequently collect data on a worldwide scale. The technique involves measuring reflected spectral radiance that results from the interaction between the green-leaf vegetation cover and incident solar spectral irradiance. These measurements can be made from any altitude above the surface (1 m, 10000 m, 1000 km) depending on the remote sensing system in question (ground-based, aircraft or satellite, respectively). Spectral estimation of green-leaf biomass generally involves the use of two wavelength regions, the red (0.6–0.7 µm) and near infrared (0.75–1.1 µm), although the specific wavelengths used often vary slightly between instruments. The 0.6–0.7 µm region correspond to the *in vivo* red region of chlorophyll absorption and is inversely related to chlorophyll density. In the 0.75–1.1 µm region, reflectance is proportional to green-leaf density. Ratio combinations of these two wavelength regions thus contain information related to the chlorophyll-green-leaf density (Tucker 1979). Use of these two bands for making plant canopy inferences by non-destructive techniques is facilitated by the proximity of the two bands in the electromagnetic spectrum. This proximity enables simple band ratios or other combinations to be used to compensate for differences in solar flux intensities. Linear or ratio combinations of red/near infrared spectral data have been used to quantify a variety of vegetation types (Tucker 1980; Curran 1983).

The choice of suitable satellites for remote sensing observations for particular applications depends upon satellite spectral, spatial, temporal and radiometric resolutions, orbital considerations and the spatial–spectral–temporal characteristics of the objects to be observed. Successful detection and monitoring of active Desert Locust breeding habitats requires a satellite that can, at any one time, detect the presence of green-leaf vegetation biomass with a suitable spatial resolution over very large areas, while radiometrically maintaining the spectral contrasts of the surface materials, and do so at frequent intervals (Tucker *et al.* 1985) (table 1). Moreover, satellite detection of vegetation cover in desert regions, which normally develops during the rainy season with frequent cloudcover, requires elimination of the latter for obtaining information on presence and amount of vegetation biomass. A technique for compositing cloudfree NOAA AVHRR vegetation index data from multiple daily NOAA orbits was successfully developed and tested by NASA Goddard Space Flight Center (GSFC) (Holben 1986).

Landsat Multispectral Scanner (MSS) data has been evaluated for mapping and monitoring Desert Locust habitats by Pedgley (1973) and Hielkema (1977, 1979, 1980), by using both visual and digital image analysis techniques. These studies showed that Landsat MSS data could accurately detect the presence of green vegetation and monitor its phytodynamics. In addition, Landsat MSS data can make an important contribution to the systematic mapping of potential Desert Locust habitats on through visual interpretation of suitably enhanced imagery at a scale of 1:250000. The imagery provides synoptic and yet detailed information on the geomorphology and soil characteristics of an area. At present, FAO is undertaking, following the 1986–1988 Desert Locust plague, a systematic mapping of potential Desert Locust habitats in the recession area, by using a methodology developed by Popov.

The Landsat Multispectral Scanner/Thematic Mapper (MSS/TM) and SPOT sensors were primarily designed to look at agricultural targets, which include many small fields. These sensors have excess resolution for repetitively detecting and monitoring the relevant phenomena of the 16 million km^2 Desert Locust recession area as required.

Moreover, the large number of scenes needed to cover this extensive area makes periodic monitoring with the Landsat and SPOT satellites logistically difficult and prohibitively expensive.

The AVHRR sensor on the NOAA satellites, providing data at 1.1 and 4 km spatial resolutions, is ideally suited for routine and cost-effective monitoring of vegetation development over large areas, such as the Desert Locust recession area. The vegetation monitoring capabilities of the AVHRR sensor in general are extensively documented in Justice (1986). Discussions of the consequences of the sensor's 56 degrees field of view on directional reflectance and atmospheric effects are given in Kimes (1983) and Holben & Fraser (1984), respectively.

The use of data from the AVHRR sensor for detecting and monitoring of active Desert Locust habitats at both resolutions, evaluated against high resolution Landsat and field data, has been developed and tested by Tucker *et al.* (1985) and Hielkema *et al.* (1986*b*) for test areas in West Africa and the Indo-Pakistan region. On the basis of the positive results of these studies, a NOAA AVHRR based vegetation monitoring technique, developed by NASA GSFC in cooperation with the FAO Remote Sensing Centre, was implemented on the operational ARTEMIS system (Hielkema *et al.* 1986*a*).

NOAA AVHRR data is operationally available at 4 km resolution on a global basis from the National Oceanic and Atmospheric Administration (NOAA) in the U.S.A. The full 1.1 km resolution data can be obtained from regional direct readout receiving stations as established in Maspalomas, Canary Islands, Spain; Niamey, Niger; Nairobi, Kenya and Islamabad, Pakistan, thus providing virtual total coverage of the Desert Locust recession area.

4. Artemis system characteristics

The ARTEMIS system, which became operational at the Remote Sensing Centre at FAO Headquarters in August 1988, was conceived and developed as a highly automated, dedicated satellite data acquisition, thematic processing and production system, based on the use of high-frequency Meteosat and NOAA AVHRR data. It consists of a number of integrated hardware and software components, which are summarized as follows (van Ingen Schenau *et al.* 1986).

(*a*) *Hardware*

(i) A Meteosat Primary Data User Station (PDUS) for direct, automated reception of high frequency (hourly), full resolution digital Meteosat data in the visible, thermal infrared and water vapour channels through a three-metre dish antenna, down converter and receiver. The

PDUS is controlled by a dedicated VME 68000 microprocessor and stores data on two 150 Mb disks before transfer of data through an ethernet link to the central system computer for mapping and thematic processing.

(ii) A computer unit consisting of a Hewlett-Packard 1000 A900 minicomputer, supported by a high-speed 6250/1600 bpi tape drive, three disks (132, 404 and 571 Mb), a clock, a system console, a printer and a modem.

(iii) An interactive image analysis workstation, consisting of a Ramtek interface to the HP 1000 A900 with a $1280 \times 1024 \times 16$ bit display memory, a 19-inch high-resolution colour monitor, a graphics terminal with touch screen capability, trackball and a printer.

(iv) An output subsystem, consisting of a hard colour copy camera for 10×15 inch polaroid and 35 mm film products, an HP Vectra microcomputer for IBM PC and ERDAS compatible output on floppy disk, the 6250/1600 bpi tape drive and a six-pen colour plotter.

The ARTEMIS system hardware was recently expanded with a multisync colour monitor and a matrox graphics card. This added capability allows the operation of the Integrated Land and Water Information System (ILWIS) geographic information system software package, which was developed by the International Institute for Aerospace Survey and Earth Sciences (ITC) (Meijerink *et al.* 1988) and implemented on the system in December 1988 for combination and manipulation of ARTEMIS digital data products with data from other sources.

ARTEMIS remote terminal connections within FAO Headquarters, through dedicated lines, linked to microcomputers of users, are currently being established in the Emergency Centre for Locust Operations (ECLO) (IBM compatible PC with IDA/ILWIS software and VGA/Matrox graphics), the FAO Global Information and Early Warning System (GIEWS) (IBM compatible PC with VGA graphics serving a network of microcomputers with IDA software and EGA graphics capability) and the Centralized Geographic Information System (GIS) Group (IBM AT with ERDAS/ArcInfo software). Two additional remote terminal ports to support other in-house users are still available on the present system configuration.

(b) Software

The ARTEMIS system software can be divided into the following groups, according to function (van Ingen Schenau *et al.* 1986; Hielkema *et al.* 1986*a*).

(i) Meteosat PDUS operating system and communications software for controlling data reception, quality control, quick-look display, data storage and transfer to HP 1000 A900.

(ii) HP 1000 A900 operating system software for operating the computer subsystem and related peripheral equipment.

(iii) Operational applications software for mapping and thematic processing of Meteosat and NOAA Global Area Coverage/Local Area Coverage/High Resolution Picture Transmission (GAC/LAC/HRPT) data and for generation of output products in specific user formats.

(iv) General applications software for image processing for operational and development purposes (NASA GIMMS software).

(v) Database management and communications software for manipulation and integration of various ARTEMIS system databases with other data and for retrieval and archival functions.

(vi) Software for definition and extraction of subarea and point data from the ARTEMIS product database and for archival of raw Meteosat data for point locations.

5. ARTEMIS DATA PROCESSING

The ARTEMIS system acquires and processes Meteosat thermal infrared data for rainfall estimation, according to the TAMSAT methodology of the University of Reading, chronologically as follows (van Ingen Schenau *et al.* 1988).

(i) Acquisition and pre-processing of Meteosat digital TIR data for Africa and the Near East on an hourly basis through pre-programmed Meteosat PDUS reception schedule and calibration data as provided by the European Space Operations Centre (ESOC), F.R.G.

(ii) At selected daily intervals, at 21h00 and 03h00, activated by the system clock, transfer of available digital Meteosat images to central system computer for processing.

(iii) Image selection for rainfall processing programme by using predefined image quality control parameters.

(iv) Computer mapping of selected images to ARTEMIS common geographic format, i.e. Hammer Aitoff equal area projection and resampling from 5 km input data resolution to 7.6 km ARTEMIS output product resolution.

(v) Execution of programme MIRPOINTS, which extracts digital Meteosat thermal infrared counts for specified point locations (some 1200 meteorological station and other locations in Africa) from available hourly images and archives extracted data in a database.

(vi) Execution of programme HRPRO, involving processing of available, in principle hourly, Meteosat images to determine cold cloud duration (CCD) for each pixel on the basis of satellite calibration data, cloudtop temperature thresholds for specific geographic regions and time intervals between successive images: product: daily cold cloud duration (CCD) images.

(vii) Execution of programme DAYPRO for determination of cumulated cold cloud duration (DCCD) and number of rainfall days (DNRFD) for 10-day periods; stores results in ARTEMIS database.

(viii) Execution of programme DECPRO, which converts 10 day CCD maps into 10-day estimated rainfall (DERF) maps, by using predetermined linear regression relations; stores result in ARTEMIS database.

(ix) Execution of the programme MNTPRO, once a month, which adds the three 10-day DERF and DNRFD maps and stores the created monthly map in the ARTEMIS database; furthermore, this programme compares the estimated monthly rainfall map (MERF) with long-term normal rainfall reference maps and generates absolute and relative rainfall anomaly maps (MRFANABS/REL); stores results in the ARTEMIS database.

The above nine processing steps are executed in a fully automated sequence, primarily during the evening and night hours, thus leaving the system free for other operational and developmental processing tasks during the day.

NOAA AVHRR data for vegetation assessment is acquired by FAO on magnetic tape from two sources:

(i) NOAA AVHRR Global Area Coverage (GAC) (4 km resolution) data, covering Africa, the Near East and southwest Asia, as received by the satellite during daily mid-afternoon overpasses and transmitted to the receiving and archiving facilities of NOAA in Washington D.C. After quality control by NASA GSFC, data are transmitted to FAO by courier and received for thematic processing generally within 7–10 days after data acquisition by the satellite.

(ii) NOAA AVHRR High Resolution Picture Transmission (HRPT) (1.1 km resolution) data, covering West Africa, as received by the ESA receiving station at Maspalomas, Canary Islands. Data are transmitted to FAO by express mail and generally received within 10–14 days after acquisition.

Processing of NOAA AVHRR GAC and HRPT data is done automatically by the ARTEMIS system as follows:

(i) Mapping of channel 1 (red), 2 (near infrared) and 5 (thermal infrared) data of daily available orbits to the ARTEMIS common geographic format projection and resampling of 4 km resolution input data to 7.6 km ARTEMIS product output resolution.

(ii) Calculation of normalized difference vegetation index: ch 2 + ch 1/ch 2 − ch 1 and identification of cloudcovered areas, by using thermal infrared channel 5 data.

(iii) Generation of 10-day vegetation index composites (DNDVI) by using a maximizing technique on all available daily data.

(iv) Storing of the resulting DNDVI images in the ARTEMIS database.

(v) Once a month, generation of a monthly vegetation index composite (MNDVI) through maximizing three DNDVI images.

(vi) Storing of the resulting MNDVI image in the database.

For processing of the NOAA AVHRR HRPT data at 1.1 km resolution, essentially the same sequence of processing steps is followed. At present, this is not an automated programme and requires operator intervention.

6. ARTEMIS OUTPUT PRODUCTS FOR SUPPORTING DESERT LOCUST MONITORING AND FORECASTING

Since August 1988, the ARTEMIS system produces the following products on a routine basis:

(i) 10-day and monthly cold cloud duration (D/MCCD) maps for the continent of Africa and the Near East, based on hourly Meteosat data.

(ii) 10-day and monthly estimated rainfall (D/MERF) maps for the Southern Sahara, Sahel, Sudanian and tropical countries of West Africa, obtained through regression relationships between CCD and ground observed rainfall.

(iii) 10-day and monthly estimated number of rainfall day (D/MNRFD) maps for Africa and the Near East.

(iv) 10-day and monthly NOAA AVHRR GAC based composite vegetation index (D/MNDVI) maps for Africa, southern Europe, the Near East and southwest Asia.
The above products all have a spatial resolution of 7.6 km and are available in both photographic and digital IBM PC compatible formats.

(v) 10-day NOAA AVHRR HRPT based composite vegetation index maps for the Desert Locust recession area of West Africa.

The above product has a spatial resolution of approximately 1.1 km and has been operationally available from mid-August 1989.

The ARTEMIS vegetation index products at 7.6 km resolution generally reveal only the major development of vegetation, following abundant rainfall, as clearly shown during the 1986–1988 plague in Sudan, Chad, Western Sahara, Morocco, Mali and Niger. This resolution is, however, too coarse for obtaining information on the basis of which detailed itineraries of field survey teams during recession periods can be planned.

The first experiences with the vegetation index product at 1.1 km resolution by the Centralized Desert Locust Reporting and Forecasting Service at FAO Headquarters has been very positive in that at this level of spatial resolution virtually all significant vegetation development in desert areas, relevant for supporting locust population development, is consistently registered, including that present in relatively narrow wadi systems. On the basis of this information, guidance can be provided for the planning of field operations.

7. Developmental research

The 1986–1988 Desert Locust plague, which affected large parts of Africa and the Near East, led to a number of initiatives by FAO, assisted by donor countries, concerning improvement of survey and control methods as well as technical and organizational infrastructures for preventing Desert Locust upsurges. One of these was the formulation and execution of a programme, financially supported by the Government of Belgium, to complete the systematic mapping of potential Desert Locust habitats in the recession area using available historical locust breeding records with Landsat imagery and to undertake a systematic evaluation of the vegetation monitoring capabilities of the NOAA satellites. For the latter programme component, the FAO Remote Sensing Centre jointly with the Plant Protection Service implemented a data collection programme between August and November 1989 in the Tamesna of Niger, consisting of the following activities.

Selection and regular ground monitoring of ecological characteristics and changes of 14 representative field sites using visual and hand-held spectral radiometer observations.

Systematic acquisition of the following satellite data:

NOAA AVHRR GAC data at 4 km resolution (2–3 times, every 9 days)
NOAA AVHRR HRPT data at 1.1 km resolution (2–3 times, every 9 days)
MOS data at 50 m resolution (August and October 1989)
multispectral video remote sensing observations from light aircraft (October 1989).

The acquired data sets are currently being processed to establish a quantitative calibration for NOAA GAC and HRPT data and to define the lower limits of the vegetation detection and monitoring capability of this satellite. The results of this research will be available in April 1990.

8. Satellite telecommunications development: the DIANA project

The ARTEMIS system, being located in Rome, serves primarily the environmental information needs of the Global Information and Early Warning System (GIEWS) and Emergency Centre for Locust Operations (ECLO) of FAO, which communicate analysed information, combined from a number of sources, to recipient and donor countries in the form of telexed summaries and GIEWS and ECLO bulletins and publications.

Rapid communication of the high volume ARTEMIS system information products to users at regional and national levels, requires special communications facilities.

Jointly with the European Space Agency (ESA), the FAO Remote Sensing Centre has formulated a project for the development, implementation and demonstration of a dedicated satellite communications system, DIANA, based on low-cost microcomputer technology and by using the commercial Intelsat satellites through the facilities of Telespazio in Fucino, Italy (Hielkema 1988).

The DIANA system will provide a capability for high speed, two-way transfer of facsimile images of documents and maps, character-coded text documents and digital images in raw or processed form from FAO Headquarters to microcomputer based receiver terminals of users at regional or national level.

This capability will enable regional organizations and national plant protection services to receive processed satellite information and other relevant communications from FAO Headquarters in near-realtime.

The DIANA system, which is funded by ESA member countries Italy, Belgium, Spain, Ireland, Norway, U.K. and Finland, is being developed industrially under ESA management.

This phase will be completed by October 1990, followed by a testing and training period. Subsequently four DIANA terminal stations, linked to the central hub station at FAO Headquarters, will be installed at selected locations in Africa and demonstrated for various applications for the period of one year. Presently confirmed DIANA terminal locations are: Nairobi, Kenya and Harare, Zimbabwe. Candidate locations are Accra, Ghana; Addis Ababa, Ethiopia, and Djibouti, Djibouti.

REFERENCES

Barrett, E. C. 1977 Mapping rainfall from conventional data and weather satellite imagery across Algeria, Libya, Morocco and Tunisia. AGP/DL/RS/1/77. Rome: FAO.

Barrett, E. C. 1980 An operational method for rainfall monitoring in Northwest Africa. Consultant Report GCP/INT/349/U.S.A. Rome: FAO.

Barrett, E. C. & Harrison, A. R. 1986 Rainfall monitoring by Meteosat in Africa. Consultants Report GCP/INT/432/NET. Rome: FAO.

Curran, P. J. 1983 Multispectral remote sensing for the estimation of greenleaf area. *Phil. Trans. R. Soc.* B **309**, 257.

England, C. E., Gombeer, R., Hechinger, E., Herschy, R. W., Rosema, A. & Stroosnijder, L. 1983 Application of Meteosat data for rainfall, evaporation, soil moisture and plant-growth monitoring in Africa. *ESA J.* **3**.

Hielkema, J. U. 1977 Application of Landsat data in desert locust survey and control. AGP/LCC/77/11. Rome: FAO.

Hielkema, J. U. 1979 Advanced training and research on satellite remote sensing techniques and applications in the United Kingdom and the United States. AGLT/RSU series 2/79. Rome: FAO.

Hielkema, J. U. 1980 Remote sensing techniques and methodologies for monitoring ecological conditions for desert locust population development. FAO/USAID Final Report. GCP/INT/349/USA. Rome: FAO.

Hielkema, J. U. 1988 Operational Environmental Monitoring and Information Dissemination by Satellite at FAO: The ARTEMIS and DIANA Systems: *USAID/NOAA International Workshop on Satellite Techniques for Estimating Precipitation, Washington, D.C., December 1988.*

Hielkema, J. U. & Howard, J. A. 1976 Application of remote sensing techniques for desert locust survey and control. AGP: LCC/76/4. Rome: FAO.

Hielkema, J. U., Howard, J. A., Tucker, C. J. & van Ingen Schenau, H. A. 1986a The FAO/NASA/NLR/ARTEMIS system: An integrated concept for environmental monitoring by satellite in support of food/feed security and Desert Locust surveillance. *Proceedings 20th ERIM Symposium on Remote Sensing of Environment, Nairobi, Kenya, December 1986.*

Hielkema, J. U., Roffey, J. & Tucker, C. J. 1986b Assessment of ecological conditions associated with the 1980/81 Desert Locust plague upsurge in West Africa using environmental satellite data. *Int. J. Remote Sensing* **6**, 1609.

Holben, B. N. 1986 Characteristics of maximum-value composite images from temporal AVHRR data: *Int. J. Remote Sensing* **7**, 1417–1434.

Holben, B. N. & Fraser, R. S. 1984 Red and near-infrared sensor response to off-nadir viewing. *Int. J. Remote Sensing* **5**, 145.

Justice, C. O. (ed) 1986 Monitoring the grasslands of semi-arid Africa using NOAA AVHRR data: *Int. J. Remote Sensing* **7**.

Kimes, D. S. 1983 Dynamics of directional reflectance factor distributions for vegetation canopies. *Appl. Opt.* **22**, 1364.

Meijerink, A. M. J., Valenzuela, C. & Stewart, A. 1988 ILWIS: The Integrated Land and Watershed Management Information System: *ITC Publication no. 7.* Enschede, Netherlands: ITC.

Milford, J. R. & Dugdale, G. 1987 Rainfall mapping over West Africa in 1986, 1987. Consultants Report GCP/INT/432/NET. Rome: FAO.
Pedgley, D. E. 1973 Testing feasibility of detecting locust breeding sites by satellite. Final report to NASA. London: COPR.
Roffey, J. 1975 Programme plan for improving desert locust survey and control by satellite remote sensing. Consultants report. Rome: FAO.
Schmugge, T. 1981 Remote sensing of soil moisture with microwave radiometers. Paper 81-4503, *Am. Soc. agr. Engng* 1981 Winter Meeting, Chicago.
Tucker, C. J. 1979 Red and photographic infrared linear combinations for monitoring vegetation. *Remote Sensing Environ.* **8**, 127.
Tucker, C. J. 1980 A critical view of remote sensing and other methods for non-destructive estimation of standing crop biomass. *Grass Forage Sci.* **35**, 177.
Tucker, C. J., Hielkema, J. U. & Roffey, J. 1985 The potential of satellite remote sensing of ecological conditions for survey and forecasting desert locust activity. *Int. J. Remote Sensing* **6**, 127–138.
van Ingen Schenau, H. A., Nicolai, R. J., Venema, J. C., van der Laan, F. B. & Versteeg, M. 1986 System definition of 'Africa real time environmental monitoring using imaging satellites' (ARTEMIS): *National Aerospace Laboratory of the Netherlands (NLR).*
van Ingen Schenau, H. A., Venema, J. C., Spaa, J. & Hielkema, J. U. 1988 ARTEMIS System Overview and Presentation of ARTEMIS Products. *Paper ITC/FAO Seminar, 14–25 November 1988, Enschede, Netherlands.*
Wetzel, P. J. & Atlas, D. 1981 Inference of precipitation through thermal measurement of soil moisture. Precipitation measurements from space. Workshop Report, April 28–May 1, 1981, NASA/Goddard Space Flight Center, pp. 170–172.

Discussion

G. B. Popov (*Food and Agriculture Organization of the United Nations, Rome, Italy*). Appropriate remote sensing techniques have great potential for making surface and aerial scouting for locusts much more efficient. This is important as national and regional control organizations are often hard pressed to scout vast, remote, and potentially infested areas with severely constrained resources. Any objective guidance on the areas most likely to be infested is most valuable for the best allocation of the scarce resources available.

J. B. Williams (*Overseas Development Natural Resources Institute, Chatham, U.K.*). Migrant pests often inhabit large, remote and inaccessible regions where few people live. Appropriate remote sensing techniques for detecting feeding and breeding and migration areas would clearly be of great advantage for planning efficient aerial and ground surveys for control. To be useful, such techniques need to be accurate, inexpensive, accessible and comprehensible to scientists and decision makers in the field.

Mr Hielkema recognized several years ago that orbiting NOAA satellite 'vegetation indices' and geostationary Meteosat satellite cold cloud/rainfall estimation techniques had much to offer to both crop production and insect pest early warning systems (especially in Africa where the ground observation network is particularly weak). Much to Mr Hielkema's credit he did something about it, and created the centralized ARTEMIS system in Rome from which it is intended that data should be disseminated to a variety of users by satellite in the DIANA project. This comment seeks to evaluate how useful, accurate, accessible, comprehensible and (in)expensive the information so produced really is, and asks to what extent the present system is sustainable.

Useful. Without doubt, prompt and reliable (to a defined degree) regional and local estimates of area rainfall and vegetation change are of the greatest value to Desert (and other) Locust, armyworm and weaver-bird control programmes. In some instances the degree of spatial and quantitative precision required is quite small, and negative (i.e. no rain) information alone can be most useful in itself.

Accuracy. (*a*) Rainfall estimation: there are reservations among the French and British research community, because the methods used by FAO (developed by ODA at Reading) do have limitations and they have not yet been validated in more than half a dozen countries in Africa. (*b*) Vegetation indices are useful for a general view, especially in the more dense biomes, but NOAA (1.1 km maximum resolution) has not yet been shown to be of much value in extremely vegetation-sparse areas where Desert Locusts may thrive. Full resolution images are not available for most of the continent yet.

Accessible. FAO dissemination plans are proceeding: DIANA intends data to reach a much wider user community but duplicates existing WMO dissemination channels. Once regional centres in Niamey, Nairobi and Harare are operational, full resolution images will be more easily available. A major question concerns the best routing of this data into national systems where operational decisions are taken.

Comprehensible. At present, few users understand the value and limitations of the images (especially the vegetation maps). When they do use them they tend to believe them as 'gospel'. Then they find that there are errors, and reject the baby with the bathwater. FAO is organizing short training courses but through Remote Sensing Centres, which tend to concentrate on mapping rather than meteorology. The images do need interpretation and workers in agricultural departments do not always have the training or experience to be able to cope with regional/synoptic scales. Meteorological departments provide an obvious route, but applied meteorology is an area where FAO and WMO have joint responsibility and where cooperation frequently leaves much to be desired.

Cost. The costs of producing satellite receiving systems are now very low (and are becoming available to departments in developing countries: portable systems will be available shortly) and running costs can be negligibly small. If the information is accurate, accessible and comprehensible it is an extremely cost effective way of estimating areal rainfall and vegetation development, and can easily be undertaken *in situ*, with the added benefits of national participation in product development. Collaboration between national, regional and continental operations is essential.

Sustainable. Continental and Regional centres (apart perhaps from DLCO-EA) do not have a happy record in Africa. Many national meteorological observing networks have never been weaker. Satellite observations are not yet sufficiently good to replace ground stations: the two methods are most powerful in their complementarity, and must be seen to be so. If the ARTEMIS/DIANA system is to be sustainable and effective it will have to focus more on improving product acceptance and use at national level in conjunction with indigenous meteorological services. Some of the difficulties experienced in establishment of the regional centres in Africa reflect the weakness of a glossy, external, top-down approach to what are sensitive national issues. More cooperation between FAO and WMO is urged.

J. HIELKEMA. In reply to Dr Williams, I can assure him that there has been contact between FAO and WMO about the developments envisaged in DIANA and there will be even greater collaboration in the future. The ARTEMIS project is preparing an active training programme for national participation through their regional centres in Nairobi and Harare.

R. W. SAUNDERS (*Meteorological Research Flight, Y46 Building, Royal Aerospace Establishment, Farnborough, Hants, U.K.*). First, I must underline the importance of the ARTEMIS system described by Mr Hielkema as being operational. It is a major step forward to have a system that can be used for monitoring meteorological and surface parameters from satellite data on a daily basis throughout the year, and this takes a great deal of effort to implement. This is reflected in the U.S. $2 million cost of the system.

However, when interpreting the parameters derived from ARTEMIS, one should bear in mind the limitations of the retrieval algorithms used to derive those parameters. For instance, for the normalized vegetation index derived from AVHRR data to be valid, a cloud free and aerosol free atmosphere between the satellite and the surface must be assumed. Other workers have shown spurious vegetation indices derived from AVHRR data due to inadequate cloud clearing of the data, viewing angle effects or high atmospheric turbidities. In addition water vapour absorption is significant in AVHRR channel 2 ($\sim 90\%$ transmittance) and this varies with total column water amount. Are you confident that your cloud clearing scheme is adequate and do you correct for viewing angle effects and water vapour absorption? Do you find the occurrence of Saharan dust limits the availability of the Normalized Difference Vegetation Index (NDVI) product over northern Africa?

Finally, I believe that the inclusion of passive microwave data from the new generation of atmospheric sounders due in 3–4 years time should be considered for the ARTEMIS system. These data have the potential at least to improve the rainfall estimates presently derived only from the Meteosat infrared data.

J. HIELKEMA. In reply to Dr Saunders' point, I accept that the limitations that you have raised are real, but they are not usually important at the operational scale generally used by ARTEMIS. We are looking at ways, such as the use of passive microwaves, to improve the system.

J. R. MILFORD (*Department of Meteorology, University of Reading, U.K.*). I congratulate Mr Hielkema on his achievement in getting the ARTEMIS system to its operational status in the face of considerable difficulties over a long period since its conception.

Mr Hielkema quoted one figure showing the success of the rainfall estimates produced by the system, and this sounded impressive. However, much more detail is needed to show whether the estimates were actually able to provide the information needed by a specific user with adequate reliability. The evaluation of such estimates (as of weather forecasts) is difficult to carry out objectively, and I feel that much more needs to be done to establish the viability of these deductions from satellite data. Each of us tends to be satisfied as soon as we have the result which we wanted to show, and truly objective assessments are rare.

Another feature of the ARTEMIS programme as reported here is that the products are still some way from the end-user. The DIANA system will provide a major step forward, but there is still a reluctance to integrate the outputs with other meteorological information, and, indeed, to work sufficiently closely with meteorological organizations generally.

PROGRESS IN THE INTERNATIONAL CONTROL OF ANOTHER MAJOR MIGRANT PEST: THE BLACKFLY VECTOR OF RIVER-BLINDNESS

Introductory remarks

By L. G. Goodwin, F.R.S.

Shepperlands Farm, Park Lane, Finchampstead, Wokingham RG11 4QF, U.K.

Onchocerca volvulus is a filarial worm, transmitted by blackflies (Diptera: Simulidae), which causes severe disease and blindness in Africa and parts of South America. The worms are threadlike; the male is usually only a few inches long, but the female grows up to 2 feet (*ca.* 0.61 m) in length. They lie paired, coiled and knotted together in nodules under the skin. The females produce millions of microscopic larvae that swim in the tissue spaces of the skin, waiting for a blackfly to pick them up and transfer them to a new host. In heavy infections they can reach a density of a hundred or more in every milligram of skin. They cause irritation, severe dermatitis and, in the eye, inflammation, fibrosis and blindness. In the past, blackflies have driven the people away from many fertile valleys of the Volta River basin. Their immature stages develop only in fast-flowing stretches of rivers, and so control methods rely on killing the larvae by aerial spraying of selected targets along river courses.

The following two papers deal, first, with the immensely successful Onchocerciasis Control Programme (OCP) of the World Health Organization (WHO), based on good science, good insecticides and a good anti-helminthic, and then with the threat to the Programme from reinvasions of blackflies from neighbouring territories into the OCP area. It was such reinvasions that led to the discovery that savanna species of the *Simulium damnosum* species complex, the principal vectors in West Africa, could migrate at least 500 km, assisted by the prevailing winds.

Phil. Trans. R. Soc. Lond. B **328**, 721–729 (1990)

Printed in Great Britain

721

The WHO Onchocerciasis Control Programme: Retrospect and Prospects

By R. Le Berre[1], J. F. Walsh[2], B. Philippon[2]†, P. Poudiougo[2], J. E. E. Henderickx[2], P. Guillet[2], A. Sékétéli[2], D. Quillévéré[2], J. Grunewald[2] and R. A. Cheke[2]

[1] *World Health Organization, 1211 Geneva 27, Switzerland*
[2] *WHO Onchocerciasis Control Programme in West Africa, B.P. 549, Ouagadougou, Burkina Faso*
Overseas Development Natural Resources Institute, Chatham Maritime, Chatham,
Kent ME4 4TB, U.K.

The history of onchocerciasis control in Africa and the genesis of the WHO Onchocerciasis Control Programme in West Africa (OCP) are briefly reviewed. The importance of experience gained in anti-locust campaigns in helping to plan the OCP is stressed. Members of the *Simulium damnosum* species complex are the vectors of onchocerciasis, which OCP is controlling with insecticide treatments on the stretches of rivers where the *Simulium* breed. Migrations of flies have been responsible for reinfestations of controlled areas and the spread of insecticide resistance. The management of these problems and related research are described, but it is emphasized that despite setbacks OCP is achieving its aims. A strategy for the future is outlined: vector control supplemented by chemotherapy is expected to continue until the year 2004.

1. Introduction

Onchocerciasis, or 'river blindness', is caused by a nematode worm (*Onchocerca volvulus* (Leuckart)), which is transmitted from person to person by its blackfly vector, *Simulium damnosum* Theobald. The World Health Organization (WHO) Onchocerciasis Control Programme (OCP) in West Africa is interrupting disease transmission by killing the fly's larvae, which inhabit fast-flowing sections of rivers.

At the first meeting on Migrant Pests (1977) several of the present authors (Le Berre *et al.* 1979) of this article described the Onchocerciasis Control Programme executed by WHO, its structure, its first results after three years and its problems, especially the reinvasion of treated areas. Twelve years later, we feel it is timely to present the results obtained so far, the problems faced, with the exception of the reinvasion, which will be discussed by Baker *et al.* (1990), and the way they were solved. We will also elaborate on the future of what can be considered as a successful venture.

2. The pre-OCP era

Before the late 1950s, onchocerciasis and its control were not considered as priorities. Several reasons accounted for this: onchocerciasis was a discrete disease at the 'end of the track' and other diseases (sleeping-sickness, leprosy) still had higher priority; there was no drug, and vector control seemed to be difficult to achieve despite the eradication of *Simulium*

† Present address: ORSTOM, 213 Rue Lafayette, Paris 75010, France.

[203]

neavei obtained by McMahon *et al.* (1958), in Kenya; WHO as a whole was deeply engaged in the 'Malaria Eradication Programme'; more simply, there were very few scientists interested in the problem, with the noticeable exceptions, on the vector side, of D. J. Lewis, M. Ovazza and R. W. Crosskey.

This initial period was followed by an era of intense research carried out by two complementary groups of scientists, one British (Medical Research Council, MRC) and the other French (Office de la Recherche Scientifique et Technique Outre-Mer, ORSTOM). The MRC group based in Kumba (Cameroun) dealt mainly with different patterns of transmission of the parasite, *Onchocerca volvulus*, epidemiology of the disease and drug trials (Duke 1957; Duke *et al.* 1975). The ORSTOM group, based in Bobo Dioulasso (then Upper Volta, now Burkina Faso), covered successively many aspects of the biology of *Simulium damnosum*, which was identified as a complex of species by Vajime and Dunbar (1975): its bionomics, behaviour, ecology in relation to disease transmission, and vector control (Ovazza *et al.* 1965; Le Berre 1966; Philippon 1977; Quillévéré 1979).

It was the huge increase in knowledge gained by these groups and other more isolated, scientists (Walsh 1970) that allowed the planning and implementation of *S. damnosum* control operations in small areas during limited periods of time (1–6 years). Two of these campaigns were successively carried out inside the present OCP area. The first (1962–1964) was sponsored by French Cooperation. It concerned a small area at the border of Upper Volta and Mali, including the Farako Valley in Mali. The second, financed by the European Communities, began in 1965. It covered successively the Farako area of Mali, the southwest zone of Burkina Faso and the Northern Bandama in Côte d'Ivoire.

As already stated by Le Berre *et al.* (1979) 'earlier campaigns all had the effect of being too limited in time and space; with perhaps one exception (Farako), the inadequate extent of the treated areas has resulted in transmission, by immigrant female flies, being maintained at too high a level to be acceptable for resettlement of the deserted valleys'.

Although showing limited success, these campaigns, often considered as experimental by the scientists involved, permitted the collection of a bulk of additional information on the vector, the *S. damnosum* species complex, the control strategies, methods and techniques. Last but not least, note was taken of earlier mistakes, not to be repeated, about the size and duration of such operations.

In 1968, aware of the progress made towards the control of what they considered as a major health and socio-economic problem, Ansari (WHO) and Richet (Organisation pour la Collaboration et la Coopération pour la lutte contre les Grandes Endémies (OCCGE)) organized a meeting in Tunis (Chairman: Janssens, Antwerp) to discuss the following questions. Is Onchocerciasis a major problem? Can it be controlled? Where, when, how, for how long, how much will it cost? Except for the very last question, all were answered positively. The Tunis meeting was really the start of the OCP.

In 1972, at the official request of seven Heads of States of West Africa, WHO organized a mission of 'preparatory assistance to Government' (the PAG mission, 1973) in charge of planning and costing a campaign now known as the OCP.

3. The OCP

Throughout its 15 years' existence, the OCP has faced many problems. Two major ones relevant to vector control have been the reinvasion by immigrating flies, sometimes infected, and resistance to insecticides. The first problem will be dealt with by Baker in his presentation to this meeting (Baker *et al.* 1990).

The second, resistance to insecticides, was detected in 1980 (Guillet *et al.* 1980) on the lower part of the Bandama river, Côte d'Ivoire. This resistance of the 'forest' species (*S. sanctipauli* and *S. soubrense*) of the *S. damnosum* complex to the initial insecticide, temephos (Abate (R)) spread rapidly, mainly in Côte d'Ivoire and Ghana. In 1983, the 'savanna' species (*S. damnosum sensu stricto* and *S. sirbanum*), those responsible for severe, blinding, onchocerciasis, also became resistant to temephos (Kurtak & Ouédraogo 1984). Resistance of these 'savannah' species also spread quickly, to reach a large part of their area of distribution, including the non-treated area located next to the OCP, i.e. the western extension area. These species also became resistant, intermittently or permanently, to another organo-phosphate insecticide, chlorphoxim. These resistances necessitated the use, under certain circumstances (low river discharge), of an insecticide of biological origin, *Bacillus thuringiensis* serotype H 14, highly selective for *Simulium* and certain mosquitoes. At present, the OCP is implementing a highly sophisticated strategy of vector control, by using alternatively five different insecticides, depending on river profiles and discharges: temephos, chlorphoxim, *B. thuringiensis* H 14, carbosulfan and permethrin. The rotational use of these insecticides, a real burden to the Vector Control Unit (VCU) of the OCP, represents the only major exercise of resistance management in the world.

The effect of insecticides on the environment has been closely monitored by independent groups under the authority of an Ecological Committee: the results have always been considered to be satisfactory.

4. Results

Despite these problems, the OCP has shown outstanding results regarding vector control, reduction of transmission and, eventually, disease and blindness. After 15 years of control, 20 million people have been protected against infection; seven million children, born within the area, have not been at risk of blindness due to onchocerciasis; 100000 people have been protected from going blind.

The socio-economic benefits have not been evaluated as completely as they should have been. However, many river valleys deserted by people are now repopulated, the density of people in some areas being so high, and the onchocerciasis prevalence so low, that vector control has been, or could have been, interrupted. To give an example, all Burkina Faso, apart from 10 km of river, has not been treated at all since 1987.

5. Research

The OCP is still regarded by the authors as an experimental operation, constantly adapting its tactics, and sometimes its strategy, to deal with forecast or unforeseen events and, last but not least, to discover new tools.

Like similar operations (i.e. locust control), one of the reasons for its success is that the OCP has produced, and provoked, a huge amount of research. The main areas of research have been, and still are: insecticides and formulations; this has always been *the* priority of the OCP. In addition to the five compounds already mentioned, several chemical insecticides of different classes (pyrethroids, organophosphates) and new formulations, both chemical and biological (*B.t.* H 14), are being tested presently. Resistance to insecticides is now being monitored through methods developed by the OCP and collaborative institutions. For operational use of insecticides, diagrams have been produced by the OCP to allow optimization procedures for dosage of each insecticide and treatment interval. Teletransmission of hydrological information, to allow precise dosages of insecticides; this method, developed by ORSTOM, has now become operational in all OCP river basins, including the extension areas. Experiments in disease transmission continue. Identification of *O. volvulus* larvae in their respective vectors is considered a priority, as it will allow a more precise delimitation of the programme boundaries. Modelling: a model has been quantified and tested by OCP in collaboration with the University of Rotterdam. Simulations of the impact of the vector control imply that 14 years of interruption of transmission will reduce the local parasite reservoir to such a low level that recrudescence of the infection is extremely unlikely if larviciding is stopped and the vector is allowed to return. Thirteen years appears to be the breakpoint and recrudescence is certain after 12 years of control. Chemotherapy: temephos, the initial insecticide had been found, not searched for. The same applies to Ivermectin, an antihelminthic compound developed for veterinary purposes. Tested on onchocerciasis patients (collaborations between OCP, the UNDP/World Bank/WHO Special Programme for Research and Training in Tropical Diseases (TDR) and many research Institutes), Ivermectin has shown to be an excellent micro-filaricide without major side-effects. This drug has been given, progressively, with extreme precautions to more than 100000 patients, with great success. It has the following advantages: very low dosage, oral, once a year to selected patients; exclusive criteria are related to pregnancy, lactation, weight and/or age. The discovery of this drug will obviously influence the strategy of onchocerciasis control inside the OCP area, especially the extension zones and outside, in those many countries that have not benefitted, and will never benefit, from another OCP.

6. TRAINING

As expected, the OCP has trained over 200 nationals in many disciplines, including entomology, hydrology, parasitology, ophthalmology, computer sciences and administration.

7. THE OCP EXTENSIONS

Success calls for success. In 1977, countries located to the west of the seven 'OCP countries' asked to be included in a larger OCP. The western extension, including Guinea, Guinea Bissau, Senegal and Sierra Leone, plus the remaining part of Mali, is now under progressive treatment with the following effects: countries are protected for themselves; the reinvasion is now restricted, and will be more so.

Similar benefits arise from the southeastern extension in Benin, Togo and Ghana.

8. THE FUTURE OF THE OCP

The central area of the initial zone (mainly Burkina Faso) is not treated any more, which concurs with the modelling results, mentioned above, concerning the period needed for protection. The regions located east and west of this core area will still be treated until they also meet the criteria required to permit a cessation of control measures.

As regards the future strategy, the last report of the OCP Expert Advisory Committee provides a good summary:

Since the inception in 1974 of the Onchocerciasis Control Programme (OCP), the control strategy has been based on larviciding aiming at virtual interruption of the transmission of the parasite and resulting eventually in the elimination of the human reservoir of the onchocercal worm within the Programme area.

When Ivermectin became available in 1987 for use in human populations, the possibility was raised that large-scale distribution of the drug could contribute to reducing transmission, and might eventually replace larviciding as a control agent. However, recent extensive field trials and epidemiological model projections demonstrate that Ivermectin is unlikely to contribute substantially to transmission control.

Consequently, the Expert Advisory Committee (EAC) recommends that the Programme should continue to rely on larviciding for the control of transmission with the use of Ivermectin to reduce morbidity in individuals, in particular to rapidly reduce high microfilarial loads and thereby avert impending blindness in heavily infected communities. Thus, in the extension areas the drug would be used widely for the treatment of infected cases.

The Original Programme area

When the original plan for the Programme was prepared, the Donors and the Participating Countries agreed in principle that OCP would operate for twenty years in the seven countries in order to reach its objective i.e., to eliminate onchocerciasis as a disease of public health and socioeconomic importance throughout the Programme area, and to ensure that there be no recrudescence of the disease thereafter.

This objective will be attained in the Original OCP area as originally scheduled, by 1994. Full control has been virtually secured throughout that seven country area. Larviciding came to an end in the northern one third of the area in 1986 and will have ceased in most of the remaining zones in the Original area by 1990.

After that date entomological surveillance will continue for another two years and larviciding will be confined to limited operations in two circumscribed zones where transmission had only been partly controlled due to operational difficulties. It is expected that such limited vector control will be completed in 1994 by which time the Programme will have withdrawn entirely from an area considerably larger than the Original area as defined in 1974. The remaining reinvaded regions in the southeast and in the west will have become operationally integrated into the Extension areas.

The Extension areas

The OCP operations planned for the Southern and Western Extension areas involving four additional countries could be seen as a consecutive programme which is distinct operationally and in its timeframe from the programme carried out in the Original OCP area, but which is, nevertheless, necessary to protect the achievements in the Original area.

In view of the fact that Ivermectin did not appear to be effective in interrupting the transmission of onchocerciasis, the Committee recommends that as larviciding was the proven method of onchocerciasis control, resulting in immediate interruption of transmission, the Programme should embark upon full-scale larviciding operations in the Extension areas.

However, recent epidemiological investigations in areas where larviciding has been conducted since 1974, supported by predictions of the epidemiological model, have now established that up to 14 years of vector control is required to eliminate the human reservoir of the onchocercal parasite as against the eleven years estimated.

Given the delay in the commencement of vector control operations in the Extension areas, and the limitations of Ivermectin in controlling transmission, OCP will need to continue vector control operations in parts of the Extension areas beyond 1997 (2004). However, by then the Programme's objective of eliminating onchocerciasis as a disease of public health and socioeconomic importance will have been achieved in more than two thirds of the eleven-country Programme area, and OCP will be actively phasing out its operations.'

9. Conclusions

The design of the OCP during the early 1970s was facilitated by the example given by the locust control operations carried out in West and East Africa: aerial treatments, radio-communication networks, and wide distributions of control and evaluation units in the field are still considered as keys to the success of the OCP, as was seen for locust control.

Locust control organizations declined after the mid 1960s, when the plague had given way to recession and international and national support progressively vanished. It has often been said that the OCP would finish the same way: what will come after? The authors strongly believe that all the information gained by the OCP regarding onchocerciasis guarantees that the 'after' will be clear and serene: vector control is efficient in almost suppressing transmission, if carried out for 14 years; vector control in extension areas will considerably reduce reinvasion; research is always considered as a vital part of the Programme; last but not least, a good and safe drug is now available for controlling the disease.

Even inside WHO, the OCP has often been regarded as a monolithic, vertical, expensive programme. A recent study by the World Bank, one of the supporting agencies of the OCP, has shown that the protection of an individual per year amounts to U.S. $0.45, compared with the miserable U.S. $1 available at the Ministry of Health 'for all other diseases'. What the World Bank's report does not say is that, for once, these U.S. $0.45 are going directly to the poorest of the poor, those living at the end of the track.

References

Duke, B. O. L. 1957 The reappearance rate of increase and distribution of the micro-filariae of *Onchocerca volvulus* following treatment with diethylcarbamazine. *Trans. R. Soc. trop. Med. Hyg.* **51**, 37–44.

Duke, B. O. L., Anderson, J. & Fuglsang, H. 1975 The *Onchocerca volvulus* transmission potentials and associated patterns of Onchocerciasis at four Cameroon Sudan-Savannah villages. *Tropenmed. Parasit.* **26**, 143–154.

Guillet, P., Escaffre, H., Ouédraogo, M. & Quillévéré, D. 1980 Mise en évidence d'une résistance au téméphos dans le complexe *Simulium damnosum* (*S. sanctipauli* et *S. soubrense*) en Côte d'Ivoire. *Cah. ORSTOM, sér. Ent. méd. Parasitol.* **18**, 291–298.

Kurtak, D. C. & Ouédraogo, M. 1984 Detection of resistance to temephos and chlorphoxim in *Simulium damnosum* s.l. by topical application to adults. *Mosquito News* **44**, 518–527.

Le Berre, R. 1966 Contribution à l'étude biologique et écologique de *Simulium damnosum* Theobald, 1903 (Diptera: Simuliidae). *Mém. ORSTOM* **17**, 204.

Le Berre, R., Garms, R., Davies, J. B., Walsh, J. F. & Philippon, B. 1979 Displacements of *Simulium damnosum* and strategy of control against onchocerciasis. *Phil. Trans. R. Soc. Lond.* B **287**, 277–288.

McMahon, J. P., Highton, R. B. & Goiny, H. 1958 The eradication of *Simulium neavei* from Kenya. *Bull. Wld Hlth Org.* **19**, 75–107.

Ovazza, M., Ovazza, L. & Balay, G. 1965 Etude des populations de *Simulium damnosum* Theobald, 1903 (Diptera: Simuliidae) en zone de gîtes non permanents. *Bull. Soc. Path. exot.* **58**, 1118–1154.

Philippon, B. 1977 Etude de la transmission d'*Onchocerca volvulus* (Leuckart, 1893) (*Nematoda, Onchocercidae*) par *Simulium damnosum* Theobald, 1903 (Diptera: Simuliidae) en Afrique tropicale. *Trav. Docums ORSTOM* **63**, 308 p.

Quillévéré, D. 1979 Contribution à l'étude des caractéristiques taxonomiques, bioécologiques et vectrices des membres du complexe *Simulium damnosum* présents en Côte d'Ivoire. *Trav. Docums ORSTOM* **109**, 304 p.

Vajime, C. G. & Dunbar, R. W. 1975 Chromosomal identification of eight species of the subgenus *Edwardsellum* near and including *Simulium (Edwardsellum) damnosum* Theobald (Diptera: Simuliidae). *Tropenmed. Parasit.* **26**, 111–138.

Walsh, J. F. 1970 Evidence of reduced susceptibility to DDT in controlling *Simulium damnosum* on the river Niger. *Bull. Wld Hlth Org.* **43**, 316–318.

Discussion

J. F. WALSH (*WHO/OCP, Kara, Togo*). Dr Le Berre has summarized the vast and long-running Onchocerciasis Control Programme (OCP) operations and has shown the considerable degree of success attained. Dr Baker will discuss the achievements in the Western Extension Area. Since the beginning of 1988, larviciding activities have also been extended into an area of 115000 km² in southern Togo, Benin and southern Ghana east of Lake Volta, which is referred to in the OCP as the Southern Extension Area. In this area we have moved from an all-out attack to a more selective approach, both temporally and spatially, in fact, from vector control to vector management. This has become feasible with our increasing knowledge of the biology of the different members of the vector complex, and is partly a necessary response to operating in an area where one of the vectors, *S. squamosum*, which is fortunately a rather inefficient vector there, breeds in small, heavily wooded montane streams. In February 1988 all known breeding sites of *S. damnosum s. str.*, *S. sirbanum*, *S. sanctipauli* (Djodji form) and *S. soubrense* (Beffa form) in the area were brought under larviciding operations. No attempt was made to cover all breeding sites of *S. squamosum*, and *S. yahense* was not considered as a target. Both the Beffa and Djodji forms have very limited dry season distributions in the OCP area and control was intended to eradicate these species, at least temporarily (Walsh *et al.* 1987).

Within eight weeks of commencing larviciding, the Djodji form disappeared from the area east of Lake Volta and no form of *S. sanctipauli* has been taken there since that time, though larviciding has been interrupted for lengthy periods. *Simulium sanctipauli*, including the Djodji form, which typically inhabits larger forest streams and rivers, was not thought to be a long distance migrant (Garms & Walsh 1987), although it is capable of considerable extension of range northwards during the rainy season (Garms *et al.* 1990). Large uncontrolled populations of *S. sanctipauli* breed to the southwest of the Southern Extension, 250 km away. This distance, which includes a 20–30 km crossing of Lake Volta, appears to be too great for recolonization to occur readily.

The elimination of the Djodji form has resulted in a reduction of the biting fly population in its original distribution area, even when control is interrupted for lengthy periods. Although *S. squamosum* rapidly recolonizes rivers that are taken off the larviciding schedule, its larval populations under these circumstances are scattered, and the high larval densities typical of the Djodji form are not reached. In addition, *S. squamosum* is a much less efficient vector than was the Djodji form (Cheke & Denke 1988), so that transmission rates have declined even more than biting rates. As against these beneficial effects of the larviciding, *S. yahense* has shown a tendency to spread into the treated rivers, once larviciding is interrupted, and though it has not yet become a serious problem it is an excellent vector in the area.

The Beffa form of *S. soubrense* was also temporarily eliminated from Benin and Togo (it did not occur in Ghana) in the 1987–1988 dry season, though uncontrolled populations breed in the Ogun and Oshun Basins of southwest Nigeria (Post 1986). Breeding sites in the Ogun River lie 120 km to the southeast of the reaches of the Okpara River, Benin, which were invaded at the start of the 1988 and 1989 wet seasons by small numbers of parous flies. This elimination of, and repopulation by, the Beffa form, is among the first, suggesting long distance movement by *S. damnosum s.l.* against the direction of the prevailing winds.

The situation following the 'intermittent eradication' of the Beffa form is epidemiologically complex: in the huge rapids of the Ouémé River, where it was invariably associated with *S. damnosum s. str.*, there is no evidence to suggest that man-biting populations remained lower after control measures were relaxed. However, in the main tributary rivers, the Okpara and the Zou, man-biting populations are slow to recover in the absence of the Beffa form, except at the end of the rainy season when there is a major influx of *S. damnosum s. str.*, probably from the east, into the Okpara valley. Also, in the lower Mono Basin in Togo where the Beffa form was certainly a major vector (Garms & Cheke 1985), control has proved extremely easy, with larviciding interrupted 50 % of the time. Unfortunately, the building of a hydro-electric dam has drastically altered the hydrological situation, and though this apparently favours greater *Simulium* production, it makes comparison of the pre- and post-control entomological data impossible.

References

Cheke, R. A. & Denke, A. M. 1988 Anthropophily, zoophily and roles in onchocerciasis transmission of the Djodji form of *Simulium sanctipauli* and *S. squamosum* in a forest zone of Togo. *Trop. Med. Parasit.* **39**, 123–127.

Garms, R. & Cheke, R. A. 1985 Infections with *Onchocerca volvulus* in different members of the *Simulium damnosum* complex in Togo and Benin. *Z. ang. Zool.* **72**, 479–495.

Garms, R., Cheke, R. A., Fiasorgbor, G. K. & Walsh, J. F. 1989 Seasonal extension of the breeding range of *Simulium sanctipauli* from forest into savanna in eastern Ghana and Togo. *Z. ang. Zool.* **76**, 457–467.

Garms, R. & Walsh, J. F. 1987 The migration and dispersal of black flies: *Simulium damnosum* s.l., the main vector of human onchocerciasis. In *Black flies, ecology, population management, and annotated world list* (ed. K. C. Kim & R. W. Merritt), pp. 201–214. University Park, Pennsylvania and London, Pennsylvania State University.

Post, R. J. 1986 The cytotaxonomy of *Simulium sanctipauli* and *Simulium soubrense* (Diptera: Simuliidae). *Genetica* **69**, 191–207.

Walsh, J. F., Philippon, B., Henderickx, J. E. E. & Kurtak, D. C. 1987 Entomological aspects and results of the Onchocerciasis Control Programme. *Trop. Med. Parasit.* **38**, 57–60.

J. B. DAVIES (*Medical Research Council and School of Tropical Medicine, Liverpool, U.K.*). I congratulate Dr Le Berre for his summary of what is a vast, complicated and long-term control exercise. The World Health Organization and the staff of the Onchocerciasis Control Programme (OCP), past and present, deserve recognition for managing to keep the Programme going for 15 years in the face of a constant onslaught of problems, both fiscal, logistic and biological. As I know from my 6-year involvement in the Vector Control Unit, one lives from day to day, never knowing what surprises and set-backs the next day will reveal. In this respect, the controllers of locusts and grasshoppers are to be envied. *Simulium* is there all the year round. There are no periods of aestivation, and no years of recession during which one can relax and take stock.

I would like to highlight some of the evidence for less dramatic movements by members of the *S. damnosum* complex than those long-distance migrations of the 'savanna' cytospecies (usually called reinvasion) that have been touched on by Dr Le Berre, and about which we will hear more from Dr Baker.

These smaller movements appear to be most evident in the forest zones adjacent to and south of the OCP which are inhabited by the 'forest' cytospecies of the *S. sanctipauli* subcomplex. Examples include: (i) In 1966 Dr Le Berre (1966) recorded what he described as 'radial dispersal' of blackflies for distances of up to 41 km westwards from the Bandama River in Côte d'Ivoire. (ii) In Sierra Leone, we have observed *S. damnosum s.l.* biting at villages in the coastal swamp grasslands only 10 km from the sea and 25–30 km south of the nearest breeding sites.

These flies only appear in December when the dry 'harmattan' wind is blowing from the north, and it seems that flies from the inland breeding sites are thus being carried southwards by these winds. *S. soubrense* 'B' is probably the species involved. Hypo–meso endemic onchocerciasis exists in these areas, and we have recorded biting rates of 70 flies per day and a Monthly Transmission Potential of over 300. Dr Walsh has just referred to a westward movement of forest flies of the 'Beffa' form from Nigeria to Benin. No doubt there are many other examples. The overall impression is one of small-scale movements of flies in all directions. Most probably wind patterns are the main vehicle for this dispersal.

It is possible that the cumulative effect of such movements over several years was responsible for the eastwards spread of temephos-resistant flies from the Bandama River in Côte d'Ivoire to the untreated rivers in Ghana between 1980 and 1983 (Meredith *et al.* 1986). The distance involved in this case was about 300 km.

I would like to remind ourselves not to be too complacent over the future of the OCP. The fact that Burkina Faso has not been treated with insecticides since 1987 is good news indeed, and we all hope that this condition will continue. However, even if all sources of 'savanna' cytospecies are eliminated, what are the risks of repopulation of areas like the headwaters of the Black Volta near Bobo Dioulasso by 'forest' species from the south (such as *S. squamosum*, which was present in the area before OCP)? We must now ask our epidemiological modellers what might happen if some of these new arrivals were to be infected. Could they start up new foci of the disease, and how would we detect such an introduction in its early stages?

References

Le Berre, R. 1966 Contribution a l'étude biologique et écologique de *Simulium damnosum* Theobald, 1903 (Diptera, Simuliidae). *Mém. ORSTOM.* **17.** (204 pages.)
Meredith, S. E. O., Kurtak, D. & Adiamah, J. H. 1986 Following movements of resistant populations of *Simulium soubrense/sanctipauli* (Diptera: Simuliidae) by means of chromosome inversions. *Trop. Med. Parasit.* **37**, 290–294.

R. LE BERRE. Regarding Dr Davies's question about modelling, I can say that Dr Remme and colleagues have shown that with 14 years of treatments, the microfilarial rates will return to pre-treatment levels after five or six years and the disease prevalence rates will follow. However, the population will not show any ocular lesions, and hence no blindness, until 30 or 40 years after the end of treatments.

L. G. GOODWIN, F.R.S. (*Shepperlands Farm, Park Lane, Finchampstead, Wokingham, U.K.*). Dr Le Berre mentioned Ivermectin and its use against the microfilariae. Is it administered once a year or more frequently?

R. LE BERRE. The drug, Invermectin, donated by M.S.D. for Onchocerciasis control was developed as a cattle anti-helminthic and after extensive trials it can now be used as a microfilaricide for people: but it does not kill the adult worms, which may live for as long as 12 years. It has to be given at least once a year, preferably before the start of the transmission season, for instance, every February. Under some circumstances, in severe foci for instance, it may need to be given at six-monthly intervals. The effects of mass treatments with Ivermectin on onchocerciasis transmission have not been as good as were expected; partly, perhaps, because there are always some people in a community who don't get treated.

Phil. Trans. R. Soc. Lond. B **328**, 731–750 (1990)

Printed in Great Britain

Progress in controlling the reinvasion of windborne vectors into the western area of the Onchocerciasis Control Programme in West Africa

By R. H. A. Baker†, P. Guillet, A. Sékétéli, P. Poudiougo, D. Boakye, M. D. Wilson‡ and Y. Bissan

Onchocerciasis Control Programme, B.P. 549, Ouagadougou, Burkina Faso

Since vector control began in 1975, waves of *Simulium sirbanum* and *S. damnosum s.str.*, the principal vectors of severe blinding onchocerciasis in the West African savannas, have reinvaded treated rivers inside the original boundaries of the Onchocerciasis Control Programme in West Africa. Larviciding of potential source breeding sites has shown that these 'savanna' species are capable of travelling and carrying *Onchocerca* infection for at least 500 km northeastwards with the monsoon winds in the early rainy season. Vector control has, therefore, been extended progressively westwards. In 1984 the Programme embarked on a major western extension into Guinea, Sierra Leone, western Mali, Senegal and Guinea-Bissau. The transmission resulting from the reinvasion of northern Côte d'Ivoire and Burkina Faso has been reduced by over 95 %, but eastern Mali has proved more difficult to protect because of sources in both Guinea and Sierra Leone. Rivers in Sierra Leone were treated for the first time in 1989 and biting and transmission rates in Sierra Leone and Guinea fell by over 90 %. Because of treatment problems in some complex rapids and mountainous areas, flies still reinvaded Mali, though biting rates were approximately 70 % lower than those recorded before anti-reinvasion treatments started. It was concluded that transmission in eastern Mali has now been reduced to the levels required to control onchocerciasis.

1. Introduction

Since vector control started in 1975, waves of *Simulium sirbanum* and *S. damnosum s.str.* the principal vectors of severe blinding onchocerciasis in the West African savannas, have reinvaded treated rivers inside the original boundaries of the Onchocerciasis Control Programme in West Africa (OCP) (Le Berre *et al.* 1979). Larviciding of potential source breeding sites has shown that these 'savanna' species are capable of travelling and carrying *Onchocerca* infection for at least 500 km northeastwards with the monsoon winds in the early rainy season (Garms *et al.* 1979; Magor & Rosenberg 1980; Walsh *et al.* 1981; Johnson *et al.* 1985; Garms & Walsh 1987). Annual Transmission Potentials (ATP), as defined by Walsh *et al.* (1978), in communities near the original western borders of the Programme have remained above tolerable levels and human infection indices have decreased much more slowly than in central areas where this reinvasion has been controlled. Vector control has, therefore, been extended progressively westwards.

Garms *et al.* (1979) documented how experimental treatments of the Marahoué and Sassandra Basins in western Côte d'Ivoire during 1977 and 1978 markedly reduced the reinvasion of the Leraba and Upper Bandama Basins in the northeast of that country. In 1979,

† Present address: Wood View, Church Fields, Stonefields, Woodstock OX7 2PP, U.K.
‡ Present address: Department of Biological Sciences, University of Salford M5 4WT, U.K.

year-round control activities were extended to these river basins and substantial reductions in biting and transmission were recorded, although the Marahoué and Sassandra continued to be reinvaded (Walsh *et al.* 1981). In 1984, the Programme embarked on a major western extension into Guinea, Sierra Leone, western Mali, Senegal, and Guinea-Bissau. The first significant impact on reinvasion occurred in 1985 when the Upper Sassandra Basin in southeastern Guinea was treated with larvicides for the first time. Biting and transmission by re-invading flies in the Sassandra, Marahoué, Bandama and Leraba Basins decreased by over 90 % (Baldry *et al.* 1985).

The Baoulé, Bagoé and Banifing tributaries of the River Niger in southeastern Mali and the extreme northwest of Côte d'Ivoire have been treated since 1977 but have proved to be more difficult to protect. The treatment of the Sankarani and Fié, the easternmost river basins in Guinea, had only a marginal effect on reinvasion (Baker *et al.* 1986). Further west, but still downwind and within 300 km of the re invaded areas, the Milo, Niandan, Kouya, Mafou and Niger Rivers were found to contain very large *S. sirbanum* breeding sites and flies collected from these sites were similar in size to those invading. Once the Inter-Tropical Convergence Zone (ITCZ) had been established to the north of the reinvasion zone, the periodicity of reinvasion could be related to changes in the height of these rivers, starting soon after their first rise at the end of the dry season and ending when they began to flood after the onset of the main rainy season. It was concluded that these rivers must be the principal sources of the reinvading flies (Baker *et al.* 1987).

This paper describes progress in discovering the sources, elucidating the migration pathways and controlling reinvasion in the western area of the OCP during 1987–1989. Progress in containing the reinvasion of the eastern area of the OCP has been presented by Garms *et al.* (1982); Cheke & Garms (1983) and Walsh (discussion after Le Berre, this symposium).

2. Study area

The area covered by this study (figure 1) extends through five countries and includes the Upper Niger River Basin in Mali and Guinea, the Sassandra, Bandama, Comoé and Black Volta River Basins in Côte d'Ivoire and Burkina Faso, together with coastal river basins in Sierra Leone such as the Great and Little Scarcies and Seli (Rokel). Descriptions of these areas have been given by Baldry *et al.* (1985) and Baker *et al.* (1986, 1987) and, for Sierra Leone, by Post & Crosskey (1985).

3. Methods

The gradual westward extension of OCP's vector control operations was preceded by extensive helicopter and ground prospections to determine the geographical and seasonal distribution of breeding sites for each vector species. A network of points (only a few of which are shown in figure 1) for 11-hour day collections of biting flies was progressively established along all the main rivers. Flies were dissected to determine parity and infection following a standard protocol (Walsh *et al.* 1979). Water-level gauges were installed or recalibrated on all the main watercourses, the most representative being fitted with Argos[R] beacons, which transmitted river heights every 20 min to the aerial bases via satellite, to enable the accurate calculation of discharge rates.

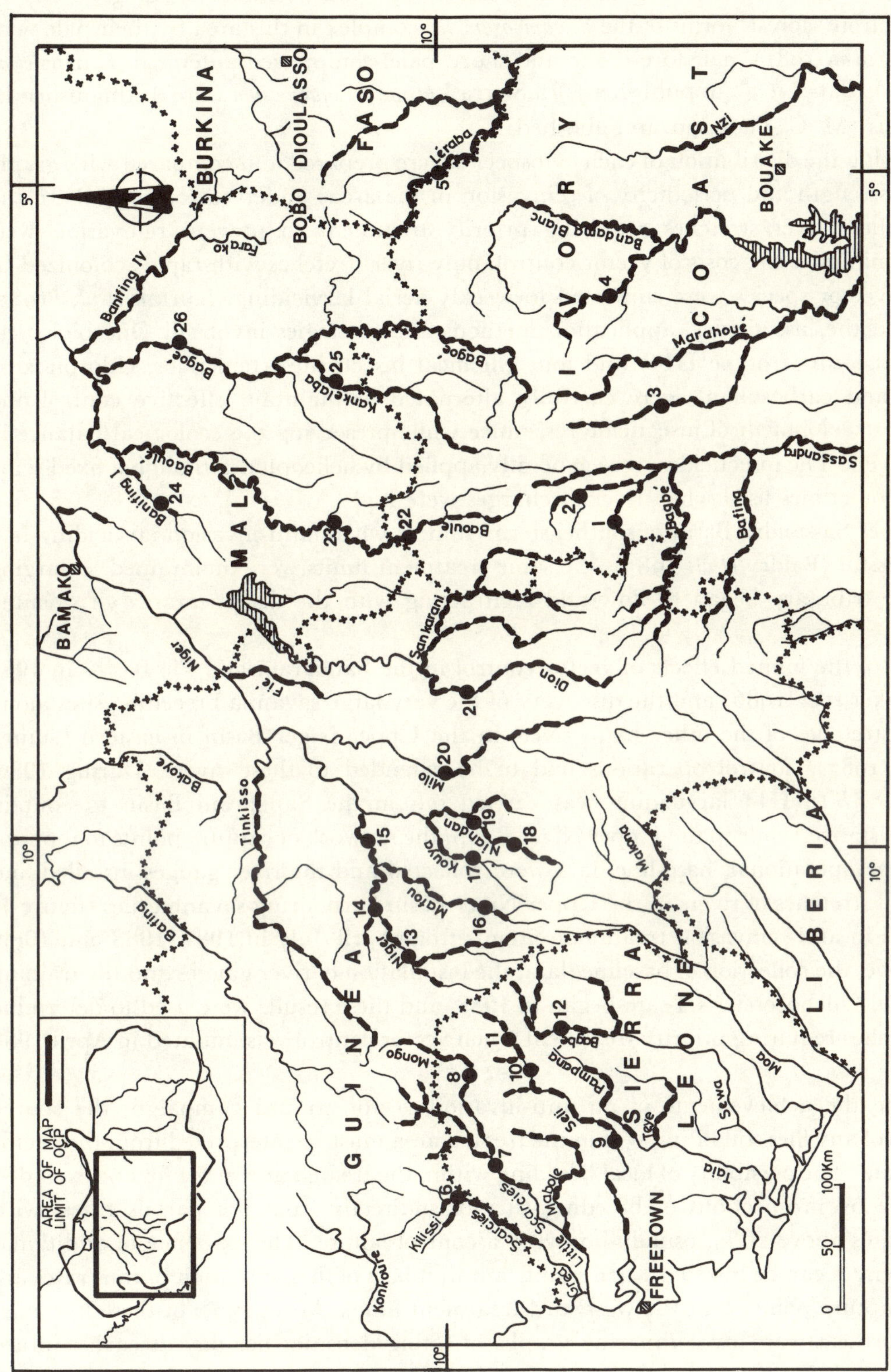

FIGURE 1. The Study Area showing: country boundaries ($+++++$), the river lengths (at maximum) treated in 1989 (thick dashed lines), the main rivers (thin lines), and the capture points mentioned in the text (1, Niamotou; 2, Massadougou; 3, Kato; 4, Gîte 3; 5, Pont Leraba; 6, Badi Kanti; 7, Kaba Ferry; 8, Musaia; 9, Makpankaw; 10, Arfanya; 11, Yirafilaia; 12, Yifin; 13, Balandougou; 14, Banfarala; 15, Diaragbela; 16, Yalawa; 17, Kouya Laya; 18, Yradou; 19, Sansanbaya; 20, Morigbedougou; 21, Téré; 22, Madina; 23, Madina Diassa; 24, Mpiela; 25, Kankela; 26, Metela).

Cytotaxonomy of the larval polytene chromosomes (Vajime & Dunbar 1975) was the only unequivocal method of identifying savanna vectors, but adult females could generally be distinguished from 'forest' forms of the *S. damnosum s.l.* complex in this area by their pale wing tufts (Kurtak *et al.* 1981), pale fore-coxae and short, pale, compressed antennae (Garms *et al.* 1982; R. H. A. Baker *et al.*, unpublished). In Sierra Leone, *S. squamosum* were distinguished by electrophoresis (M. C. Thomson, unpublished).

Maps showing the distribution of each cytospecies were prepared and compared with graphs showing the strength and periodicity of reinvasion in the areas under vector control. In this way, the critical river stretches requiring priority treatment to prevent reinvasion were selected. To minimize the costs of vector control, only river stretches with rapids colonized by the savanna vector species were subjected to weekly aerial larviciding. Kurtak *et al.* (1987) have reviewed the insecticides, application methods and strategies involved. One biological (*Bacillus thuringiensis* serotype H14) and four chemical insecticides (temephos, chlorphoxim, permethrin, and carbosulfan) were carefully alternated to maintain effective control and minimize the development of insecticide resistance while preserving the ecological balance in the treated rivers. The insecticides were generally applied by helicopter, although a fixed-wing aircraft was sometimes used when river discharges were high.

In the Upper Sassandra Basin of southeastern Guinea, where anti-invasion larviciding had been so successful (Baldry *et al.* 1985), the same treatment limits were maintained, enlarging as each large tributary began to flow and contracting with the disappearance of savanna species.

As a result of the limited effects of vector control in the Sankarani and Fié Basins in 1984 and 1985 (Baker *et al.* 1986) and the discovery of the very large savanna breeeding sites along the lowland stretches of the other main rivers in the Upper Niger Basin in eastern Guinea (Baker *et al.* 1987), control operations had to be extended to these rivers. During 1986, *B. thuringiensis* (*B.t.*) H14 larviciding was carried out in the Sankarani Basin to contain insecticide resistance, while in the Upper Niger Basin, the network of capture points to monitor control became operational, base-line data were collected and the river gauges installed and calibrated. All stretches of rivers in the Upper Niger Basin harbouring savanna flies (figure 1) were included in anti-reinvasion treatment circuits from April–July in 1987, 1988 and 1989. In Sierra Leone, the collection of baseline data, the installation of river gauges and the training of the National Onchocerciasis Team began in 1988, and these results were used to determine the river stretches requiring priority treatment when vector control was initiated in April 1989 (figure 1).

To evaluate the achievements of an anti-invasion vector control campaign, the source breeding sites of any flies still biting within the treated area must be interpreted from the vector monitoring data. The possibility of local breeding within the treated area must first be excluded either directly by prospections of breeding sites or indirectly from the parous rates, with nulliparous rates above 20%, usually implying a control failure. The existence of additional re-invasion sources can be inferred when significant numbers of flies with high parous rates are collected at capture points near the westward treatment limits. Movements of these flies may be tracked northeastward by comparing graphs of biting densities per day at each capture point on their path (Johnson *et al.* 1985). Our incomplete knowledge of the factors initiating and terminating such long-distance movements, coupled with uncertainty as to the times of day and heights of travel in the air stream (reviewed by Garms & Walsh (1987)) make it difficult to back-track reinvasions accurately to their source and to predict their strength from year to year. Detailed interpretation was also limited by the incompleteness of the meteorological data from Guinea and Sierra Leone. These difficulties were compounded in this study by the limited sampling programme, with captures at most points restricted to one day per week. Capture frequencies were increased at heavily reinvaded sites but some points on the apparent reinvasion path, particularly those not situated close to large breeding sites, rarely recorded the passage of migrant flies.

4. RESULTS

(a) *Reinvasion of Côte d'Ivoire* 1986–1989

Table 1 shows that since 1985, the treatments of the Upper Sassandra in southeastern Guinea have maintained their impact on biting and transmission during the critical April–June reinvasion period with 75–95% reductions in Monthly Biting Rates (MBRs) and 95–99% reductions in Monthly Transmission Potentials (MTPs) in comparison with the same period in 1982–84, when the treatment strategy in the reinvaded zones was most comparable. Annual Transmission Potentials (ATPs) (figure 2) were first reduced in 1979 when year-round vector control was undertaken in the Marahoué and Sassandra Basins and again when the Upper Sassandra in Guinea was added in 1985. The community microfilarial load in man, the geometric mean number of *O. volvulus* microfilariae per skin snip in adults of 20 years or older, for Massadougou was dropping rapidly (figure 3), and was approaching the trend predicted by a computer model (Remme *et al.* 1990).

(b) *Reinvasion of southeastern Mali* 1987–1988

(i) *Biting and transmission in Guinea and Mali*

Vector control in both 1987 and 1988 greatly reduced biting and transmission in Guinea at those points where collections were made in 1986 (table 2). Over 80% reductions in biting rate were recorded at Téré and Morigbedougou in the Sankarani and Milo Basins, and over 60% at Sansanbaya on the River Niandan. Despite these reductions, considerable numbers of parous flies were caught. At Sansanbaya an MBR of 24 830 was obtained in May–July 1986, but 7579 and 10 436 were still recorded in 1987 and 1988.

The 1987–1988 reinvasion in Mali, as represented by cumulative May–July MBRs, constituted a 72% reduction at Madina Diassa and a 36% reduction at Mpiela, but only a 10% reduction at Kankela and Madina, when compared with 1977–1983 when no treatments were conducted in Guinea (table 3). At Metela there was a 35% increase. However, estimates of the reductions in biting and transmission achieved by spraying source breeding sites can only be very approximate, because MBRs and MTPs have varied considerably from year to year (figures 4 and 5) and it has not been possible to predict the strength of the annual reinvasion.

(ii) *Effectiveness of treatments in Guinea and Mali*

Parous rates at capture points beside treated river stretches in Guinea and Mali remained very high in both 1987 and 1988. In 1987, during the period when flies were invading Mali, however, some local breeding was reported in several of the complex rapid systems on the River Niandan and River Niger in Guinea, which had been treated with *B.t.* H14. Formulations of this insecticide could have a relatively short 'carry' and treatment failures

TABLE 1. Cumulative April–June monthly biting rates and cumulative monthly transmission potentials in the Côte d'Ivoire reinvasion Zone 1982–1989, together with mean values for 1982–1984 and 1985–1989 for both indices and percentage reductions between the two periods

capture point and map number (figure 1)	1982	1983	1984	1985	1986	1987	1988	1989	mean 1982–84	mean 1985–89	percentage reduction
					Monthly biting rates (MBR)						
Niamotou (1)	16892	14536	6919	808	230	270	1082	516	12782	581	95.5
Massadougou (2)	12372	3762	4613	209	1372	1068	711	173	6916	707	89.8
Kato (3)	13782	3540	7905	189	224	1094	946	1387	8409	768	90.9
Gîte 3 (4)	7102	1233	2761	111	288	744	1106	1599	3699	770	79.2
Pont Leraba (5)	2423	259	705	20	459	269	351	323	1129	284	74.8
					Monthly transmission potentials (MTP)						
Niamotou (1)	808	986	551	20	0	0	15	0	782	7	99.1
Massadougou (2)	476	66	179	2	8	15	6	0	240	6	97.4
Kato (3)	410	154	211	1	0	17	5	2	258	5	98.1
Gîte 3 (4)	452	30	177	11	0	22	13	9	220	11	95.0
Pont Leraba (5)	161	18	36	0	4	3	0	0	72	1	98.0

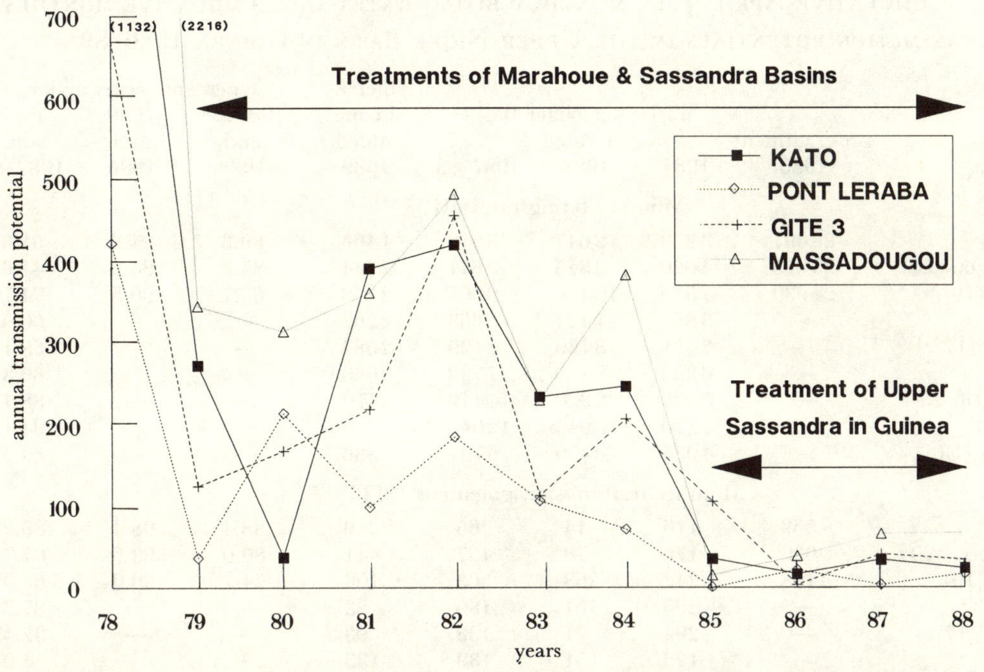

Figure 2. The effect of insecticide treatments of the Marahoué and Sassandra Basins followed by the Upper Sassandra in Guinea, on Annual Transmission Potentials at four capture points in the Côte d'Ivoire reinvasion zone.

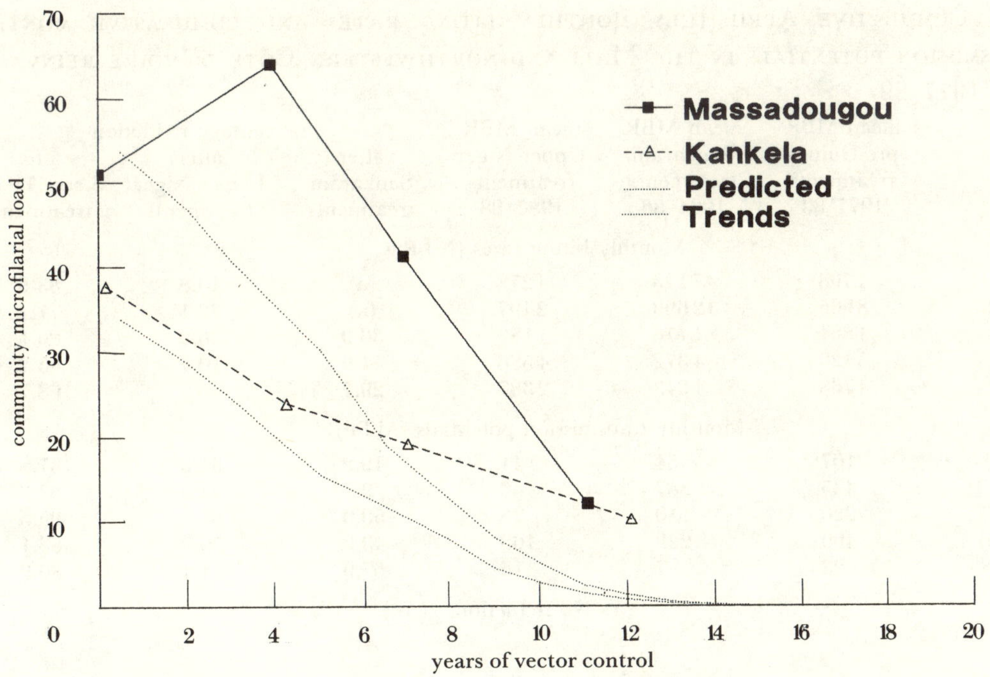

Figure 3. Epidemiological trends in the reinvasion zones. The predicted trends are those of Remme et al. (1990).

TABLE 2. CUMULATIVE APRIL–JULY MONTHLY BITING RATES AND CUMULATIVE MONTHLY TRANSMISSION POTENTIALS IN THE UPPER NIGER BASIN IN GUINEA 1986–89

capture point and map	pre-treatment 1986	all Upper Niger Basin treated			Sierra Leone added 1989	percentage reduction		
		1987	1988	1987–88		1987–88 and 1986	1989 and 1986	1989 and 1987–88
Monthly biting rates (MBR)								
Tere (21)	28097	3373	2647	3010	1466	89.3	94.8	51.3
Morigbedougou (20)	23675	4026	3315	3671	2144	84.5	90.9	41.6
Sansanbaya (19)	24830	7579	10436	9008	2276	63.7	90.8	74.7
Yradou (18)	—	3881	7424	5653	2262	—	—	60.0
Kouya Laya (17)	—	7471	8769	8120	2587	—	—	68.1
Yalawa (16)	—	6986	8489	7738	3069	—	—	60.3
Diaragbela (15)	—	5978	2320	4149	470	—	—	88.7
Banfarala (14)	—	8140	15945	12043	716	—	—	94.1
Balandougou (13)	—	4955	8459	6707	1336	—	—	80.1
Monthly transmission potentials (MTP)								
Téré (21)	559	116	14	65	9	88.4	98.4	86.2
Morigbedougou (20)	636	176	78	127	41	80.0	93.6	67.7
Sansanbaya (19)	1168	242	363	303	106	74.1	90.9	65.0
Yradou (18)	—	205	154	180	23	—	—	87.2
Kouya Laya (17)	—	204	71	138	93	—	—	32.4
Yalawa (16)	—	126	151	139	133	—	—	4.0
Diaragbela (15)	—	370	102	236	14	—	—	94.1
Banfarala (14)	—	420	347	384	0	—	—	100.0
Balandougou (13)	—	39	407	223	69	—	—	69.1

TABLE 3. CUMULATIVE APRIL–JULY MONTHLY BITING RATES AND CUMULATIVE MONTHLY TRANSMISSION POTENTIALS IN THE MALI AND NORTHWESTERN CÔTE D'IVOIRE REINVASION ZONE 1977–89

capture point and map number	mean MBR pre-Guinea treatments 1977–83	mean MBR Sankarani treatments 1984–86	mean MBR Upper Niger treatments 1987–88	percentage reduction		
				after Sankarani treatments	after Upper Niger treatments	after Sierra Leone treatments
Monthly biting rates (MBR)						
Madina (22)	4798	7115	4278	(a)	10.8	83.6
M-Diassa (23)	8965	12699	2497	(a)	72.1	71.1
Mpiela (24)	1864	1176	1189	36.9	36.2	86.1
Kankela (25)	5120	4372	4586	14.6	10.4	65.2
Metela (26)	1768	1243	2387	29.7	(a)	68.7
Monthly transmission potentials (MTP)						
Madina (22)	107	54	41	49.2	61.6	87.8
M-Diassa (23)	425	387	66	9.1	84.6	83.3
Mpiela (24)	220	110	28	50.0	87.5	95.5
Kankela (25)	409	229	168	43.9	59.0	83.1
Metela (26)	23	7	15	67.9	34.4	86.9

(a) No reduction.

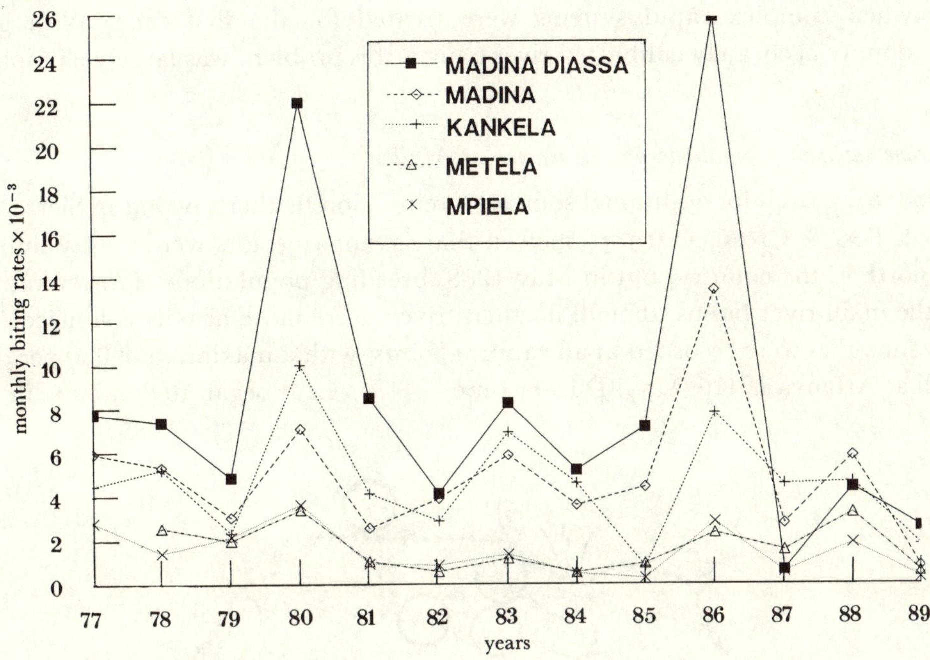

FIGURE 4. May–July cumulative Monthly Biting Rates at five capture points in the
Mali/northwestern Côte d'Ivoire reinvasion zone from 1977–1988.

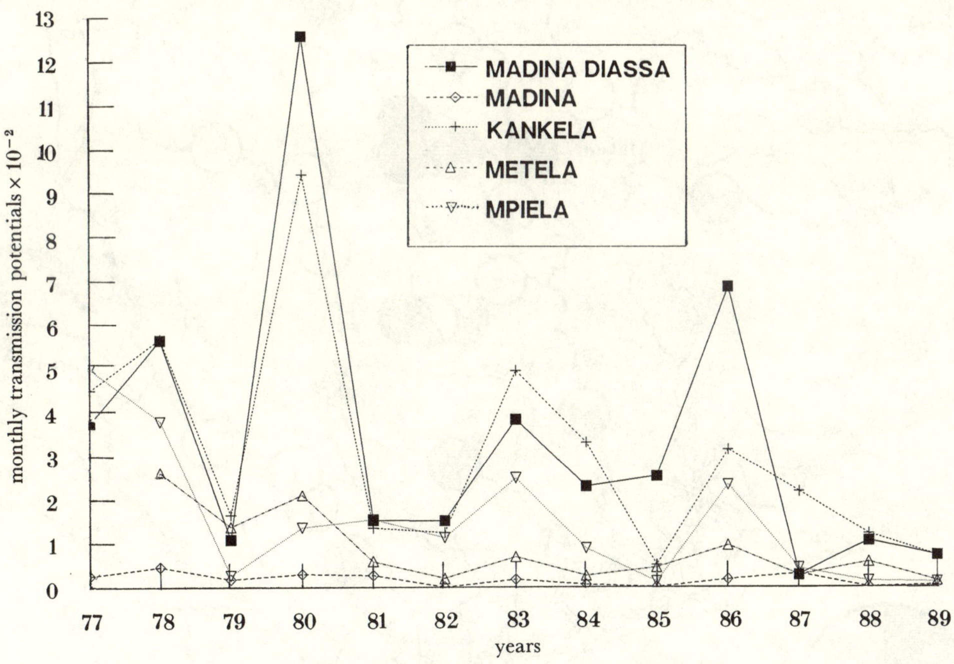

FIGURE 5. May–July cumulative Monthly Transmission Potentials at five capture points in the
Mali/northwestern Côte d'Ivoire reinvasion zone from 1977–1989.

could occur when complex rapid systems were treated for the first time. With greater experience and more accurately calibrated river gauges this problem was largely eliminated in 1988.

(iii) *Sierra Leone savanna populations and reinvasion of Mali*

During 1988, the search for additional sources of reinvasion further upwind in Sierra Leone was intensified. Post & Crosskey (1985) showed that savanna vectors were mostly limited to the extreme north of the country, but in May 1988, breeding populations of *S. sirbanum* were found in all the main river basins, though northern rivers were more heavily colonized (figure 6). Biting savanna flies were recorded at all capture points, with a maximum of 650 per day on the River Seli at Arfanya in late May. During June, *S. soubrense* B began to dominate in biting

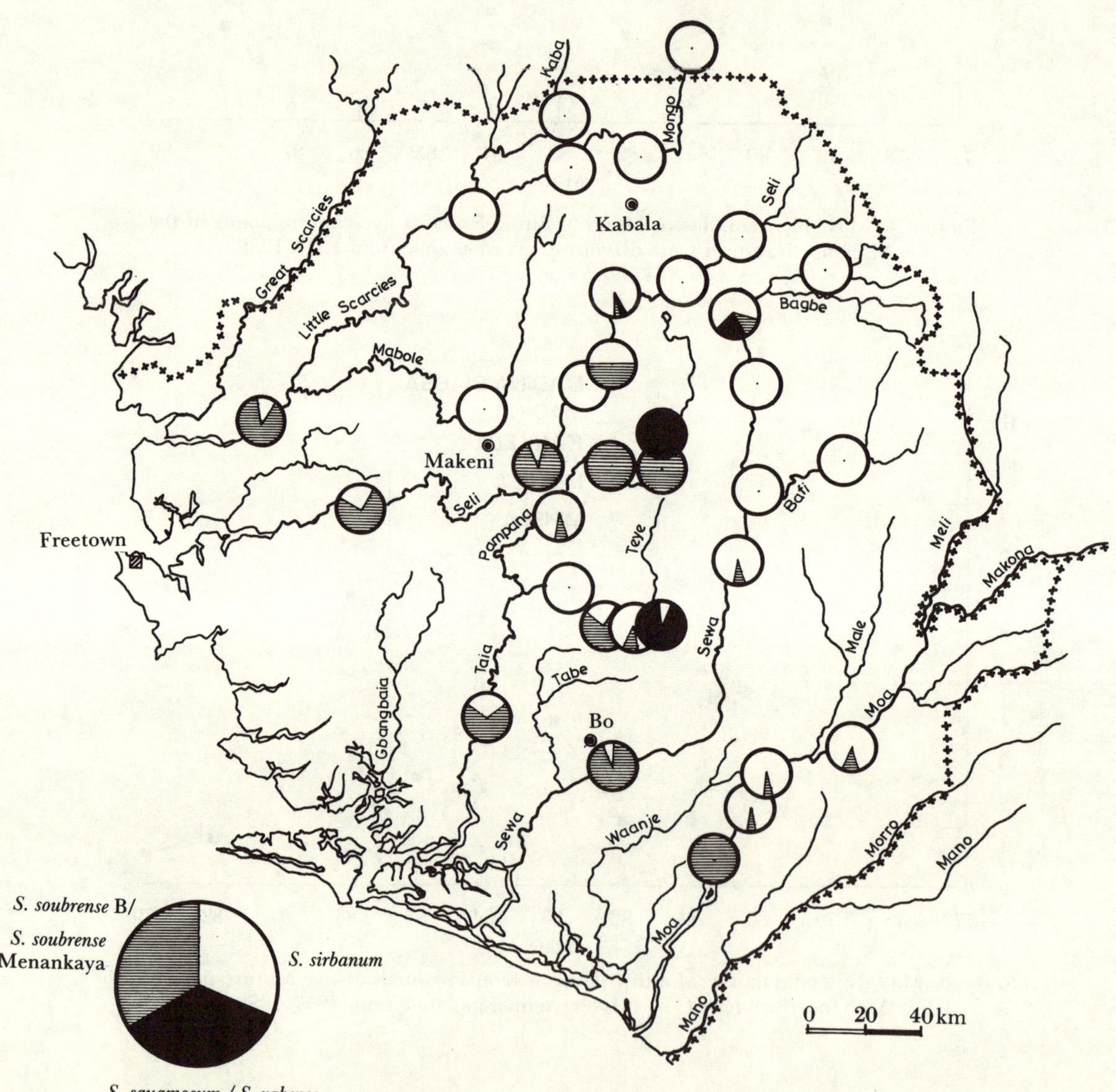

FIGURE 6. The distribution of different members of the *Simulium damnosum* species complex in Sierra Leone in April–May 1988, identified by larval cytotaxonomy.

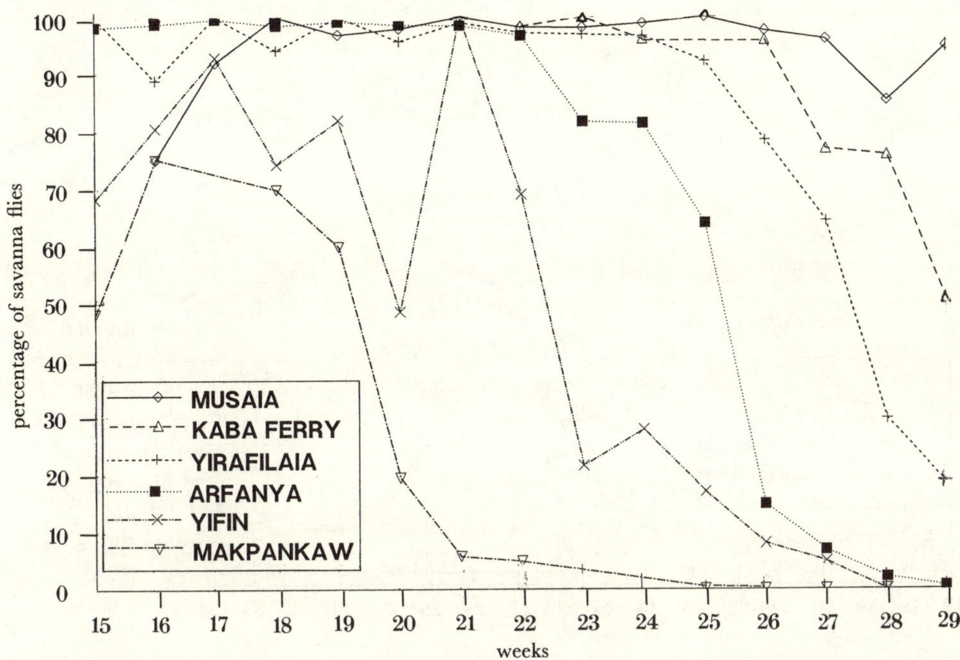

FIGURE 7. Percentage of savanna biting flies in weekly catches at six capture points in Sierra Leone during calendar weeks 15–29, 1988.

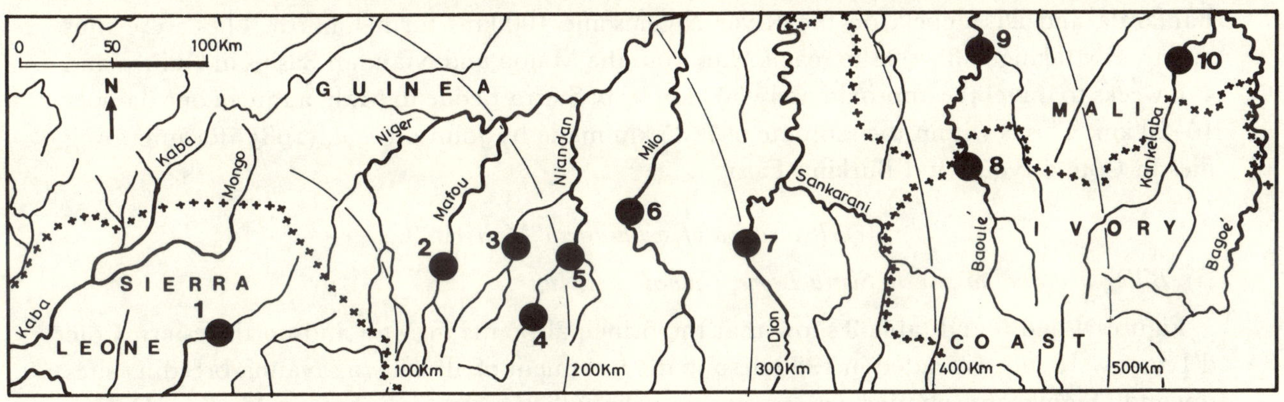

FIGURE 8. Map situating the capture points (1, Arfanya; 2, Yalawa; 3, Kouya Laya; 4, Yradou; 5, Sansanbaya; 6, Morigbedougou; 7, Téré; 8, Madina; 9, Madina Diassa; 10, Kankela) on the reinvasion axis between Sierra Leone and Mali as used in figures 9 and 10.

collections until, by the end of July, biting savanna flies were virtually restricted to the Little and Great Scarcies Basins in either side of the border with Guinea (figure 7).

When the 1988 mean daily biting rates per week at capture points in Sierra Leone, Guinea and Mali (figure 8) were plotted on the same graph (figure 9), a wave of savanna flies could be followed northeastwards from Arfanya through Yalawa (101–150 km) on the R. Mafou, Kouya Laya on the River Kouya and Yradou on the River Niandan (151–200 km), Sansanbaya on the R. Niandan and Morigbedougou on the River Milo (201–250 km), Téré on the R. Dion (251–300 m) and reaching Madina in northwestern Côte d'Ivoire (401–450 km), Madina Diassa (451–500 km) and Kankela in Mali (501–550 km) in calendar week 24–25. Another wave was observed two weeks later passing through Balandougou,

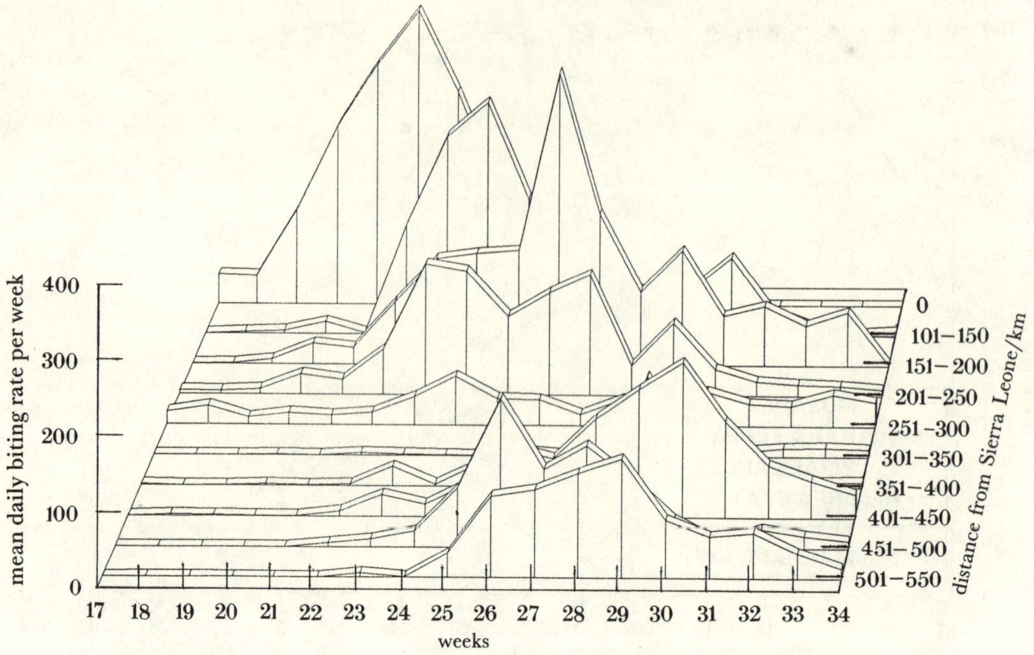

FIGURE 9. Three-dimensional representation of the mean daily biting rate per week at increasing distances from Sierra Leone in 1988.

Banfarala and Diaragbela on the River Niger some 100 kms further north. Flies were thus apparently taking 2–3 weeks to reach Mali from the Mafou and Niandan Basins in Guinea and 4–5 weeks to travel the approximately 500 km from Sierra Leone to Mali, a rate of one day per 15–20 km. This is within the estimate of 7–35 km made by Johnson *et al.* (1985) for migrating flies in Côte d'Ivoire and Burkina Faso.

(c) Reinvasion of southeastern Mali 1989

(i) Biting and transmission in Sierra Leone, Guinea and Mali

Figures 4 and 5 and table 3 show that the principal points in Mali and northwestern Côte d'Ivoire were still reinvaded in 1989 despite the treatment of all known savanna breeding sites upwind. Significant numbers of parous invading flies were recorded at Madina Diassa, Kankela and Mpiela with maximum mean daily biting rates of 145, 91 and 94, respectively. However, very few flies were caught at Madina (maximum mean daily biting rate per week of 31, and a May–July MBR of 788), the best result ever recorded. In 1989, May–July MBRs were 65–86 % and MTPs 83–95 % less than means for 1977–1983, before any control was undertaken in the Western Extension.

The 1989 treatments also further reduced biting and transmission in the Upper Niger Basin (table 2). This effect was most noticeable at Sansanbaya where the cumulative May–July MBR of 2276 was 78 % lower than that recorded in 1988, 70 % lower than 1987 and 91 % lower than 1986. MBRs on the River Niger were reduced by over 80 % compared to the 1987–1988 mean and, at all points where 1986 data were collected, MBRs in 1989 were also reduced by at least 80 %. On the Kouya, Upper Niandan and Mafou, MBRs remained at between 60–70 % of the 1987–1988 mean.

MTPs followed the same trend. At Sansanbaya, the MTP of 106 was 70 % lower than

1988, 55% lower than 1987 and 91% lower than 1986. Over 65% reductions compared to the 1987–1988 mean were recorded in the Sankarani, Milo, Niandan and Niger Basins, but there was little or no change on the Kouya and Mafou. In May, ivermectin, a microfilaricidal drug potentially capable of reducing transmission by up to 75% in isolated endemic communities (Remme *et al.* 1989), was delivered to hyperendemic communities in the Milo, Niandan, Mafou and Niger Basins, but no consistent effect on vector infectivity was recorded.

Vector control also greatly depressed biting and transmission rates in northern Sierra Leone. At Arfanya, the May–July MBR dropped by 97% (from 25420 to 680, with forest vectors making up 27% of the residual MBR) and the MTP by 98% (from 1810 to 1833). Low savanna vector biting rates and high parous rates were observed at all points except those in the Upper Great and Little Scarcies Basins near or north of the Guinea border.

(ii) *Effectiveness of treatment in Sierra Leone, Guinea and Mali*

Parous rates remained high at virtually all capture points beside treated river stretches in Sierra Leone, Guinea and Mali. However, some local breeding may have occurred for a short time in Mali near Madina Diassa because on the two days when the highest numbers of flies (238 and 200) were caught, only 76% and 67% were parous although by the next day the parous rate had reverted to 98%. This suggests that reinvasion at Madina Diassa was less important than the biting levels imply. Parous rates at Kankela and Madina were very high throughout and these two points may be better reflections of the reinvasion events.

Savanna vector breeding was not, however, completely eliminated in Sierra Leone. This was again partly because of difficulties in applying *B.t* H14 to complex breeding sites for the first time, but mainly to the problems encountered in treating tributaries of the Great and Little Scarcies which, at a critical time, were dry near their confluence with the main river, but flowing productively upstream. Biting rates on the Great Scarcies River at Badi Kanti increased rapidly from week 23, well before the river began to flow. Substantial breeding by *S. sirbanum* was found two weeks later in the River Kilissi, a major tributary in Guinea, which

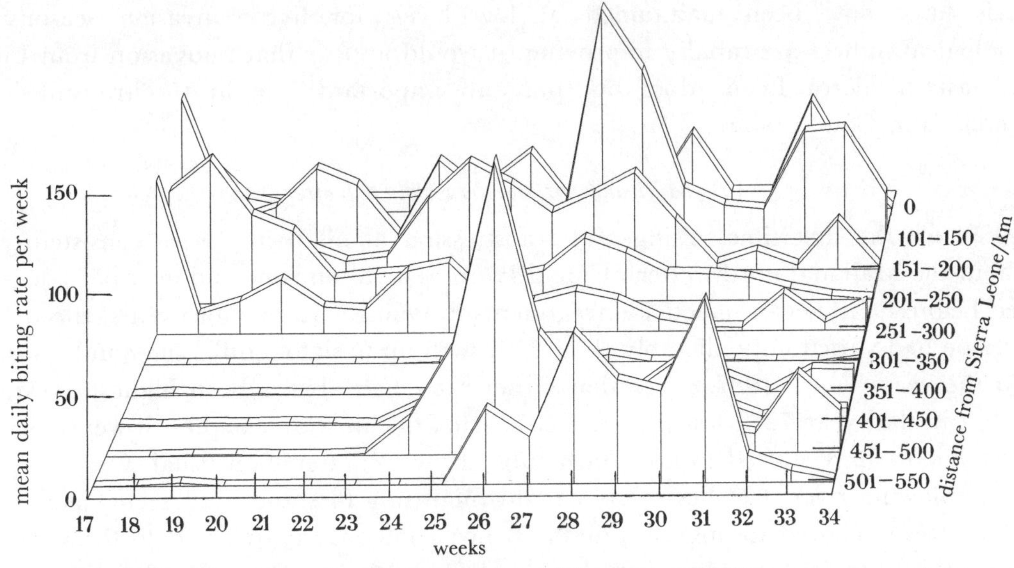

FIGURE 10. Three-dimensional representation of the mean daily biting rate per week at increasing distances from Sierra Leone in 1989.

was then included in the treatment circuit. However, catches at Badi Kanti peaked at 230 per day and were clearly an important potential source of invading flies.

(iii) *Sources of reinvasion in 1989*

A graph of biting rates along the reinvasion axis for 1989 (figure 10) was much less clear cut than for 1988 (figure 9). In week 22 the situation looked excellent with about 10 flies per day at all points (25 at Sansanbaya). Then in weeks 23 and 24 daily biting trebled or quadrupled at Yalawa, Yradou, Kouya Laya and Téré, though remaining stable at Sansanbaya and Morigbedougou. Such rises could be correlated with the increases in Mali during weeks 25 and 26, but it is not clear why they were not detectable at Sansanbaya and Morigbedougou, both points where reinvasion had been very important in previous years. An additional rise in weeks 27 and 28 can, however, be detected at all points on the southern Guinea axis.

Along the Niger River fly numbers gradually increased at Balandougou in weeks 25–27, peaking at Balandougou, Banfarala and Diaragabela in week 27. A second peak at Balandougou was observed in week 31.

5. Discussion

(a) *Côte d'Ivoire*

Since treatments began in the Upper Sassandra Basin in Guinea, May–July MBRs in the Sassandra, Marahoué, Bandama and Leraba Basins in Côte d'Ivoire have been maintained at 4–25% and MTPs at 1–5% of pre-1985 levels. Epidemiological indices (figure 3) have fallen rapidly and, with ATPs below 100 since 1985 (figure 2), this trend should continue.

Potential reinvasion sources still exist to the southwest of the Côte d'Ivoire reinvasion zone. Garms (1987) and discussion after this paper has collected adults and larvae of both savanna vector species from several rivers in Liberia. A large population of breeding and biting savanna flies (maximum 335 per day in late May 1988) was found on the River Moa in southeastern Sierra Leone in both 1988 (figure 6) and 1989. However, since biting rates and transmission potentials have now been maintained at low levels for five reinvasion seasons and epidemiological indices are rapidly improving, it would appear that reinvasion from Liberia or southeastern Sierra Leone does not play an important role in the transmission of onchocerciasis in Côte d'Ivoire.

(b) *Mali and northwestern Côte d'Ivoire*

In 1989, for the first time, biting and transmission at all points were consistently low throughout the Mali and northwestern Côte d'Ivoire reinvasion zone and over 65% less than recorded before OCP began insecticide treatments in Guinea. In previous years, the strength of reinvasion had varied unpredictably (figure 4), with up to sixfold differences in biting rates between the years. Until 1987, reinvasion biting rates had always been highest at Madina Diassa, but in that year 72% less than the 1977–1983 mean were caught. However, in 1987 the River Baoulé only started to flow on 6 July, six weeks later than usual. Capture points beside slow flowing river stretches rarely caught migrating flies and if, as seems likely, these flies, which were believed to migrate gravid (Bellec 1976), were attracted to the vicinity of rapids then this behaviour could explain the low catch at Madina Diassa in 1987. If years with unusually late flow by the capture points are excluded (together with Kankela in 1985, which may have been affected by treatments of the Sankarani Basin (Baker *et al.* 1986)), the 1989 reinvasion was lower than ever before at all points. In 1989, cumulative May–July MTPs (figure 5) were below 100 for the first time. Epidemiological indices, which have only been slowly decreasing in this area (figure 3 for Kankela) should improve rapidly if this trend of low MTPs is maintained.

(c) *Sources of reinvasion in* 1989

The key questions would appear to be to determine the source of the flies invading Mali during the main peak, which started in week 24 and was at a maximum 2–3 weeks later. If we accept that flies do take at least 4 weeks to travel the 500 km from northeastern Sierra Leone to Mali, then the flies breeding in the poorly treated tributaries of the Great and Little Scarcies, 100–150 km further west, should take even longer. This makes it unlikely that these flies, which first appeared in week 23, were the main contributors to this invasion.

An alternative hypothesis is based on the increase in biting rates observed in the Mafou, Kouya, Upper Niandan and Dion Basins during weeks 23–24, but this is confounded by the stability of the biting rates at Sansanbaya and Morigbedougou during the same period. Could the occasional poorly treated breeding site in Sierra Leone and Guinea have been responsible for the 30–40 flies per day in the southern part of the Upper Niger Basin and could these flies have produced 150 per day at Madina Diassa? If we assume that the Madina Diassa peak was inflated by local breeding, because of the relatively low parous rate in week 26, and if wind conditions favoured long distance movement on particular days, thus increasing densities in the reinvaded zones, we may have a partial explanation. There were no other obvious potential source rivers.

Breeding sites in the Upper Great and Little Scarcies were probably responsible for the wave of invading flies that passed through capture points in the Niger River in weeks 26–27 and maintained relatively high biting rates in Mali and northwestern Côte d'Ivoire until weeks 34–35.

With the experience obtained in treating complex breeding sites and partially flowing rivers in Sierra Leone during 1989 and with the evidence for the existence of source areas there, future reinvasions of Guinea and Mali should be further minimized by the appropriate insecticide treatments.

In a programme of the scale of OCP it is impossible to acknowledge all the individuals who contributed to the work described here. Special thanks are extended to: Mr I. Sesay, Entomologist of the Sierra Leone National Onchocerciasis Team and his staff; Messrs R. Lama, A. Sagno and Dr K. Kaba, entomologists of the Guinea National Onchocerciasis Team; Dr A. Akpoboua, OCP Bouaké, Sector Chief, his Sub-Sector Chiefs and their staff; Messrs I. Diallo and Y. Diarra, OCP Sub-sector Chiefs at Bamako and Bougouni, and their staff, Mr M. Kassambara, OCP Bobo Dioulasso, and his staff; Mr. S. Dramane, Dr K. Doucouré, Dr L. Toé, and Mr B. Tele for detailed surveys in Sierra Leone; Dr M. C. Thomson for electrophoretic identifications; Mr A. Soumbey and Dr S. Doumbia for data analysis; Dr C. Back for comments on the manuscript; Drs G. De Sole, K. Y. Dadzie and J. Remme for epidemiological information; Messrs A. Sib and Y. Coulibaly for cytotaxonomic assistance; Mr A. Belli and the administrative services for logistic support; the pilots of Viking Helicopters Ltd. and Evergreen Helicopters Inc. and the OCP aerial operations staff.

We acknowledge the key roles played by Dr B. Philippon and Dr D. Quillévéré, past and present Chiefs of the Vector Control Units; Dr D. A. T. Baldry and Mr P. Kaboré, past and present Chief Aviation Officers; Dr J. Grunewald, Research Coordinator; D. G. Zerbo, Chief Entomological Evaluation and Dr H. Agoua, Chief Central Zone.

Finally, we are grateful to Dr E. M. Samba, Programme Director for his continuing encouragement and permission to publish.

REFERENCES

Baker, R. H. A., Baldry, D. A. T., Boakye, D. & Wilson, M. 1987 Measures aimed at controlling the invasion of *Simulium damnosum* Theobald s.l. (Diptera: Simuliidae) into the Onchocerciasis Control Programme Area. III. Searches in the Upper Niger Basin of Guinea for additional sources of flies invading southeastern Mali. *Trop. Pest Mgmt.* **33**, 336–346.

Baker, R. H. A., Baldry, D. A. T., Pleszak, F. C., Boakye, D. & Wilson, M. 1986 Measures aimed at controlling the invasion of *Simulium damnosum* Theobald s.l. (Diptera: Simuliidae) into the Onchocerciasis Control Programme Area. II. Experimental aerial larviciding in the Sankarani Basin of eastern Guinea in 1984 and 1985. *Trop. Pest Mgmt.* **32**, 148–161.

Baldry, D. A. T., Zerbo, D. G., Baker, R. H. A., Walsh, J. F. & Pleszak, F. C. 1985 Measures aimed at controllng the invasion of *Simulium damnosum* Theobald s.l. (Diptera: Simuliidae) into the Onchocerciasis Control Programme Area. I. Experimental aerial larviciding in the Upper Sassandra Basin of south-eastern Guinea in 1985. *Trop. Pest Mgmt.* **31**, 255–263.

Bellec, C. 1976 Captures d'adultes de *Simulium damnosum* Theobald, 1903 (Diptera, Simuliidae) à l'aide de plaques d'aluminium, en Afrique de l'Ouest. *Cah. ORSTOM, Sér. Entomol. med. Parasitol.* **14**, 209–217.

Cheke, R. A. & Garms, R. 1983 Reinfestations of the southeastern flank of the Onchocerciasis Control Programme area by windborne vectors. *Phil. Trans. R. Soc. Lond.* B **302**, 471–484.

Garms, R. 1987 Occurrence of the savanna species of the *Simulium damnosum* complex in Liberia. *Trans. R. Soc. trop. Med. Hyg.* **81**, 518.

Garms, R., Cheke, R. A., Vajime, C. G. & Sowah, S. 1982 The occurrence and movements of different members of the *Simulium damnosum* complex in Togo and Benin. *Z. angew. Zool.* **69**, 219–236.

Garms, R. & Walsh, J. F. 1987 The migration and dispersal of black flies: *Simulium damnosum* s.l., the main vector of human onchocerciasis. In *Black flies: ecology, population management and annotated world list* (ed. K. C. Kim & R. W. Merritt), pp. 201–214. University Park and London: Pennsylvania State University.

Garms, R., Walsh, J. F. & Davies, J. B. 1979 Studies on the reinvasion of the Onchocerciasis Control Programme in the Volta river basin by *Simulium damnosum* s.l. with emphasis on the south-western areas. *Tropenmed. Parasit.* **30**, 345–362.

Johnson, C. G., Walsh, J. F., Davies, J. B., Clark, S. J. & Perry, J. N. 1985 The pattern and speed of displacement of females of *Simulium damnosum* Theobald s.l. (Diptera: Simuliidae) across the Onchocerciasis Control Programme area of West Africa in 1977 and 1978. *Bull. ent. Res.* **75**, 73–92.

Kurtak, D. C., Grunewald, J. & Baldry, D. A. T. 1987 Chemical population management of *Simulium* vectors of onchocerciasis in Africa. In *Black flies: ecology, population management and annotated world list* (ed. K. C. Kim & R. W. Merritt), pp. 341–362. University Park and London: Pennsylvania State University.

Kurtak, D. C., Raybould, J. N. & Vajime, C. G. 1981 Wing tuft colours in the progeny of single individuals of *Simulium squamosum* (Enderlein). *Trans. R. Soc. trop. Med. Hyg.* **75**, 126.

Le Berre, R., Garms, R., Davies, J. B., Walsh, J. F. & Philippon, B. 1979 Displacements of *Simulium damnosum* and strategy of control against onchocerciasis. *Phil. Trans. R. Soc. Lond.* B **287**, 277–288.

Magor, J. I. & Rosenberg, L. J. 1980 Studies of winds and weather during migrations of *Simulium damnosum* Theobald (Diptera: Simuliidae), the vector of onchocerciasis in west Africa. *Bull. ent. Res.* **70**, 693–716.

Post, R. J. & Crosskey, R. W. 1985 The distribution of the *Simulium damnosum* complex in Sierra Leone and its relation to onchocerciasis. *Ann. trop. Med. Parasit.* **79**, 169–194.

Remme, J., Baker, R. H. A., De Sole, G., Dadzie, K. Y., Walsh, J. F., Adams, M. A., Alley, E. S. & Avissey, H. S. K. 1989 A community trial of ivermectin in the onchocerciasis focus of Asubende, Ghana. I. Effect on the microfilarial reservoir and the transmission of *Onchocerca volvulus*. *Trop. med. Parasit.* **40**, 367–374.

Remme, J., De Sole, G. & van Oortmarssen, G. J. 1990 The predicted and observed decline in onchocerciasis infection during 14 years of successful vector control with reference to the reproductive lifespan of *Onchocerca volvulus*. *Bull. Wld Hlth Org.* (In the press.)

Vajime, C. G. & Dunbar, R. W. 1975 Chromosomal identification of eight species of the subgenus *Edwardsellum* near and including *Simulium* (*Edwardsellum*) *damnosum* Theobald (Diptera: Simuliidae). *Z. Tropenmed. Parasit.* **26**, 111–138.

Walsh, J. F., Davies, J. B. & Garms, R. 1981 Further studies on the reinvasion of the Onchocerciasis Control Programme by *Simulium damnosum* s.l.: the effects of an extension of control activities into southern Ivory Coast during 1979. *Tropenmed. Parasit.* **32**, 269–273.

Walsh, J. F., Davies, J. B. & Le Berre, R. 1979 Entomological aspects of the first five years of the Onchocerciasis Control programme in the Volta River basin. *Tropenmed. Parasit.* **30**, 328–344.
Walsh, J. F., Davies, J. B., Le Berre, R. & Garms, R. 1978 Standardisation of criteria for assessing the effect of *Simulium* control in onchocerciasis control programmes. *Trans. R. Soc. trop. Med. Hyg.* **72**, 675–676.

Discussion

R. Garms,[1] R. A. Cheke[2] and R. Sachs[1] ([1] *Bernhard Nocht Institute for Tropical Medicine, Bernhard-Nocht-Strasse* 74, *D-2000 Hamburg* 36, *F.R.G.*; [2] *Overseas Development Natural Resources Institute, Chatham, U.K.*). Baker *et al.* described movements in a northeastward direction of savanna members of the *Simulium damnosum* species complex from uncontrolled areas. Little is known about whether savanna flies also move in the opposite direction from north to south, from the savanna into the forest. Some recent observations in Liberia are therefore of interest.

Savanna species were known to occur and occasionally breed in northern Liberia (Garms & Vajime 1975), but they were not found elsewhere in the country until the dry season of 1985, when a few flies were caught biting man at sites beside the St Paul River in the evergreen rainforest zone. They were thought to have arrived from northern sites assisted by the northeasterly harmattan winds (Garms 1987). Three years later, during the dry season of 1988, residents of the Bong iron ore mine complained about a serious nuisance caused by blackflies biting them within the mine's concession area as much as 10 km away from the St Paul River, where flies had been extremely rare in previous years. Fly catches in May and June 1988 confirmed that biting densities were very high (more than 500 in 7 h). Morphological identifications showed that practically all of these flies were savanna members of the *S. damnosum* species complex.

This phenomenon of mass biting did not recur in the dry season of 1989 when regular fly catches were performed at four sites within and five sites outside the Bong Mine concession area. However, some savanna flies were present, but they were only caught in small numbers within the concession area. Their source could be traced to the Yea Creek, a small stream emerging from the Bong Mine's tailings pond. Larvae collected on 20 March, 21 and 22 April were cytologically identified as 39 *S. damnosum s.str.*, 19 *S. sirbanum* and 2 *S. soubrense* (G. K. Fiasorgbor, personal communication). Interestingly, they were associated with *S. adersi*, another savanna species, which had not been recorded from the area before. It is probable that the savanna species had invaded the region from the north with the harmattan winds, had encountered favourable breeding conditions and were able to establish themselves in the area for at least a few generations. They were not found in any of six streams and rivers around the concession area which were regularly checked.

It is not known why the outlet of the tailings pond, a unique and highly artificial environment, is such an attractive breeding site for the savanna species but not for the local forest species. Most of the water originates from the St Paul river from where it is pumped through a pipe line *ca.* 10 km long to the concentrator of the Bong Mine. From there the water, heavily loaded with inorganic material, is passed to a large impondment area, the tailings pond, where the solid wastes settled prior to the release of clean water to the river. The water leaving the tailings pond is characterized by constant high temperatures of 30 °C and several chemical characteristics, in particular hardness, which are quite distinct from those of the natural watercourses. It is very rich in microorganisms, probably produced by the drowning forests and decaying trees within the impondment area.

[229]

It is not yet clear whether or not the invasion by savanna species and their colonization of the Bong Range will have any epidemiological consequences.

References

Garms, R. 1987 Occurrence of the savanna species of the *Simulium damnosum* complex in Liberia. *Trans. R. Soc. trop. Med. Hyg.* **81**, 518.

Garms, R. & Vajime, C. G. 1975 On the ecology and distribution of the species of the *Simulium damnosum* complex in different bioclimatic zones of Liberia and Guinea. *Trop. med. Parasit.* **26**, 375–380.

R. H. A. BAKER. Despite the potential threat of savanna flies re-invading the Sassandra, Marahoué and Bandama Basins from isolated seasonal breeding sites in Liberia, described by Dr. Garms *et al.* and southern Sierra Leone (figure 6), biting and transmission rates in the reinvasion zones have been maintained at low levels for the past five years (since the treatment of the Upper Sassandra in southeastern Guinea began). This was true even in 1988, when the savanna flies were most numerous at the Bong Mine in Liberia. Rivers in southern Sierra Leone will be included in treatment circuits from 1990, further diminishing this threat.

R. A. CHEKE[1], M. A. HOWE[2], M. J. LEHANE[2], A. L. MILLEST[2], T. KONE[3] R. H. A. BAKER[3] (*[1]Overseas Development Natural Resources Institute, Chatham, U.K.*; *[2]School of Animal Biology, University College of North Wales, Bangor, Gwynedd U.K.*; *[3]WHO Onchocerciasis Control Programme. B.P. 549, Ouagadougou, Burkina Faso*). Baker *et al.* suggested that the migrant *Simulium sirbanum* reaching Madina Diassa in Mali during the 1988 wet season were 4–5 weeks old. The basis for their conclusions were analyses of the time sequences of waves of flies traversing a series of sites between Madina Diassa and their likely sources, 450–500 km away in Sierra Leone. We have recently estimated the ages of a sample of *S. sirbanum*, caught near Madina Diassa in June 1988, by analysis of their pteridine concentrations. Our results, based on this biochemical technique, agree very well with the conclusions of Baker *et al.*

Pteridine concentrations are known to increase with age and fly size in different members of the *S. damnosum* species complex (Cheke *et al.* 1987, 1989). Such variation was established for Malian populations of *S. sirbanum* by experiments with flies emerged from pupae, collected at Tienfala (12° 44′ N, 08° 00′ W; 40 km E of Bamako) in the Niger River. The flies were maintained at different temperatures (20–33 °C), killed at daily intervals, measured for size, and pteridine concentrations estimated by fluorescence spectrometry. Although, surprisingly, temperature had no effect ($P = 0.99$), a significant multiple regression was obtained relating pteridine concentrations in the flies' heads to fly age and size, as follows: Log PTHD = 0.0113 age (days) + 0.2158 fly thorax length (mm) + 0.53 ($P < 0.0001$), where PTHD is the pteridine concentration in the head capsule. This equation was then used to estimate the ages of 199 flies caught at Kanibougoula, 10 km upstream of Madina Diassa, on 22 and 23 June 1988. A histogram of the results is shown in figure 1. The mean of this distribution was 29.08 days (s.d. = 8.99), within the range of 4–5 weeks suggested by Baker for his flies caught nearby during the same period. A few flies were estimated as being less than two weeks old and some others more than six weeks, with a maximum of 67 days. This maximum is more than twice the previous longevity record, estimated by pteridines, of 27 days for a female *S. damnosum s.str.* (Cheke *et al.* 1989). However, it is recognized that conclusions about individual flies must be treated with caution as the method is best restricted to populations (Cheke *et al.* 1989). As such

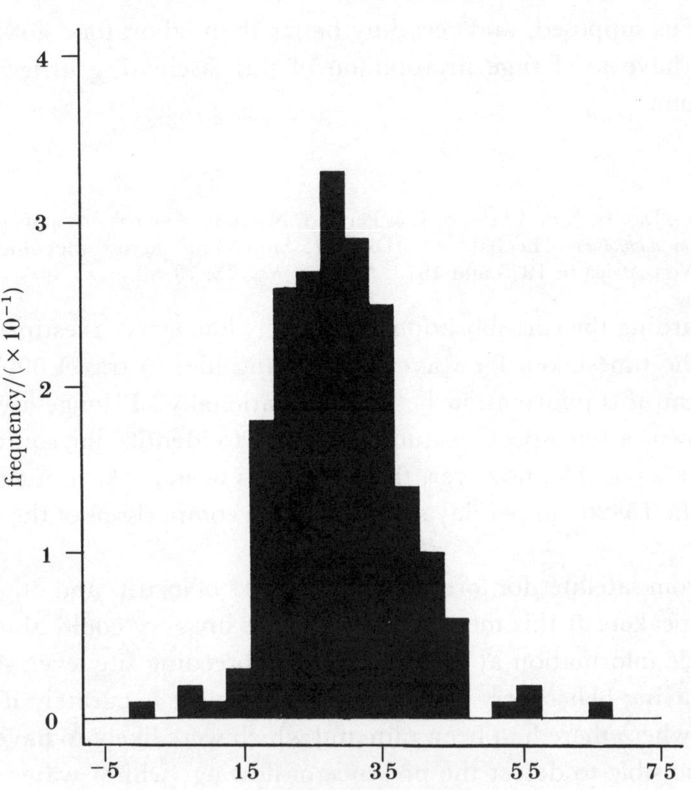

FIGURE 1. A frequency histogram of the estimated ages of *Simulium sirbanum*, caught at Kanibougoula on 22–23 June 1988. (Average age, $\bar{x} = 29.08$ days; s.d. = 8.99.)

a population analysis, the results in figure 1 provide good supportive evidence for the conclusion of Baker *et al.* that *S. sirbanum* take about a month to migrate 500 km.

References

Cheke, R. A., Garms, R., Howe, M. A. & Lehane, M. J. 1987 Possible use of pteridine concentrations for determining the age of adult *Simulium damnosum* s. l. *Trop. Med. Parasit.* **38**, 346.

Cheke, R. A., Dutton, M., Avissey, H. S. K. & Lehane, M. J. 1989 Age and size dependency of pteridine concentrations in different members of the *Simulium damnosum* species complex. *O-Now!: symposium on onchocerciasis: recent developments and prospects for control, September 20–22, 1989* (ed. H. J. van der Kaay), p. 101. Leiden, the Netherlands: The Institute of Tropical Medicine, Rotterdam.

J. B. DAVIES (*Medical Research Council and School of Tropical Medicine, Liverpool, U.K.*). I am extremely interested in Dr Baker's observation that flies were taking 4–5 weeks at average speeds of 15–20 km per day, to travel 500 km or so, and note how this is supported by Dr Cheke's estimates of age from pteridine concentrations and how closely both sets of results agree with earlier estimates for similar migrations (Johnson *et al.* 1985). I have always found Dr Johnson's earlier estimates of survival difficult to accept, although I have never questioned the fact that the flies were covering these distances. However, although the data showed the journeys took 30–60 days to accomplish, I found it hard to conceive that sufficient numbers of flies were surviving so long to provide such high biting rates at the furthest destinations. With this new corroborating evidence it seems that under the right conditions *S. damnosum* is a better survivor than some of us supposed, and certainly better than laboratory survival experiments show. I for one, will have to change my opinion of this fascinating little insect. This new evidence is very welcome.

Reference

Johnson, C. G., Walsh, J. F., Davies, J. B., Clark, S. J. & Perry, J. N. 1985 The pattern and speed of displacement of females of *Simulium damnosum* Theobald s.l. (Diptera: Simuliidae) across the Onchocerciasis Control Programme area of West Africa in 1977 and 1978. *Bull. ent. Res.* **75**, 73–92.

R. H. A. BAKER. Regarding the corroboration between fly longevity, as estimated by pteridine concentrations, and the time taken for waves of invading flies to travel 500 km from Sierra Leone to Mali, how can this information be used operationally? If longevity can be directly related to distance flown, a retrospective study could try to identify the source breeding sites of the invading flies. It is possible, however, that some flies would take a much shorter time to fly this distance than the 15–20 km per day suggested from comparisons of the waves of moving flies.

The value of data from satellites for forecasting outbreaks of locusts and other pests has been discussed by various speakers at this meeting, but satellite imagery could also be of benefit to OCP if it could provide information at the river or even breeding site level. At the beginning of the wet season, spraying helicopters could be targetted more accurately if it were possible to identify tributaries where there had been rain and which were likely to have started to flow. Ideally, it would be possible to detect the presence of flowing (white) water.

M. D. WILSON (*Department of Medical Entomology, School of Tropical Medicine, Liverpool, U.K.*). There is a lot of deforestation taking place in the lowland rainforest outside the Onchocerciasis Control Programme area of West Africa. Will this phenomenon result in the creation of habitats permanently suitable for colonization by the dangerous savanna vector species? This would have obvious implications for the future control strategy and boundary of the Programme.

R. H. A. BAKER. This is a very important point. In recent years, the savanna vector species have begun to colonize deforested parts of the Programme area and this situation can only deteriorate.

Phil. Trans. R. Soc. Lond. B **328**, 751–755 (1990)
Printed in Great Britain

General discussion

K. A. BROWNING, F.R.S. (*Meteorological Office, Bracknell, U.K.*). I would like to recall Mr Boulahya's advocacy of an operational Locust Control Watch and suggest that in parallel we need a properly resourced research programme integrating the physical and biological aspects of the problem. Today we heard about the advances in our observational capability from satellite, radar and aircraft and about our understanding of aspects of the weather that affect insect migration. We seem to have a physically attractive hypothesis, that convergent wind patterns in the boundary layer play an important role in concentrating insects; but the hypothesis is still of unproven generality. There are a few case studies and a lot of circumstantial evidence, but to find out whether meteorological convergence events are abundant enough to be really important we need a research programme with much more comprehensive coverage in space and time of both meteorology and insects. Assuming we are able to prove that the convergence events are important, we would also need more research into the best methods to detect them and to control the insects, perhaps by using a combination of satellite, radar, aircraft and mesoscale numerical models as part of an integrated system. The question is who should run such an integrated research experiment bringing migrant pest, remote sensing and meteorological communities together. I would suggest that a joint FAO/WMO Programme is required.

D. RIJKS (*World Meteorological Organization, Geneva, Switzerland*). Given the available knowledge about relationships between meteorology and the physiology and behaviour of at least some insects, and the possibilities to use this knowledge to monitor and combat outbreaks of migrant pests, am I right to think that there is, in this meeting, a consensus that this knowledge should be used now, to install a permanent preventive migrant pest watch, using among others the available meteorological infrastructure and reinforcing it, where necessary?

P. M. SYMMONS (*Food and Agriculture Organization of the United Nations, Rome, Italy*). My response concerns the relevance of what we have been talking about for the past couple of days. I suppose I stand as near to the customer as anyone can here. I've been involved with locust forecasting for twenty years. I ran the Desert Locust Information Service, at the Anti-Locust Research Centre and its successor, the Centre for Overseas Pest Research, in London for a half a dozen years and I am now responsible for Desert Locust forecasting with FAO.

On the subject of research there is a coordinating group for Desert Locust research, under UNDP auspices, consisting of a group of donors supported by a joint FAO/UNDP scientific advisory committee for which I am the secretary. This has the job of reviewing and attempting to coordinate research in general with a specific role of looking after, advising and managing any UNDP-financed projects or ones operated by FAO with funds entrusted to it. So this body could consider any initiative raised at this meeting which comes within the committee's general terms of reference, which are indeed very wide.

Some of the suggestions made here about general monitoring are unrealistic. A lot of what we have been discussing pre-supposes the existence of organizations in the field, but this is in doubt. It is depressing the way in which, within a few months of the disappearance of locust

[233]

swarms, reporting has gone down to a very low level and it appears that the will is not there, either in the donors or in the countries affected, to maintain organizations in the field on a long-term basis. I thing a pre-requisite for anything of this sort is a technical unit in FAO itself. In six months time, after my retirement, the current unit will cease to exist unless something is done in the interim. Without this unit, or something that can step in with advice and help on the appropriate scale if there is an emergency, I do not see locust organizations in the field having the central motivating core to keep them going. The only realistic hope is to try to create and maintain a relatively small unit (and I emphasise small) that can be properly and gainfully occupied during a recession. There are some schemes afoot involving very large scale, grandiose, preventative activities, but to my mind they are wholly unrealistic: they are talking in terms of tens of millions of dollars and I don't see this sort of thing being either sensible or continuously funded. We have got to look to donors to support operations on a small scale on a long-term basis. Donors don't like putting in money on a continuing basis, but that is the only way it can be done; vehicles have to be replaced and equipment has to be kept going, things that need hard currency to buy.

On the specific points of Desert Locust forecasting, I think we get a bit mixed up on the timescales, as I said earlier after Mr Boulahya's talk. What we try to do is forecast on three timescales. One of these scales is six months or more ahead, which can only be done in a very vague way, but if you are going to get extra money out of donors or even bring in pesticides on a large scale you have got to look that far ahead. The standard forecasting is done on a four to six weeks ahead basis, which is what has been done with the bulletin over many years, and is done now from FAO in Rome. Then there is the tactical forecasting on hours and days, but very little has been done at this scale. However I do feel, having operated at the sharp end in the field, that you can get away in a large measure without it. What is needed when operating in the field is a good information system; it is like war, as many people have said, and what is needed is intelligence. If you can get people out there looking and get the information back you can build on that very quickly. In other words you are forecasting only a very short time ahead and basically forecasting on the assumption that what has happened is going to continue to happen. That is a pretty good basis in meteorological forecasting as well as with locusts.

What one is trying to do during a plague is essentially to try to track the swarms. Your main interest here, as Mr Pedgley has pointed out, is to forecast for invasions, which means translation over very long distances and is certainly not related to convergence zones, though convergence on a large scale may be important in keeping swarms in a particular area where one can get at them. But on the whole, what people want to know is whether they are going to get some swarms from somewhere else and that means a large scale migration, often a thousand kilometres or more.

During a recession period, as Mr Boulahya implied, our main problem is where the locusts are breeding and for that we need to know where rain has fallen and where vegetation is likely to be growing: the rain providing both the necessary conditions for the eggs to mature and the green feed for the hoppers to come through. I am afraid that the products we have to do this, from the user end, are really not very satisfactory. One of the problems is that we are looking at a very narrow band in a critical area where it is not really very wet; the cold-cloud analysis does not work except in the west African Sahelian area and even then it has to be looked at it with very considerable care. I am very unhappy with the greenness index maps: the first product is the 1 km data and the big advance of that is that some sort of topographical structure underneath can be seen that can be related to it and wadis, for example, can be identified. But there are a lot of spurious responses: for instance, outcrops and suchlike which show up as things in the middle of the desert, which one knows very well as not real. The confidence from the users' end is simply not there at the moment and it will take quite a lot of use and testing to build it up.

S. G. Cornford (*Meteorological Office, Bracknell, U.K.*) The most notable progress in the field of migrant pests since the Society's 1977 meeting has been in operational meteorology. Data from meteorological satellites and analysed and predicted fields of wind and weather from mathematical models run on supercomputers can be available operationally.

Mr Hielkema has said ARTEMIS will process meteorological satellite data into a form useful to agriculturalists in Africa and DIANA will send the results to potential recipients there.

The size, lifetimes and speeds of movement of weather systems are such that the national meteorological services of the world must be interdependent. Those in Europe cannot be efficient unless those in Africa are, and vice versa. National meteorological services (NMSs) in Africa provide observations to Europe and those in Europe provide satellite data and the results from the numerical prediction models. It is therefore essential that any products coming from ARTEMIS and DIANA are complementary with those which will be distributed by the Meteorological Data Distribution (MDD) mission of the European meteorological satellite Meteosat. It is also essential that DIANA routes ARTEMIS products in ways that are agreed with the NMSs of Africa. If not, the NMSs will be undermined and that is against the interests of the national meteorological services of Europe, which control and fund Meteosat on which ARTEMIS depends.

At the intergovernmental level it is essential that cooperation between FAO and WMO be cordial and mutually understanding. In the context of this meeting, a possible framework for the cooperation is provided by the United Nations' International Decade for Natural Disaster Reduction.

A. G. Gatehouse (*University of Wales, Bangor, U.K.*). The papers and discussion at this meeting have, for the most part, emphasized the implications of advances in meteorological knowledge and remote sensing techniques for our understanding of the origin and development of outbreaks, upsurges and plagues of migratory insects. While these advances are undeniably important, there have also been significant advances in recent years in our knowledge and understanding of biological aspects of insect migration which have a direct bearing on the subject of this meeting.

Most important, perhaps, is the growing evidence from a wide range of species that the migratory potential of individuals and thus of populations is highly variable. While this variability is often modulated by environmental cues, particularly in temperate-zone species, there is now good evidence for primarily genetic determination of migratory potential in several subtropical/tropical insects. In these, which include some noctuid moths and one acridoid, *Melanoplus sanguinipes* (Fabricius), variations in the frequencies of persistent fliers within and between populations and/or seasons can be related to variations in the distribution of suitable habitats in time and space (Gatehouse 1989, and references therein). In the Desert Locust (and perhaps other gregarizing species), the distribution of low-density populations among the scattered and ephemeral habitats of the recession areas depends on nocturnal

[235]

migration of individual solitary-phase locusts. Thus these movements must make a major contribution to the spatial and temporal dynamics leading to gregarization and the onset of upsurges. In spite of this, we know virtually nothing of the variation in migratory potential of solitary-phase locusts (although anecdotal accounts suggest it exists; G. Popov, personal communication). Neither do we know anything of the factors regulating their migratory behaviour. I suggest that these are questions of central importance in understanding the development of upsurges of the Desert Locust. What is more, I would argue strongly that the potential of the impressive meteorological and remote sensing techniques we have heard about to allow us to monitor and predict developments leading to upsurges will not be fully realised until these questions are answered.

Reference

Gatehouse, G. 1989 Genes, environment and insect flight. In *Insect flight* (ed. G. J. Goldsworthy & C. H. Wheeler), pp. 115–138. Florida: CRC Press.

R. J. V. JOYCE (*Cranfield Institute of Technology, Bedfordshire, U.K.*). Dr Symmons fears that with the impending retirement from FAO of so many locust specialists, the ability of that organization to continue its work of coordination of locust control will be greatly impaired. Captain Kitenda has already drawn attention to the serious loss of operational capability for locust control in the Regions. Yet, it is near certain that, sometime, there will be a new upsurge of Desert Locusts. Moreover, as Dr Riley and others have pointed out, grasshoppers continue to be a serious annual threat to crops throughout the Sahel and no satisfactory, or cost effective methods are yet available to combat them.

In the absence of action now, the next upsurge of locusts will undoubtedly be accompanied by the need to provide emergency help, and this will inevitably be again of a kind which cannot fail to be less efficient than a well-planned operation. The supply of insecticides and formulations unsuitable for locust control, the employment of inappropriate and wasteful crop spraying techniques for controlling swarms and hopper bands, the execution of spraying by crop-spraying pilots inexperienced in locust control and without proper directions with regard to targets, or means of finding targets or of checking the results of control, the use of aircraft with navigation aids quite inadequate for work in featureless desert areas, these are among the factors which have resulted in an unacceptable waste of resources, especially insecticide, and grave environmental hazards. In my view, the banning of dieldrin and the absence of a substitute with adequate persistence, makes it impossible to expect any significant contribution to the regulation of Desert Locust numbers from the control of hoppers. The barrier spraying techniques which used dieldrin at a rate of 5–10 g active ingredient per hectare, permitted up to 500 000 ha per day per aircraft to be treated for hopper infestations. The substitute (and more expensive) insecticides have to be directed at individual bands and applied at about 500 g ha^{-1} so that work rates are unlikely to reach 10 000 ha per aircraft per day. This means that Desert Locust control must rely even more heavily on swarm control and this dictates the use of aircraft adequately instrumented to permit the crew to find their targets, spray them and return to them again to complete the job and assess results.

It is evident from the contributions to our discussions from Captains Kitenda and Kinvig, that skills and experience in locust control are not absent in the regions. What is lacking is a means of retaining and employing them, particularly during periods of recession and it is this to which I wish to address myself.

Skaf *et al.* have emphasized the need for 'preventative control' to avoid the kind of emergency which otherwise results. Mr Popov has drawn attention to the type of habitat in the Sahel and elsewhere where gregarization and swarm formation occurs. Mr Hielkema and Mr Zick have graphically shown the irreplaceable role which satellite imagery can play in defining areas where recent rainfall makes such areas a current risk. These areas can be vast, many of them are difficult or impossible of access by ground-based teams, but they must be searched for infestations in about 4–6 weeks. Search in such areas can be economically achieved in the time available only by airborne crews, and systematic quantitative search only by aerial traverses by using aircraft fitted with suitable navigation and insect detecting equipment, namely of the type described in the paper by Dr Rainey and me. Such an aircraft, flown by pilots experienced in locust control, such as Captain Kitenda, could locate and search not only zones of wind discontinuities for flying swarms, but also areas where gregarization may have occurred, and so apply to infestations from either mechanism, control on the scale needed and, finally, assess the results quantitatively.

I would like to see funds made available to FAO for the provision of such an aircraft, and that organization, by using an experienced crew, undertake the responsibility for searching for swarm and hopper infestation in any areas which information from all sources indicate at possible risk. The same experienced aircrew would also have a training role to update the staff of Regional Control Organizations, and in an upsurge, any new emergency staff, on the best methods of survey and control. An annual expenditure of about \$600000 for operating such a '*search and strike*' procedure could well avoid emergency aid estimated, during the last upsurge, at \$300000000.

L. G. Goodwin, F.R.S. (*Shepperlands Farm, Park Lane, Wokingham, U.K.*). In drawing this discussion meeting to a close I would stress that the key to its success has been the bringing together of physical and biological communities. This must be a feature of our future approach to the important problem of migrant pests. We have not made as much progress as we should over the past quarter century. Given better resources and the determination to go forward together, perhaps we can do better in the future. So, with thanks to our speakers, chairmen and organizers, and special thanks and best wishes to our absent friend and lead organizer, Reg Rainey, I close this meeting.